Preface

Historically, network theory and system theory developed in parallel through the 40's and 50's. During the 60's, however, the two fields began to diverge with network theorists motivated by the problems of integrated circuits, directing their efforts toward the study of interconnected configurations of large numbers of relatively simple (RLC) components. System theorists, on the other hand, were motivated by problems in control and estimation and directed their efforts toward the analysis and design of relatively simple configurations of complex components. Fortunately, in the past decade, the two fields have, with the increasing emphasis on the area of large scale systems, begun once again to merge. Indeed, for the analysis of a large system the only alternative to going to ever larger computers is to efficiently exploit the decoupling inherent in the system's connectivity structure. System theorists have thus begun to adopt the practice of viewing a large system as a complex interconnection of complex components. At the same time the network theorist has been steadily increasing the degree of complexity of the components with which he deals within his traditional interconnected structure. Thus the two fields have, by different routes, come to a common point: the analysis and design of large systems from the point of view of complex interconnections of complex components.

The goal of *Interconnected Dynamical Systems* is to meet the need for a textbook at the beginning graduate level in this "new" field at the intersection of network and system theory. The book has been written so that it can either follow or subsume the standard linear systems course now taught in the first semester in most EE graduate programs. At Texas Tech, we are teaching a two semester course, the first semester of which replaces the linear systems course, while at Purdue we are teaching a one semester course that follows the classical linear systems course. By the end of the two semesters the student has seen all of the material usually taught in the linear system course and, in addition, the new material on interconnected systems. The advantage of the integrated course over the classical linear systems course is that the student is given a background that simultaneously prepares him for advanced courses in both networks and systems, as well as the emerging area of large scale systems.

The text is directed primarily towards teaching the student analytical tools for the analysis and design of interconnected networks and systems. However, since the networks and systems to which these techniques are applied are typically very large and highly nonlinear it is recognized that these analytic tools must ultimately be computer implemented. Therefore, each chapter contains at least one section devoted to a discussion of the computational problems encountered in implementing the theory. The purpose of these sections is not to teach computation but simply to familiarize the student with computational reality and thus motivate the student towards the development of analytic techniques which are amenable to the computer. For instance, the section on sparse matrix inversion serves primarily to motivate the formulation of analytic formulae in terms of an LU factorization rather than direct matrix inversion, etc.

The text is divided into seven chapters dealing, respectively, with Components, Connections, Systems, Simulation, Sensitivity, Stability, and Control. In each of the first five chapters both linear and nonlinear techniques are developed and the numerical methods required for their implementation are discussed. Although the last two chapters are restricted to linear systems (not enough is presently known about the nonlinear control problem, while the literature on nonlinear stability theory would require an entire text to summarize), the numerical emphasis is continued. The entire text is illustrated by several hundred detailed examples wherein both analytic and numerical techniques are presented. Although some of the numerical examples are quite long, all of the examples can be done with the aid of a hand calculator, thereby allowing the student to follow the example without going to a computer.

Unlike most technological activities which are motivated by the requirements of a single industry, the field of interconnected dynamical systems evolved in response to the simultaneous requirements of several distinct industries: aerospace, semiconductor, computer, electric power, communications, etc., each of which has fostered a research effort in the field. As such, the topics presented in the present text are derived from a wide variety of sources and represent the work of a myriad of researchers too numerous to identify. We are, however, particularly indebted to our teachers who influenced our activity in the area and our students who supported it. Finally, we would like to acknowledge prodigious efforts of Mrs. Flores Myers who prepared and edited a series of manuscripts which eventually evolved into the present text.

Interconnected Dynamical Systems

Raymond A. DeCarlo

School of Electrical Engineering
Purdue University
West Lafayette, Indiana

Richard Saeks

Department of Electrical Engineering
Texas Tech University
Lubbock, Texas

MARCEL DEKKER, INC.   New York and Basel

Library of Congress Cataloging in Publication Data

DeCarlo, Raymond A., [date]
    Interconnected dynamical systems.

    (Electrical engineering and electronics; v. 10)
    Bibliography: p. 485
    Includes index.
    1. Electric networks.   2. System theory.   I. Saeks, R.
II. Title.   III. Series: Electrical engineering and
electronics; 10.
TK454.2.D35   620.7′2   81-9784
ISBN 0-8247-6639-3   AACR2

MARCEL DEKKER, INC.
270 Madison Avenue, New York, New York 10016

Current printing (last digit):

10 9 8 7 6 5 4 3 2 1

PRINTED IN THE UNITED STATES OF AMERICA

D
620·72
DEC

Contents

I. COMPONENTS

1. INTRODUCTION

The components of an interconnected dynamical system may take on a wide variety of physical forms characterizing electrical, mechanical, chemical, social, and economic phenomena; and/or hybrids thereof. Intuitively, any such system component is representable by a black box as shown in figure 1.1. Here the *component input*, a, is a vector of time

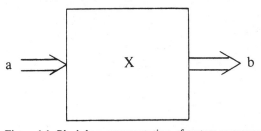

Figure 1.1. Black box representation of system component.

functions depicting the signals applied to the various component input terminals or ports. Similarly, b denotes the *component output* vector and x is a vector of internal signals termed the *state vector*, which may or may not be explicitly included in the component model. The dimensions and units

1

of these vector variables are quite arbitrary. In particular, mixed units are allowed and are quite common in practice. For instance, in a model of an electric *power system* one normally encounters electro-mechanical devices and occasionally it is necessary to include ecomomic factors in the model. Of course, whenever components are interconnected, the appropriate vectors must be conformable, both with respect to dimension and units.

The primary purpose of the present chapter is to fabricate mathematical models for the components of an interconnected dynamical system. Since one desires to encompass a wide variety of physical and socio-economic phenomena within the theory, these models are necessarily mathematical in nature. Thus the models are independent of the physical significance of the device being characterized. On the other hand, our goal is to study systems, not mathematical models. Consequently no attempt is made at achieving generality with respect to either the class of permissible input/output signals or with respect to the *component-models* themselves. Indeed, we assume the input signals are both smooth and bounded in a manner commensurate with the existence of a well behaved Laplace transform. Similarly, the relationship between component inputs and component outputs will always take the form of an *ordinary differential equation*. Although it is possible to lift both of these assumptions, the resultant gain in generality fails to justify the required jump in the complexity of the exposition. Furthermore, the components and signals in most real world systems can be viably approximated by models satisfying these constraints when they cannot be modeled exactly. [6,12]

In the following section we formalize the class of allowed input/output signals for our components and then discuss their Laplace transformation. System components modeled by rational transfer functions and/or transfer function matrices are presented in section 3, followed by a development of the linear (time-invariant) state model and its relationship to the transfer function model in section 4. In section 5, state models for nonlinear components are introduced. Here special emphasis is placed on the piecewise linear case. Finally the composite component model for an interconnected dynamical system is introduced in the last section of the chapter.

2. SIGNALS

The sequel assumes that all component inputs, outputs and internal (state) signals lie in the following class of admissible signals.

DEFINITION 2.1: A real valued function of a real variable is an *admissible signal* if it is *smooth, exponentially bounded* and has *support bounded on the left*. That is, the function, f, is admissible if:

(i) it is *infinitely differentiable* at every point.

(ii) there exist constants, M and c, such that

$$|f(t)| < M \exp(ct) \qquad (2.2)$$

(iii) there exists a real constant, r, such that

$$f(t) = 0 \text{ for all real } t < r. \qquad (2.3)$$

The above criteria insure that the admissible signals possess a viable *two-sided Laplace transformation*. Unfortunately, the criteria exclude some of the classic signals of system theory such as the *Heavyside step function*. However, such discontinuous signals are usually idealizations of admissible "real world" signals, so there is no real loss of applicability.

At any rate, condition (i) formalizes the physical observation that measured signals are smooth (i.e., C^∞); (ii) is a weak bound which allows signals to "explode" provided they remain bounded by an exponential; (iii) abstracts the physical concept that all signals are "turned on" at a finite point in time. Note that since we allow "c" to be positive and "r" to be negative, (ii) and (iii) are both minimal restrictions.

Now by virtue of the fact that we consider only *one-sided functions* (i.e., axiom (iii) of 2.1) one may advantageously work with the two-sided Laplace transform. This alleviates many of the usual difficulties associated with questions of continuity at zero when employing the *one-sided Laplace transform*. Recall that the *two-sided Laplace transform* of a function, f, is defined by the integral

$$\int_{-\infty}^{\infty} f(q)e^{-sq}dq \qquad (2.4)$$

for those complex values of s for which the integral exists.[13] Although nonstandard and an abuse of notation we find it convenient to use the same symbol, f, to denote both the Laplace transform of a function and the function itself. This causes little difficulty since either the context or the argument of the function (t or s respectively) will clearly indicate the use of the *time* or *frequency domain*. The notation also has the benefit of yielding a significant simplification of the many arguments which are identical in both domains.

The *existence theorem for the Laplace transform* of an admissible signal is as follows:

THEOREM 2.5: Let f be an admissible signal, then there exists a real constant c such that the two-sided Laplace transform of f exists and is analytic for $\text{Re}(s) > c$.

PROOF: Since f is admissible, there exist constants r, M and c such that (ii) and (iii) of definition 2.1 are satisfied. Let $s = \sigma + j\omega$. Then

$$\left| \int_{-\infty}^{\infty} f(q)e^{-sq}dq \right| = \left| \int_{r}^{\infty} f(q)e^{-sq}dq \right|$$

$$\leq \int_{r}^{\infty} |f(q)| e^{-\sigma q} dq \leq \int_{r}^{\infty} Me^{cq}e^{-\sigma q}dq \qquad (2.6)$$

$$= \frac{1}{\sigma - c} e^{-(\sigma - c)r} < \infty$$

whenever $\sigma > c$. Thus the integral exists for σ strictly greater than c. Moreover, f(s) is analytic in its region of existence ($\sigma > c$) since s enters into the integrand, which defines the Laplace transform, analytically.

Notice that we have not used the smoothness of f in the above argument. Consequently, the theorem is applicable to any *Lebesque integrable function* satisfying (ii) and (iii) of definition 2.1. Thus in the sequel we are free to apply the theorem to any *piecewise continuous function.*[13]

The above defined transform has all the usual properties of the classical one-sided Laplace transform such as *linearity* and the *scaling property*. In several respects the properties of the two-sided transform, when considering functions in the admissible signal space, are less complex than their one-sided counterparts. In particular, the *transform of the derivative* of a function, f, is $s[f(s)]$ rather than $[f(s)] + f(0)$. This fact follows from 2.4 via integration by parts and exploitation of the fact that both f and its derivative are zero in a neighborhood of $-\infty$.[13] Also, the *inversion formula* for the two-sided transform is simpler. The two-sided inversion formula is derivable from the inversion formula for the *Fourier transform*. To see this again, let $s = \sigma + j\omega$. Observe that the Laplace transform of f at s is just the Fourier transform at $j\omega$ of the function f_σ where

$$f_\sigma(t) = f(t)e^{-\sigma t} \qquad (2.7)$$

Now, since f is continuous, it may be computed via the Fourier inversion formula for all t. This yields

$$f(t) = e^{\sigma t}f_\sigma(t) = e^{\sigma t}\int^{\infty} f(\sigma + j\omega)e^{j\omega t}d\omega \qquad (2.8)$$

$$= \int_{-\infty}^{\infty} f(s)e^{st}d\omega$$

Note that unlike the inversion formula for the one-sided Laplace transform, the integral of 2.8 is with respect to a real variable, the imaginary part of s with the real part of s fixed at any $\sigma>c$. An immediate corollary to the above inversion formula is the following *uniqueness theorem*:

THEOREM 2.9: Let f and g be admissible signals whose Laplace transforms both exist and coincide for $\text{Re(s)}>c$. Then $f=g$.

PROOF: Since f and g have the same transforms for $\text{Re(s)}>c$ one can apply equation 2.8 to both functions to compute

$$f(t) = \int_{-\infty}^{\infty} f(s)e^{st}d\omega = g(t) \tag{2.10}$$

Although the use of admissible signals greatly simplifies the Laplace transformation theory, a number of difficulties remain. In particular, even though f(s) is only defined for $\text{Re(s)}>c$, the function f(s) may have an *analytic extension* into a larger region. For instance, the Heavyside step function (an inadmissible signal since it is discontinuous at zero) has a Laplace transform equal to $1/s$ in the region $\text{Re(s)}>0$. Yet the function $1/s$ is well defined for all $s\neq0$. Although it is quite possible and legitimate to deduce information about f(s) (in its region of existence) from this analytic extension, one must realize that f(s) equals the Laplace transform integral only for $\text{Re(s)}>c$. In practice, we often characterize f(s) for $\text{Re(s)}>c$ with $j\omega$-axis information even though c may be greater than zero.

3. TRANSFER FUNCTIONS

Recall that a complex valued function of a complex variable is *rational* if it is the ratio of two *polynomials*.[7] Such a function is *strictly proper* if the *order* of its numerator polynomial is less than that of the denominator polynomial; equivalently if the function has a *zero at infinity*. A rational function is *proper* if the order of its numerator is less than or equal to the order of its denominator—i.e., the rational function does not have a *pole at infinity*. Finally, if the order of the numerator exceeds the order of its denominator, then the rational function has a pole at infinity and it is said to be *improper*. The *degree* of a rational function is the maximum of the orders of the numerator and denominator polynomials. Thus the degree of a rational function is also equal to either the number of its poles or the number of its zeros (counting those at infinity and multiple poles or zeros by their *multiplicity*). A *rational matrix* is a matrix whose entries are rational functions. Such a matrix is (*strictly*) *proper* if all its entries are (strictly) proper.

Now the *components* considered in our theory of interconnected dynamical systems have natural physical interpretations. However, from the point of view of their *mathematical models*, they are simply mappings from their *input signals* to their *output signals* as symbolically illustrated in Figure 1.1. The input is a vector of *admissible signals* denoted by "a" and the output is a vector of admissible signals denoted by "b". In general these vectors have different dimensions and units. For instance, the output characterizing a *generator* in an *electric power system* may be composed of a mechanical variable characterizing the angular velocity of the machine rotor, an electrical variable describing the output voltage of the machine, and an economic variable accounting for the fuel consumption of the machine.

So by a *component model* we will mean a formula with which one computes the *component output vector* "b" given its *input vector* "a". Probably the most common component model is the *transfer function* and hence we will consider it first.

DEFINITION 3.1 A *transfer function matrix* for a system component is a rational matrix, Z, satisfying the equality

$$b(s) = Z(s)a(s) \tag{3.2}$$

for all s such that $\text{Re}(s) > c$, some real constant.

Before conjuring up the myriad possible examples of rational transfer functions we demonstrate (through a series of lemmas) that *every rational matrix is a well defined transfer function*—i.e., any rational matrix defines a unique mapping from vectors of admissible signals, "a", to vectors of admissible signals, "b". To this end, we have the following lemmas.

LEMMA 3.3: Let f be an admissible signal and let g be a real valued function of a real variable such that $g(t) = 0$ for $t \leq 0$, $g(t)$ is smooth for $t > 0$, and $g(t)$ is exponentially bounded. Then the *convolution* of f and g is a well defined admissible signal.

PROOF: Since $g(t) = 0$, $t < 0$, and $f(t) = 0$ for $t < r$, the infinite limits on the convolution formula

$$h(t) = \int_{-\infty}^{\infty} f(t-\lambda)g(\lambda)d\lambda \tag{3.4}$$

may be replaced by the finite limits

$$h(t) = \int_0^{t-r} f(t-\lambda)g(\lambda)d\lambda \tag{3.5}$$

This assures the existence of $h(t)$, since both f and g are smooth in the range of integration. Moreover, if $t < r$, the region in which the integrand is non-zero becomes trivial. This verifies that $h(t) = 0$ for $t < r$. Furthermore, the n-th derivative of h may be computed by the formula[13]

$$\frac{d^n h(t)}{dt^n} = \int_{-\infty}^{\infty} \frac{d^n f(t-\lambda)}{dt^n} g(\lambda)d\lambda \tag{3.6}$$

which is also well defined via the same argument used in showing that h is well defined. Hence h is *smooth*.

To show that h is exponentially bounded use the fact that both f and g are exponentially bounded as:

$$|f(t)| < Me^{ct} \tag{3.7}$$

and

$$|g(t)| < Ne^{dt} \tag{3.8}$$

Assume without loss of generality, that $c > d$. If this were false, the fact that the support of f is bounded on the left would allow one to obtain a new bound with $c > d$, simply by increasing both M and c. Thus

$$|h(t)| = |\int_{-\infty}^{\infty} f(t-\lambda)g(\lambda)d\lambda| = |\int_0^{\infty} f(t-\lambda)g(\lambda)d\lambda|$$

$$\leq \int_0^{\infty} |f(t-\lambda)| |g(\lambda)| d\lambda \leq \int_0^{\infty} Me^{c(t-\lambda)} Ne^{d\lambda} d\lambda \tag{3.9}$$

$$= \frac{MN}{c-d} e^{ct}$$

which is an exponential bound for h. Thus h is an admissible signal and the proof is complete.

LEMMA 3.10: Let f and g satisfy the hypotheses of lemma 3.3 and h be their convolution. Then $h(s) = f(s)g(s)$ in the region where all three are defined.

PROOF: Note that h is admissible by lemma 3.3 and hence has a well defined Laplace transform by theorem 2.5. As such the equality $h(s) = f(s)g(s)$ is well defined in a *half plane* whose boundary is the largest of the boundaries defining the domains (the *regions of convergence*) of $h(s)$, $f(s)$, and $g(s)$.

Now we must verify that this equation is indeed valid. From the convolution formula 3.4 and the definition of the Laplace transform 2.4, we have

$$h(s) = \int_{-\infty}^{\infty} \exp(-st) \left[\int_{-\infty}^{\infty} f(t-\lambda)g(\lambda)d\lambda \right] dt \qquad (3.11)$$

$$= \int_{-\infty}^{\infty} \exp(-s(\mu-\lambda)) \left[\int_{-\infty}^{\infty} f(\mu)g(\lambda)d\lambda \right] d\mu$$

$$= \left[\int_{-\infty}^{\infty} f(\mu)\exp(-s\mu)d\mu \right] \left[\int_{-\infty}^{\infty} \exp(-s\lambda)g(\lambda)d\lambda \right]$$

$$= f(s)g(s)$$

Here we have used the change of variables $\mu = t-\lambda$. Also the interchange of the order of integration is legitimate in the region where all three integrals exist. [10]

At last with the aid of this lemma, we obtain our *main* theorem.

THEOREM 3.12: Every proper rational matrix uniquely defines a mapping from vectors of admissible signals, "a", to vectors of admissible signals, "b".

PROOF: Clearly, it suffices to consider the scalar case, since *linearity* insures the extension of the result to the matrix case. Now, if "a" is admissible, its Laplace transform is well defined and analytic in the half plane, $Re(s) > c$. Hence for any rational function, $Z(s)a(s)$ is well defined and analytic in the half plane, $Re(s) > d$. Here d is equal to max $[c, \sigma_1, ..., \sigma_n]$ and the σ_i are the real parts of the poles of $Z(s)$. Note that the only points at which $Z(s)$ is not analytic are its *poles*. Moreover, since $Z(s)$ is a rational function, it has only a finite number of poles.[13] Therefore, the problem is to show that $Z(s)a(s)$ is the Laplace transform of a unique admissible function in the region $Re(s) > d$.

By theorem 2.9, there can be at most one such admissible function. Consequently, it suffices to construct any admissible function whose Laplace transform coincides with $Z(s)a(s)$ on the appropriate half plane. Towards this end first consider the case where $Z(s)$ is strictly proper. Here

Z(s) always has the *partial fraction expansion*

$$Z(s) = \sum_{i=1}^{k} \sum_{j=1}^{m^i} \frac{r_{ij}}{(s-p_i)^j} \qquad (3.13)$$

where the p_i are the (possibly complex) poles of Z(s), and the r_{ij} are the appropriate *residues*. By direct computation, Z(s) is the Laplace transform of the function

$$Z(t) = \begin{cases} \sum_{i=1}^{k} \sum_{j=1}^{m^i} \frac{r_{ij}}{(j-1)!} t^{j-1} \exp(p_i t) & t \geq 0 \\ 0 & t < 0 \end{cases} \qquad (3.14)$$

in the region $Re(s) > \max[Re(p_i)] = \max[\sigma_1,...,\sigma_n]$. Now Z(t) and a(t) satisfy the hypotheses of lemma 3.3 and 3.10. Thus their convolution is an admissible signal whose Laplace transform exists and equals Z(s)a(s) in the appropriate region. This proves the theorem for the case of a strictly proper Z(s).

For the general case expand Z(s) via *continued fractions* into the form

$$Z(s) = R(s) + P \qquad (3.15)$$

where R(s) is a strictly proper rational function and P is a constant. Therefore

$$Z(s)a(s) = R(s)a(s) + Pa(s) \qquad (3.16)$$

where R(s)a(s) is the Laplace transform of an admissible signal by the above argument, and Pa(s) is the Laplace transform of Pa(t). Linearity then guarantees that Z(s)a(s) is the Laplace transform of an admissible signal in the region $Re(s) > d$.

In other words for any admissible a(t), a proper rational transfer function uniquely defines an *admissible* b(t). Equivalently, a proper transfer function uniquely specifies a mapping between admissible input signals and admissible output signals. Therefore every proper rational matrix is a well defined transfer function for a component of an interconnected dynamical system.

With these technicalities resolved, let us muse over some specific examples of *system components*, modeled by *transfer functions*.

EXAMPLE 3.17: The most common example of a system component is the *integrator* whose transfer function is $Z(s) = 1/s$. This formula results from a computation of the Laplace transform of an "indefinite integral". Since the input and output terminals are stipulated, the *integrator* is dubbed a *unilateral component*. Of course, the units of the input signal may be arbitrary so long as the output signal has units of time multiplied by the units of the input signal.

A second common unilateral system component is the *2nd-order Butterworth filter* whose transfer function is

$$Z(s) = \frac{1}{2s^2 + 2s + 1} \tag{3.18}$$

Typically the input and output signals of the filter are both voltage, although other variables may be employed.

As a final example of unilateral components, consider the *linearized dynamics of a rocket* in the pitch plane. Here, the horizontal thrust at the nozzle of the rocket in the pitch plane is the only input. Denote this thrust by u. There are two outputs, the horizontal position of the rocket, y, and the pitch of the rocket, ϕ. The transfer function for this system component is a 2x1 rational matrix as follows.

$$Z(s) = \begin{bmatrix} \dfrac{mL/2}{[(m+M)I + mML^2/4]s^2 + m(m+M)g} \\[4ex] \dfrac{(m+M)}{[(m+M)I + mML^2/4]s^2 + m(m+M)g} \end{bmatrix} \tag{3.19}$$

where m is the mass of the rocket, M is the mass of its gimbel assembly, L is the length of the rocket, and I is its moment of inertia.

EXAMPLE 3.20: Unlike the above unilateral system components, the *passive components* in an *electrical network* and the components of many fluidic and economic systems are *bilateral*. The physics of a bilateral component does not prescribe which measurable signals are inputs and which are outputs. in such cases, one is more or less free to pick both the input and the output signals. For instance, a one farad *capacitor* is a bilateral component. It has three equivalent representations, each with a different set of input-output signals: its *impedance description* with transfer function $1/s$; its *admittance description* with transfer function s; and its *scattering description* with transfer function $s-1/s+1$. Similarly, *transistors* are bilateral components commonly characterized by a *hybrid matrix*. For these components, the choice of input and output signals used in defining the component's transfer function is up to the system analyst. In practice, the placement of the component within the interconnected system often fixes the choice of input and output signals. In any event, the notation Z(s) always depicts the component transfer function model and "a" and "b" denote its inputs and outputs, independent of their units. Indeed, even our notation is hybrid in nature since Z(s) is the usual notation for impedance in circuit theory, and "a" and "b"are the classical symbols for the scattering variables.[7]

Before concluding the section, we note that one can suitably expand the concept of a transfer function by allowing for *non-rational* Z(s). For example, the *shift theorem* of Laplace transform theory implies that e^{-sT} is the transfer function of a delay. While such components certainly arise in "real world" systems, their inclusion would complicate our theory without significantly adding to our understanding. Therefore our exposition drops any further discussion of such objects.

Finally observe that a rational function is analytic everywhere in the complex plane except at its poles. This fact notwithstanding, however, a rational function only coincides with the Laplace transform of a linear combination of exponentials a-la 3.14, in a *half plane*. Thus even though our theory deals with rational functions, whose properties are completely characterized by their values at a finite set of points (poles and zeros), the Laplace transform theory is valid only in a half plane.

Before jumping to the notion of *state*, a few remarks are in order. Hopefully these remarks will dispel any misgivings of the engineering student at the spectre of theorems, proofs, and theory.

The preceding sequence of lemmas and theorems has constructed an avenue of analysis, equivalent, in some sense, to the real world system. In particular, with each real world time function, one uniquely associates a complex valued function called its Laplace transform. Essentially this is an equivalence between time domain analysis and frequency domain analysis. Moreover, these theoretical statements have shown that a transfer function (matrix) can legitimately model a real world system. In addition, this transfer function model maps admissible input signals to admissible output signals. This is an obvious prerequisite if the formulation is to be of any value. Thus our immersion into the realm of theory is justifiable, for only a reasonable theoretical argument can legitimatize a methodology. Running a thousand "correct" examples only serves to verify that a formulation works for those specific examples. With these thoughts in mind let us begin consideration of the extremely important concept of state.

4. STATE MODELS

The *state model* of a *system component* is *a vector first order, ordinary differential equation* relating the component's input and output signals. However, rather than directly differentiating the output variables, it is convenient to deal with a vector of intermediary variables termed the *state variables* of the component. x will denote these state variables.

DEFINITION 4.1: A *state model* for a system component is a set of four matrices, (A,B,C,D) such that

$$\dot{x} = Ax + Ba$$

$$(4.2)$$

$$b = Cx + Da$$

where a and b are the vectors of the component input and output signals respectively, and x is the aforementioned intermediary (state) vector of admissible signals.

Observe that an *initial condition* need not be explicitly specified for the state vector x since if x is admissible it is zero in a neighborhood of $-\infty$, an implicit initial condition.

In many applications the state vector of a system component often has a generic physical interpretation, such as capacitor voltage or inductor current. However in a *passive electrical circuit*, capacitor charges and inductor fluxes may easily replace the capacitor voltage and inductor current of the other model.

EXAMPLE 4.3: The most natural state model for *the integrator* is

$$\dot{x} = [0] x + [1] a$$

$$(4.4)$$

$$b = [1] x$$

Here the state is the output. Equally valid, however, are the state models

$$\dot{x} = [0] x + [k] a$$

$$(4.5)$$

$$b = [1/k] x$$

wherein x = ka or

$$\begin{bmatrix} \dot{x}_1 \\ \dot{x}_2 \end{bmatrix} = \begin{bmatrix} 0 & 0 \\ 1 & 2 \end{bmatrix} \begin{bmatrix} x_1 \\ x_2 \end{bmatrix} + \begin{bmatrix} 1 \\ 0 \end{bmatrix} a \tag{4.6}$$

$$b \Big] = \begin{bmatrix} 1 & 0 \end{bmatrix} \begin{bmatrix} x_1 \\ x_2 \end{bmatrix}$$

These models demonstrate that not only x, but even its dimension is non-unique.

EXAMPLE 4.7: As with the integrator a *Butterworth filter* has several possible state models. A little algebra reveals that the second order filter has the state model

$$\begin{bmatrix} \dot{x}_1 \\ x_2 \end{bmatrix} = \begin{bmatrix} 0 & 1 \\ -1/2 & -1 \end{bmatrix} \begin{bmatrix} x_1 \\ x_2 \end{bmatrix} + \begin{bmatrix} 0 \\ 1/2 \end{bmatrix} a \tag{4.8}$$

$$b \Big] = \begin{bmatrix} 1 & 0 \end{bmatrix} \begin{bmatrix} x_1 \\ x_2 \end{bmatrix}$$

Paralleling the terminology used for transfer functions a state model is *strictly proper* if D = 0 and *proper* otherwise. The concept of an *improper state model* is discussed later. Similarly one may define the notion of *degree* of a state model. Here, if one lumps all the different state models of a given system into a single group, then the degree of any of these models is equal to the smallest (minimum) dimension of any of the state vectors in the group. Clearly this definition implicitly recognizes that the state model for a given system component is non-unique.

Now two state models are *equivalent* if each model defines the same mapping between the input "a" and the output "b"(though the x's may differ). Clearly equivalent models have identical degrees.

Before developing a kinship between the state and transfer function models of a component, we must confirm that *the state model is well defined*. Precisely, for each admissible "a", there must exist unique vectors of admissible signals, x and "b", satisfying equation 4.2. The method of attack first gives an explicit solution for x and "b" as per 4.2. Then we show that the resultant signal vectors are admissible. For this undertaking, it is necessary to define a *"matrix valued exponential function"*as follows.

DEFINITION 4.9: For any square matrix, A, its exponential is defined by the infinite series

$$\exp(At) = e^{At} = 1 + At + \frac{A^2 t^2}{2!} + \frac{A^3 t^3}{3!} + \frac{A^4 t^4}{4!} + \ldots$$

$$(4.10)$$

$$= \sum_{i=0}^{\infty} \frac{A^i t^i}{i!}$$

Mathematically speaking, a series is well defined (valid) if it "converges". By *convergence*, we mean the "distance" between partial sums of the series becomes smaller with higher order partial sums. Here, the analytic concept of *norm* is a measure of "distance". Thus, in order to show that the defining series of the matrix exponential converges it is necessary to define a *matrix norm*. For an nxn matrix A we define its norm to be

$$|A|_\infty = n \max_{i,j} |A_{ij}| \qquad (4.11)$$

where A_{ij} denotes the i,j entry of the matrix A.

The following proposition illustrates some properties of the matrix norm which are useful in showing that *the matrix exponential series converges*.

PROPOSITION 4.12: For nxn matrices A and B and a scalar c, the matrix norm $|\cdot|_\infty$ (there are other norms[4]) satisfies the following conditions:

(i) $|M|_\infty \geqslant 0$ with equality if and only if M = 0.

(ii) $|cM|_\infty = |c||M|_\infty$

(iii) $|M + N|_\infty \leqslant |M|_\infty + |N|_\infty$

$$(4.13)$$

(iv) $|MN|_\infty \leqslant |M|_\infty |N|_\infty$

The proof of these properties follows trivially from the fact that $|a + b| \leqslant |a| + |b|$ and $|ab| = |a||b|$ for scalars a and b, together with an appropriate manipulation of the entries of the matrices M and N. The proof[4] then is left as an exercise for the reader.

LEMMA 4.14: The *matrix exponential converges* for all nxn matrices, A, and real t and has the following *properties*:

(i) $\exp(At)\exp(As) = \exp(A(t+s))$; in particular
$$\exp(-At) = (\exp(At))^{-1}$$

(ii) $\dfrac{d(\exp(At))}{dt} = A\exp(At) = \exp(At)\dot{A}$

(iii) $|\exp(At)|_{\infty} \leqslant \exp(|At|_{\infty})$

PROOF: To show that the series of 4.10 converges, it suffices to show that the sequences S_{ij}^{k} corresponding to the ij-th entry in the matrices

$$S^{k} = \sum_{i=1}^{k} \frac{A^{i}t^{i}}{i!} \tag{4.15}$$

are each *Cauchy*. That is, $|S_{ij}^{k} - S_{ij}^{m}|$ goes to zero (uniformly) as k and m go to infinity. Now, 4.12 implies that

$$|S_{ij}^{k} - S_{ij}^{m}| \leqslant |S^{k} - S^{m}|_{\infty} \tag{4.16}$$

Thus it is sufficient to show that $|S^{k} - S^{m}|_{\infty}$ approaches zero as k and m go to infinity. Assuming without loss of generality, that $k > m$ we have

$$|S^{k} - S^{m}|_{\infty} = |\sum_{i=m+1}^{k} \frac{A^{i}t^{i}}{i!}|_{\infty} \leqslant \sum_{i=m+1}^{k} \frac{|At|_{\infty}^{i}}{i!} \tag{4.17}$$

Now

$$\sum_{i=m+1}^{k} \frac{|At|_{\infty}^{i}}{i!}$$

is the difference of partial sums for the expanison of the scalar function $\exp(|At|_{\infty})$. This must converge to zero as k and m go to infinity, since the scalar exponential is convergent.[10] This supplies an upper bound, which converges to zero, in the terms $|S^{k} - S^{m}|_{\infty}$. Consequently $|S^{k} - S^{m}|_{\infty}$ converges to zero as k, m go to infinity. Thus the matrix exponential series is convergent.

To verify property (i) multiply the series for $\exp(As)$ term by term by the series for $\exp(At)$ and then regroup terms with the same factorials in the denominator. Property (ii) follows from a term by term differentiation of the series for exp(At).

Finally to verify property (iii) observe

$$|\exp(At)|_\infty = \sum_{i=0}^{\infty} \frac{A^i t^i}{i!}{}_\infty \leq \sum_{i=0}^{\infty} \frac{|At|_\infty^i}{i!} \qquad (4.18)$$

$$= \exp(|At|_\infty)$$

Thus (iii) is verified.

With these preliminaries swept away, the following theorem legitimizes the state model as a characterization of a component. More precisely, this formulation maps *admissible inputs* to *admissible outputs*.

THEOREM 4.19: Let (A,B,C,D) be the state model for a system component. Then the vectors

$$x(t) = \int_{-\infty}^{t} \exp(A(t-q))Ba(q)dq \qquad (4.20)$$

and

$$b(t) = \int_{-\infty}^{t} C\exp(A(t-q))Ba(q)dq + Da(t) \qquad (4.21)$$

exist and are the unique admissible solutions to the state equations 4.2 for a given admissible input a.

PROOF: Assume "a" and "b" are scalars while the state x, remains a vector. The extension to "a" and "b" as vectors follows from *linearity*. As motivation, consider the scalar equation x(t) μ x(t). Clearly x(t) has the form $x(t) = e^{\mu t}$. With this in mind, define a function, g, by

$$g(t) = \begin{cases} C\exp(At)B & t \geq 0 \\ 0 & t < 0 \end{cases} \qquad (4.22)$$

Step 1: We show that b is admissible. For this we substitute g(t) into equation 4.21, yielding

$$b(t) = \int_{-\infty}^{\infty} g(t-\lambda)a(\lambda)d\lambda + Da(t)$$

$$= \int_{-\infty}^{\infty} g(\mu)a(t-\mu)d\mu + Da(t) \qquad (4.23)$$

where $\mu = t-\lambda$. By assumption the input $a(t)$ is admissible. By construction $g(t) = 0$ for $t < 0$. Also $g(t)$ is smooth for $t > 0$ since it is a uniformly convergent sum of polynomials (in t) each of which is smooth. Finally $g(t)$ is exponentially bounded by lemma 4.14.

Observe that equation 4.21 is a *convolution* formula. Therefore by lemma 3.3, the convolution $a * g = b$ exists and is admissible.

Step 2: Now we show x exists and is admissible, by demonstrating that each entry of x exists and is admissible. Now, the i-th entry in x is computed by the same formula as was $b(t)$ with the matrix, D, in 4.21 set to zero and C of 4.21 set equal to a row vector. This row vector has all of its entries zero except the i-th which is one.

Now the arguments of step 1 were independent of the specific properties of C and D. Using the same arguments as above we thus conclude that all entries of x exist and are admissible.

Step 3: Here we show x satisfies the differential state equation.

$$\dot{x} = Ax + Ba \qquad (4.24)$$

For this step use the formula for the derivative of a product of two functions and the fact that the derivative of an integral with respect to its upper limit is the integrand evaluated at this upper limit.[9] This yields

$$\dot{x} = \frac{d}{dt}\left[\int_{-\infty}^{t} e^{A(t-\lambda)} Ba(\lambda)d\lambda\right] = \frac{d}{dt}\left[e^{At} \int_{-\infty}^{t} e^{-A\lambda} Ba(\lambda)d\lambda\right]$$

$$= Ae^{At} \int_{-\infty}^{t} e^{-A\lambda} a(\lambda)d\lambda + e^{At}e^{-At}Ba(t) \qquad (4.25)$$

$$= Ax + Ba$$

where the last equality is legitimate by the identities of lemma 4.14. Finally by substitution into 4.20 and 4.21 we have $b = Cx + Da$.

Step 4: To show that x is the unique solution to 4.24, assume that z is an arbitrary solution (x is one such solution) satisfying

$$\dot{z} = Az + Ba \qquad (4.26)$$

Move the Az term to the left side of the equation and multiply both sides of the equations through by $\exp(-At)$. The non-singularity (lemma 4.14) of $\exp(-At)$ assures that it does not affect the solution space of 4.26. Employing the formula for the derivative of a product and lemma 4.14 we have

$$\frac{d}{dt}\left[e^{-At}z\right] = e^{-At}\dot{z} - e^{-At}Az = e^{-At}Ba \qquad (4.27)$$

Integrating both sides of 4.27 over the interval $r < q < t$ and evaluating the lefthand integral yields

$$e^{-At}z(t) - e^{-A}z(r) = \int_r^t \frac{d}{dq} e^{-Aq}A(q)dq$$

$$= \int_r^t e^{-Aq}B\dot{a}(q)dq \qquad (4.28)$$

Upon transferring the $e^{-Ar}z(r)$ term to the right side of 4.28 and multiplying through by e^{At}, we find that 4.28 is equivalent to

$$z(t) = e^{A(t-r)}z(r) + \int_r^t e^{A(t-q)}Ba(q)dq \qquad (4.29)$$

Invoking the identities of lemma 4.14, we observe that equation 4.29 holds independently of the choice of r. Consequently it holds in the limit as r goes to $-\infty$. Now since z is admissible, $z(r)$ is zero in a neighborhood of $-\infty$, accordingly the limiting equality is

$$z(t) = \int_{-\infty}^t e^{A(t-\lambda)}Ba(\lambda)d\lambda = x(t) \qquad (4.30)$$

showing that an arbitrary admissible solution of 4.24 coincides with x.

Finally since x is unique and b is explicitly dependent on x, b is also unique. This completes the proof.

As can be seen in the above proof the matrix exponential is fundamental to many manipulations of the state model. For this reason it is often given a special name, the *state translation matrix*, and denoted by $\Phi(t)$. When evaluating equation 4.20, one usually computes $\exp(At)$ numerically. However, analytic computation proves less tedious with the aid of the identity

$$\exp(At) = T\exp(Bt)T^{-1} \qquad (4.31)$$

whenever $A = TBT^{-1}$.

To verify this identity, one need only substitute $A = TBT^{-1}$ into equation 4.10, the matrix exponential series. Notice that all of the T's cancel out except two: one of these factors out to the left side of the series and one factor out to the right side, leaving the exponential series for exp(Bt) in the middle.

The specific computational procedure is to diagonalize A by a *similarity transformation* into the form $A = TDT^{-1}$ where D is the diagonal matrix of the *eigenvalues* of A—i.e. $D = diag(\lambda_1, \lambda_2, ..., \lambda_n)$ where λ_i are the n eigenvalues of A.[4] Consequently, exp(At) may be computed analytically via 4.31 and the observation that $exp(Dt) = diag(exp(\lambda_1 t), ..., exp(\lambda_n t))$. As an illustration of this method, consider the following example.

EXAMPLE 4.32: Regard the state model

$$(A,B,C,D) = \left(\begin{bmatrix} -1 & 0 \\ 2 & -2 \end{bmatrix}, \begin{bmatrix} 1 \\ 1 \end{bmatrix}, \begin{bmatrix} -3 & 0 \\ 1 & -2 \end{bmatrix}, \begin{bmatrix} 2 \\ 4 \end{bmatrix} \right) \quad (4.33)$$

A straight forward manipulation of the eigenvalues and *eigenvectors* of A will yield the diagonalization

$$A = \begin{bmatrix} -1 & 0 \\ 2 & -2 \end{bmatrix} = \begin{bmatrix} 1 & 0 \\ 2 & 1 \end{bmatrix} \begin{bmatrix} -1 & 0 \\ 0 & -2 \end{bmatrix} \begin{bmatrix} 1 & 0 \\ -2 & 1 \end{bmatrix} \quad (4.34)$$

Using the aforementioned identity,

$$exp(At) = Texp(Dt)T^{-1} = \begin{bmatrix} 1 & 0 \\ 2 & 1 \end{bmatrix} \begin{bmatrix} e^{-t} & 0 \\ 0 & e^{-2t} \end{bmatrix} \begin{bmatrix} 1 & 0 \\ -2 & 1 \end{bmatrix}$$

$$\quad (4.35)$$

$$= \begin{bmatrix} e^{-t} & 0 \\ 2(e^{-t}-e^{-2t}) & e^{-2t} \end{bmatrix}$$

Upon subsitution of 4.35 into 4.21, we obtain

$$\begin{bmatrix} b_1(t) \\ b_2(t) \end{bmatrix} = \int_{-\infty}^{t} \begin{bmatrix} -3e^{-(t-q)} \\ -3e^{-(t-q)}+2e^{-2(t-q)} \end{bmatrix} a(q)dq + \begin{bmatrix} 2 \\ 4 \end{bmatrix} a(t) \quad (4.36)$$

This is then the desired formula for computing the output of the system component characterized by the state model of 4.33.

At this point let us consider the derivation of a *frequency domain* description for a system component from its state space formulation. From theorem 4.19, if the input of a system component characterized by a state model is admissible, then both its output and state vector are uniquely defined admissible functions. Therefore the Laplace transform of the state equations defined by 4.2 is well defined. In equation form we have

$$s x(s) = Ax(s) + Ba(s)$$
$$b(s) = Cx(s) + Da(s) \qquad (4.37)$$

Solving the top equation in 4.37 for x(s) and substituting the result into the bottom equation yields:

$$b(s) = [C(sI-A)^{-1}B + D]a(s) \qquad (4.38)$$

Since (sI−A) is a matrix of polynomials in s, it's inverse is rational (verify this by expanding the inverse in determinants[4]). Clearly then

$$Z(s) = C(sI-A)^{-1}B + D \qquad (4.39)$$

is a *transfer function matrix for the component with state model* (A,B,C,D). In conclusion, the transfer function of a system component is readily computable from its state space model. *Converting from the transfer function to a state model* is far more complex. In the following we delineate one of several algorithms for achieving this end.

From a homework problem 7.22, the matrix $(sI-A)^{-1}$ is *strictly proper* implying that $C(sI-A)^{-1}B$ is strictly proper. This property guarantees that the limit of $C(sI-A)^{-1}$ as s approaches infinity is the zero matrix. In addition the transfer function $Z(s) = C(sI-A)^{-1}B + D$ is a *proper* rational matrix. Therefore the strict properness of $C(sI-A)^{-1}B$ and the properness of Z(s) insures that

$$Z(\infty) = \lim_{s \to \infty} [C(sI-A)^{-1}B + D] = D \qquad (4.40)$$

Clearly the D matrix of a state model may be computed trivially from the transfer function matrix.

To find the A,B, and C matrices observe that every proper *transfer function matrix* has the *decomposition*

$$Z(s) = \frac{N(s)}{d(s)} + Z(\infty) \qquad (4.41)$$

where d(s) is the least common multiple of the denominators of the various entries of Z(s) (assume d(s) is of order n); N(s) is a matrix of polynomials whose *order* is strictly less than n; and Z(∞) is the constant matrix as per 4.40. Since D = Z(∞), the task dwindles to finding matrices A,B, and C that

$$\frac{N(s)}{d(s)} = C(sI-A)^{-1}B \qquad (4.42)$$

For this endeavor expand both sides of equation 4.42 about infinity and choose matrices A,B, and C so that coefficients of like powers of $1/s$ coincide. In particular, let N(s)/d(s) have the *Laurent expansion*

$$\frac{N(s)}{d(s)} = \frac{M_1}{s} + \frac{M_2}{s^2} + \frac{M_3}{s^3} + \ldots$$
$$= \sum_{i=1}^{\infty} \frac{M_i}{s^i} \qquad (4.43)$$

Since the degree of the polynomial entries in N(s) is always less than n, such an expansion can invariably be carried out without positive or zero powers of s occurring, by *synthetic division*. The coefficients, M_i, of 4.43 are termed the *Markov parameters* of the system component.

In general a system component may have a series representation wherein the Markov parameters completely characterize the component. This point of view indicated that the set of Markov parameters is an alternate *system-component model*. Moreover since a rational function is a special form of a series representation, the *Markov parameter model* encompasses a broader class of components. However our specific intent is to use the concept as an intermediary step in computing a component's *state model*.

A *rational function* possesses a finite amount of information in that it has a finite number of coefficients. Thus the infinite set of Markov parameters, M_i , born of N(s)/d(s) as per 4.43 brandish some kind of degenerate behavior. The degeneracy takes the form of a recursive relationship among the series coefficients, the Markov parameters, as formalized in the following lemma.

LEMMA 4.44: Let N(s), d(s), and the M_i, $i \geqslant 1$, be as in 4.43 where d(s) is an n-th order polynomial

$$d(s) = \sum_{j=0}^{n} d_j s^j. \tag{4.45}$$

Then for all $k > n$

$$M_k = \sum_{r=1}^{n} \frac{-d_{n-r}}{d_n} M_{k-r} \tag{4.46}$$

PROOF: First multiply both sides of 4.43 by d(s), then group like powers of s to obtain

$$N(s) = d(s) \sum_{i=1}^{\infty} M_i s^{-i} = \sum_{j=1}^{n} d_j s^j \sum_{i=1}^{\infty} M_i s^{-i} \tag{4.47}$$

$$= \sum_{k=1}^{\infty} \left(\sum_{r=1}^{n} d_{n-r} M_{k-r} \right) s^{n-k}$$

with $M_i = 0$) for $i \leqslant 0$. Clearly the left side of 4.47 is a polynomial in s. Hence it has no negative powers of s present. The right side has potentially both positive and negative powers of s. Equality occurs only if the coefficients of all negative powers of s in the right hand side of 4.47 are zero. In other words

$$\sum_{r=0}^{n} d_{n-r} M_{k-r} = 0 \quad \text{for } k > n \tag{4.48}$$

which is equivalent to 4.46.

This lemma, knowledge of d(s), and the first n Markov parameters allows complete determination of the remaining Markov parameters, verifying their degeneracy.

The desired algorithm uses the lemma and the Markov parameters to construct A,B, and C matrices from the transfer function Z(s). In this vein define a *Hankel matrix*, H_i, to be a block matrix bred from the Markov parameters as follows:

$$H_i = \begin{bmatrix} M_i & M_{i+1} & M_{i+2} & \cdots & M_{i+n-1} \\ M_{i+1} & M_{i+2} & M_{i+3} & \cdots & M_{i+n} \\ \vdots & & & & \vdots \\ M_{p+n-1} & & \cdots & & M_{i+2(n-1)} \end{bmatrix} \qquad (4.49)$$

Also define A to be a block matrix in *rational cononical form* constructed from the coefficients of d(s) as below:

$$A = \begin{bmatrix} 0 & I & 0 & & 0 \\ 0 & 0 & I & & 0 \\ \vdots & & & & \vdots \\ 0 & 0 & 0 & & I \\ (\frac{-d_0}{d_n})I & (\frac{-d_1}{d_n})I & (\frac{-d_2}{d_n})I & & (\frac{-d_{n-1}}{d_n})I \end{bmatrix} \qquad (4.50)$$

where the sub-blocks are square and have dimension equal to the number of rows of each M_i. Direct computation [1,2] coupled with lemma 4.44 implies

$$AH_i = H_{i+1} \qquad (4.51)$$

Iterating this equation r times we conclude that

$$A^r H_i = H_{i+r} \qquad (4.52)$$

These statements complete the preliminary groundwork. The following theorem pieces these puzzle parts together resulting in the state model for a system component given its transfer function matrix.

THEOREM 4.53: Let Z(s) be a proper transfer function matrix for a system component. Then a state model for the component is (A,B,C,D) where A is defined by equation 4.50, $B = \text{col}(M_1, M_2, M_3,..., M_n)$,

$C = row(I, [0], [0], ..., [0])$, and $D = Z(\infty)$.

PROOF: By 4.39, the transfer function associated with (A,B,C,D) is

$$C(sI-A)^{-1}B + D = D + \sum_{i=1}^{\infty} CA^{i-1}Bs^{-i} \qquad (4.54)$$

where the expansion of the inverse of $(sI-A)$ follows from the identity $(1-X)^{-1} = \Sigma X^i$ which is valid for small X (i.e. for s in a neighborhood of infinity[10]). To prove the theorem it suffices to compare the expansion of $C(sI-A)^{-1}B+D$ as per 4.54 with the similar expansion of $Z(s)$ obtained from 4.41 and 4.43. Recall

$$Z(s) = Z(\infty) + \sum_{i=1}^{\infty} M_i s^{-i} \qquad (4.55)$$

By definition $D = Z(\infty)$, hence to complete the proof we show that $CA^{i-1}B = M_i$ for all $i \geq 1$. First note that $B = H_1 E$ where $E = col(I, 0, 0, ..., 0)$, and the square blocks in E have dimension equal to the number of columns in M_i. Consequently

$$CA^{i-1}B = CA^{i-1}H_1E = CH_iE = M_i \qquad (4.56)$$

Here "$CH_iE = M_i$" follows because pre-multiplication of a block matrix by C and post-multiplication by E picks out the $l-1$ block entry of the center matrix, H_i. By equation 4.49 this is equal to M_i. The proof is thus complete.

The derivation of this theorem is not nearly as complex as other procedures such as the ND^{-1} algorithm.[8] Moreover its implementation is quite straightforward: simply expand $Z(s)$ by synthetic division to compute $Z(\infty)$ and the first m Markov parameters, then form the required A,B,C, and D matrices.

EXAMPLE 4.57: Recall that the transfer function of a *Butterworth filter* is

$$Z(s) = \frac{1}{2s^2+2s+1} \qquad (4.58)$$

Via synthetic division, it has the expansion

$$\frac{1/2s^2 - 1/2s^3 + 1/4s^4}{2s^2+2s+1 \ \big| \ 1}$$

$$\frac{1+1/s+1/2s^2}{\big| \ -1/s-1/2s^2}$$

$$\frac{-1/s-1/s^2-1/2s^3}{\big| \ +1/2s^2+1/2s^3}$$

yielding $Z(\infty) = M_1 = 0$, $M_2 = 1/2$, $M_3 = -1/2$, $M_4 = 1/4$. Since $n = 2$ only the first two Markov parameters are needed. According to 4.53, the state model becomes

$$\left(\begin{bmatrix} 0 & 1 \\ -1/2 & -1 \end{bmatrix} , \begin{bmatrix} 0 \\ 1/2 \end{bmatrix} , \begin{bmatrix} 1 & 0 \end{bmatrix} , \begin{bmatrix} 0 \end{bmatrix} \right) \tag{4.59}$$

This model is identical to the state model for the Butterworth filter given in 4.8.

Theorem 4.53 implies that D is zero if and only if $Z(\infty) = 0$. Equivalently a system component's *transfer function is strictly proper if and only if its state model is strictly proper*. Our terminology, then, is compatible for both types of component models. This duality permits us to view the component itself as being "*strictly proper*" rather than the component's models; similarly for *proper* models. The terminology for *degree* is also dual with respect to the transfer function and state model formulations.

COROLLARY 4.60: For a single-input single-output proper component, the degree of its transfer function model (deg_{TF}) and the degree of its state model (deg_{SM}) coincide.

PROOF: For a single-input single-output proper transfer function the deg_{TF} is the order (degree) of its denominator polynomial. This is precisely the dimension of the A matrix fabricated in theorem 4.53, since the sub-blocks of A are lxl in this case. Consequently by definition, the deg_{SM} must be less than or equal to the degree of the transfer function. Thus

$$deg_{SM} \leq deg_{TF} \tag{4.61}$$

Conversely the definition of the degree of a state model implies there exists an A matrix with dimension (the number of rows or columns) equal to $\deg_{SM}$. Observe that each entry in $(sI-A)$ is either a 1-st or 0-th order polynomial. Upon computing

$$Z(s) = D + C(sI-A)^{-1}B \tag{4.62}$$

via expansion of the inverse by determinants[4], it becomes clear that $Z(s)$ has a denominator whose order is less than or equal to the dimension of A (equal to $\deg_{SM}$) implying

$$\deg_{TF} \leq \deg_{SM} \tag{4.63}$$

which in combination with 4.61 completes the proof.

Note that in the process of proving corollary 4.60 we have demonstrated

COROLLARY 4.64: For single-input single-output components the state model constructed in theorem 4.53 is of *minimal* possible *dimension*.

For multiple-input multiple-output components, the state model manufactured via theorem 4.53 is not of minimal dimension. Hence, an analog of corollary 4.60 fails to exist in such cases. It is for this very reason that we have avoided defining the degree of a multiple-input multiple-output transfer function. Such a degree concept, termed *McMillan degree*,[8] is formulated in terms of the orders of the various minors of the transfer function matrix and does coincide with the degree of the corresponding state model. Although the concept is extraneous to the sequel we state without proof[8] a theorem characterizing the degree of a multiple-input multiple-output component in terms of its Markov parameters. The actual degree is easily computed from either its transfer function or its state model.

THEOREM 4.65: For any proper multiple-input multiple-output component $\deg_{TF} = \deg_{SM} = \text{rank}(H_1)$ where $\deg_{TF}$ is taken to be the McMillan degree.

Before concluding this section we briefly outline some generalizations of the state model. First it is possible to define an *improper state model* (A, B, C, D, D_1, D_2 ,..., D_r) which characterizes a system component via the *ordinary differential equations*

$$\dot{x} = Ax + Ba$$

$$b = Cx + Da + D_1 \frac{da}{dt} + D_2 \frac{d^2a}{dt^2} + \ldots + D_r \frac{d^ra}{dt^r} \quad (4.66)$$

which are well defined when the appropriate derivatives of a are admissible. Taking Laplace transforms on both sides of equation 4.66 yields

$$C(sI-A)^{-1}B + D + D_1s + D_2s^2 + \ldots + D_rs^r \quad (4.67)$$

This *transfer function is improper* if the D_i, $i \geq 1$, are non-zero. Conversely, given an improper transfer function, it has the *continued fraction expansion*

$$Z(s) = P(s) + M_{-1}s + M_{-2}s^2 + \ldots + M_{-r}s^r \quad (4.68)$$

where $P(s)$ is proper. By identifying the matrices M_{-i} with D_i and computing A, B, C, and D from $P(s)$ one can now determine the improper state model corresponding to the improper transfer function. In practice the concept of an improper state model is more confusing than useful, so the sequel assumes that all state models are proper.

A final component model which finds occasional usefulness in our theory is the *Rosenbrock model*. Formulated in the *frequency domain*, the Rosenbrock model subsumes both the transfer function and the improper state models. This model, like the state model, uses an intermediate *frequency domain "state" variable*. The equations characterizing the component's input-output signals are

$$W(s)x(s) = U(s)a(s)$$
$$b(s) = V(s)x(s) + Y(s)a(s) \quad (4.69)$$

where $W(s)$, $U(s)$, $V(s)$, and $Y(s)$ are polynomial matrices.

Clearly this model embodies the improper state model. To verify this, simply take the Laplace transform on both sides of 4.66 and obtain $W(s) = (sI-A)$, $U(s) = B$, $V(s) = C$ and $Y(s) = D + D_1s + \ldots + D_rs^r$. Also to verify that 4.69 incorporates the improper transfer function component model simply factor a least common denominator out of $Z(s)$. Thus $Z(s) = N(s)/d(s)$ where unlike 4.41, $N(s)$ potentially has polynomial entries of order greater than the order of d(s). By letting $W(s) = d(s)I$, $U(s) = N(s)$, $V(s) = I$ and $Y(s) = 0$ one secures a Rosenbrock model for a system component from its transfer function.

Although the Rosenbrock Model is the foundation for an entire system theory,[8] this book works with the more familiar transfer function and state models. On occasion, however, the Rosenbrock model proves to be especially convenient for the study of interconnection phenomena. In such instances we freely cultivate its usefulness.

5. NONLINEAR STATE MODELS

For *nonlinear system components*, the Laplace transform enjoys little value, except for components which are *almost linear*.[3] Thus the *state model* is the dominant mode of analysis. Yet even the state model furnishes little analytical information about nonlinear system components. This fact forces an orientation towards their *numerical* analysis to permeate our investigation.

DEFINITION 5.1:   The *state model for a nonlinear system component* is a pair of functions, (f,g), characterizing the component's *input, output, and state variables* via the *ordinary differential equations:*

$$\dot{x} = f(x,a)$$
$$b = g(x,a) \tag{5.2}$$

In general the existence of admissible solutions, x and b, given the admissible input, a, requires extremely tight smoothness conditions on f and g.[5] Further complications arise from the lack of a viable global existence and uniqueness theory for nonlinear ordinary differential equations.[5] As such we postulate no definitive conditions on f and g to assure that the equations of 5.2 are well defined. Since our primary goal is the numerical analysis of 5.2, the smoothness of these functions plays a minor role in our theory. This grows from the methods of numerical analysis in which discrete samples are manipulated, thereby reducing the *smoothness requirements* between samples to a less significant role. Moreover, in practice, it is convenient to approximate the well defined f and g of a "real world" component by functions which are easier to manage, say *piecewise linear* but not smooth.

Following our notation for the linear case, a *nonlinear state model* such as in 5.2 is termed *proper* whereas a state model of the form

$$\dot{x} = f(x,a)$$
$$b = g(x)$$

(5.3)

is *strictly proper*.

EXAMPLES 5.4: Consider the *charge controlled capacitor* whose voltage is a function of the capacitor charge, $v = C(q)$. As usual, current is the rate of change of charge, $i = \dot{q}$. This yields the first order state equations

$$\dot{q} = [0]\, q + i$$
$$v = C(q)$$

(5.5)

with capacitor charge serving as the state variable. Similarly a *diode* is normally modeled by the exponential function

$$i = I(e^{kv} - 1)$$

(5.6)

This is a well defined *0-th order state equation*. Specifically the diode is a *memoryless* component characterized by a nonlinear function

$$b = g(a)$$

(5.7)

Essentially this is a state equation in which the zero order state vector is not explicitly written. Of course, there are moments when one desires to exploit the memoryless character of a component. Hence one would disregard the above viewpoint. In these cases system components are classified into two categories, *dynamical components* with non-trivial state equations and *memoryless components* with characterizations such as 5.7. Recall the linearized model of the pitch plane dynamics of a rocket given in 3.19. Small angular deviations from the vertical is all that normally occurs since most rockets could not survive the structural stresses resulting from large ϕ. Thus a *linearized model* is often satisfactory. However, elementary mechanics provides an exact fourth order nonlinear model for the *rocket dynamics* as follows.

$$
\begin{bmatrix} \dot{x}_1 \\ x_2 \\ x_3 \\ x_4 \end{bmatrix} = \begin{bmatrix} x_2 \\ A\sin(x_1) + Bx_2^2\sin(x_1) + C\cos(x_1)u \quad \bigg| \cos(x_1)u \\ x_4 \\ D\cos(x_1)\sin(x_1) + Ex_2^2\sin(x_1) + Fu \end{bmatrix}
$$

$$
= \quad f(x,u) \tag{5.8}
$$

$$
\begin{bmatrix} \phi \\ y \end{bmatrix} = \begin{bmatrix} x_1 \\ x_3 \end{bmatrix} = g(x,u)
$$

Here y is the horizontal position of the rocket in the *pitch plane*; ϕ is the pitch angle; and u is the horizontal thrust applied in the pitch plane. In this particular state model, the state variables represent y and ϕ and their derivatives. As in the linear case, other state variables are also possible.

Clearly such a set of nonlinear equations is difficult to analyze. Another example is the smooth resistance function of 5.6 which accurately models the diode. In practice it is often convenient to use the less accurate but more easily analyzed *piecewise linear model*

$$
i = \begin{cases} r_f v & v \geq 0 \\ r_b v & v < 0 \end{cases} \tag{5.9}
$$

Unlike the *diode model* of 5.6, the piecewise linear model of 5.9 fails to map admissible voltages to admissible currents since di/dv does not exist at zero. However, computationally speaking, it is far superior to the smooth model. Thus the practitioner often chooses to work with the piecewise linear model in spite of the fact that it is not well defined in our formal setting. For a single-input single-output component, piecewise linearity is a straightforward concept. For multi-input multi-output components, it is a less straightforward notion. Since we will be manipulating multi-input components, it is necessary to formalize the concept of a *piecewise linear function* mapping $\mathbf{R}^n$ to $\mathbf{R}^m$.

By a *hyperplane* in $\mathbf{R}^n$, we mean a set of vectors, x in $\mathbf{R}^n$, satisfying the equality

$$
h^t x = c \tag{5.10}
$$

where h is an arbitrary vector in $\mathbf{R}^n$ and c is a real constant. In $\mathbf{R}^1$ the hyperplanes are points; in $\mathbf{R}^2$ they are straight lines; in $\mathbf{R}^3$ they are planes; and in general hyperplanes are $(n-1)$ dimensional *linear manifolds* in $\mathbf{R}^n$. Essentially a set of hyperplanes partitions $\mathbf{R}^n$ into *polygonal regions*, say $S_1, \ldots, S_k$. Assuming such a partitioning of $\mathbf{R}^n$ via a set of hyperplanes, a *piecewise linear function* from $\mathbf{R}^n$ to $\mathbf{R}^m$ is defined to be a function of the form

$$y = A_i x + w_i \qquad (5.11)$$

for those x's in the region S_i, where A_i is an mxn matrix and w_i is an m-vector.

EXAMPLE 5.12: Although somewhat degenerate the concept of piecewise linearity is most easily illustrated in the one dimensional case. Consider the *hard limiter* whose input–output characteristic is shown in figure 5.13. Here there

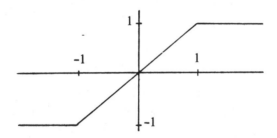

Figure 5.13. Input-output curve for hard limit.

are two hyperplanes corresponding to the points 1 and -1 of $\mathbf{R}^1$. The equalities, $1x = 1$ and $1x = -1$ ($h = 1$ and $c = 1$ for the point 1 and $h = 1$ and $c = -1$ for the point -1), characterize these two hyperplanes. They divide $\mathbf{R}^1$ into three regions: $\{S_1 = x: x \leqslant -1\}$, $S_2 = \{x: -1 \leqslant x \leqslant 1\}$, and $S_3 = \{x: 1 \leqslant x\}$. Thus for x in the appropriate region, equation 5.14 suitably characterizes the limiter.

$$y = \begin{cases} [0]x + -1 & \text{for x in } S_1 \\ [1]x + 0 & \text{for x in } S_2 \\ [0]x + 1 & \text{for x in } S_3 \end{cases} \qquad (5.14)$$

EXAMPLE 5.15: A more interesting if less applicable example of a piecewise linear function is illustrated in figure 5.20. This function maps $\mathbf{R}^2$ to $\mathbf{R}^2$ and is partitioned into 16 regions defined by six hyperplanes as shown in figure 5.16−c. Formally, the hyperplanes are defined by the equalities

$$[1 \ -1] \begin{bmatrix} x_1 \\ x_2 \end{bmatrix} = 1 \ ; \ [1 \ -1] \begin{bmatrix} x_1 \\ x_2 \end{bmatrix} = -1 \ ; \ [1 \ \ 0] \begin{bmatrix} x_1 \\ x_2 \end{bmatrix} = 0]$$

$$[1 \ \ 1] \begin{bmatrix} x_1 \\ x_2 \end{bmatrix} = 1 \ ; \ [1 \ \ 1] \begin{bmatrix} x_1 \\ x_2 \end{bmatrix} = -1 \ ; \ [0 \ \ 1] \begin{bmatrix} x_1 \\ x_2 \end{bmatrix} = 1] \tag{5.16}$$

whereas the piecewise linear function is defined by

$$\begin{bmatrix} y_1 \\ y_2 \end{bmatrix} = \begin{cases} \begin{bmatrix} -1 & -1 \\ 1 & 1 \end{bmatrix} \begin{bmatrix} x_1 \\ x_2 \end{bmatrix} + \begin{bmatrix} 1 \\ -1 \end{bmatrix} & \text{for x in } S_1 \\ \begin{bmatrix} 1 & -1 \\ -1 & 1 \end{bmatrix} \begin{bmatrix} x_1 \\ x_2 \end{bmatrix} + \begin{bmatrix} 1 \\ -1 \end{bmatrix} & \text{for x in } S_2 \\ \begin{bmatrix} 1 & 1 \\ -1 & -1 \end{bmatrix} \begin{bmatrix} x_1 \\ x_2 \end{bmatrix} + \begin{bmatrix} 1 \\ -1 \end{bmatrix} & \text{for x in } S_3 \\ \begin{bmatrix} -1 & 1 \\ 1 & -1 \end{bmatrix} \begin{bmatrix} x_1 \\ x_2 \end{bmatrix} + \begin{bmatrix} 1 \\ -1 \end{bmatrix} & \text{for x in } S_4 \end{cases} \tag{5.17}$$

and is zero in the regions S_5 through S_{16}.

Out of curiosity, compare the above A_i matrices, including those for which A_i is zero, for any two regions. One discovers that their difference is either zero or a rank one matrix. In general this is not true. However, if the piecewise linear function is *continuous* and the A_i matrices correspond to adjacent regions, this property does hold true, since *continuity* requires the value of the maps to be equal on their common boundary, an $(n-1)$ dimensional hyperplane. This fact greatly simplifies the analysis of *piecewise linear continuous functions,* making it a fundamentally important property.

THEOREM 5.18: For any continuous piecewise linear function, f mapping $\mathbf{R}^n$ to $\mathbf{R}^m$, let S_i and S_j be continuous regions with a common boundary characterized by the hyperplane $h^t x = c$. Then there exists an m-vector, r, such that

$$(A_i - A_j) = \underline{r h^t} \tag{5.19}$$

PROOF: Without loss of generality we may assume that h is a *unit vector*

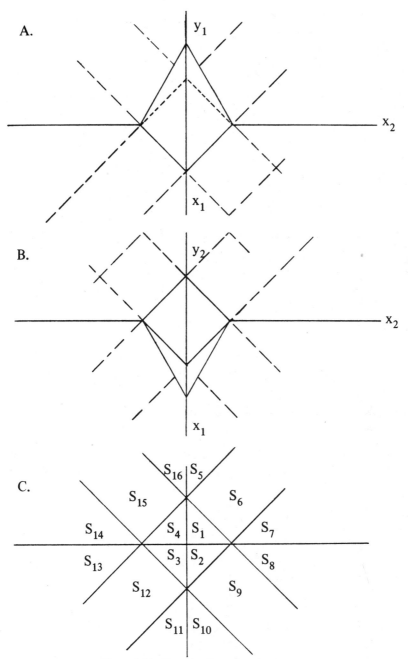

Figure 5.20. Piecewise linear function.

i.e. it has a Euclidean length of one. Let x_1 and x_2 be any two vectors in the hyperplane separating S_i and S_j. More precisely

$$h^t x_1 = h^t x_2 = c \qquad (5.21)$$

Since f is continuous the representations for $f(x)$ in S_i and S_j must coincide for both x_1 and x_2. Mathematically

$$A_i x_1 + w_i = A_j x_1 + w_j \qquad (5.22)$$

and

$$A_i x_2 + w_i = A_j x_2 + w_j \qquad (5.23)$$

Subtracting 5.23 from 5.22 yields

$$(A_i - A_j)(x_1 - x_2) = 0 \qquad (5.24)$$

Observe from 5.21 that

$$h^t(x_1 - x_2) = 0 \qquad (5.25)$$

Now consider an arbitrary vector z satisfying $h^t z = 0$. By letting

$$x_1 = z + \frac{ch}{h^t h} \quad \text{and} \quad x_2 = \frac{ch}{h^t h} \qquad (5.26)$$

we observe that z can be written as the difference of two vectors in the hyperplane separating S_i and S_j. As such the arguments leading to equation 5.24 imply that

$$(A_i - A_j) z = 0 \qquad (5.27)$$

whenever

$$h^t z = 0. \qquad (5.28)$$

Since the set of z in $\mathbf{R}^n$ satisfying 5.28 is an $n-1$ dimensional *subspace* of vectors perpendicular to h, we can construct an *orthonormal basis* for $\mathbf{R}^n$ composed of vectors $v_1...,v_n$ where $v_n = h$. Now every row of $(A_i - A_j)$ is the transpose of an n-vector, so every row is a linear combination of the transposes of the v_k, $k = 1, \ldots, n$. Thus $(A_i - A_j)$ may be expressed as

$$(A_i - A_j) = \sum_{k-1}^{n} c_k v_k^t \qquad (5.29)$$

where the c_k are column vectors. Since the basis v_p, $p = 1, \ldots, n$ is orthonormal and since $v_n = h$, then

$$h^t v_p = 0 \text{ for } p = 1, \ldots, n-1 \qquad (5.30)$$

$$0 = (A_i - A_j)v_p = \left[\sum_{k-1}^{n} c_i v_t^i \right] v_p = c_p, \ p = 1, \ldots, n-1$$

By 5.27

$$A_i x_1 + w_i = A_j x_1 + w_j \qquad (5.22)$$

where all the column vectors c_p, $p = 1, \ldots, n-1$ are zero. Letting $c_n = r$, equation 5.27 reduces to the desired result

$$(A_i - A_j) = rh^t. \qquad (5.31)$$

The proof is now complete.

With these preliminary notions in mind, define a *state model to be piecewise linear* if it is characterized by a pair of state equations (as per definition 5.2) in which both f and g are piecewise linear. As an example suppose that a capacitor has the magnitude of its voltage self limited to prevent breakdown. A possible piecewise linear characterization is

$$\dot{q} = i$$
$$v = C(q) \qquad (5.32)$$

where $C(\cdot)$ is the hard limiter function shown in Figure 5.13.

In addition to the classes of nonlinear components discussed here there are an abundance of other components arising in "real world" systems which are incapable of being modeled by a pair of state equations such as 5.2. These include *time varying* components, *improper* components, *distributed* components, *logical* components, *sampled data* and *discrete* time components, etc. Components modeled by nonlinear state equations, however, are sufficiently general to permit the exposition of our theory. Fortunately generalizations to these other classes of components are straightforward.[7,11] As such, we rarely deal with any component which fails to have a nonlinear state model such as 5.2, or its special cases: linear state equations and transfer functions.

6. COMPOSITE COMPONENT MODELS

An interconnected dynamical system is composed of myriad components, each of which has a mathematical model as discussed in the preceding sections. If n components have *transfer functions models*, then we may index the models as

$$b_i = Z_i a_i \quad i = 1, 2, \ldots, n \qquad (6.1)$$

where a_i is the vector of *input signals* for the i-th component; b_i is the vector of *output signals* for the i-th component; and Z_i is the transfer function (matrix) for the i-th component. Similarly, if the system components have a linear or nonlinear *state equation model*

$$\begin{aligned} \dot{x}_i &= A_i x_i + B_i a_i \\ b_i &= C_i x_i + D_i a_i \end{aligned} \quad i = 1, 2, \ldots, n \qquad (6.2)$$

or

$$\begin{aligned} \dot{x}_i &= f_i(x_i, a_i) \\ b_i &= g_i(x_i, a_i) \end{aligned} \quad i = 1, \ldots, n \qquad (6.3)$$

For numerical reasons it is often profitable to handle the n component equations of 6.1, 6.2, or 6.3 as separate entities. This permits one to store the different component models separately in a computer memory and to analyze them one at a time. From a theoretical point of view, however, it is convenient to lump the n component equations together forming a single *composite component model*. For composite component transfer function models we have

$$b \cdot = Za \qquad (6.4)$$

where $b = \text{col}(b_1, \ldots, b_n)$ is obtained by stacking the n different vectors of component outputs. Thus one forges a single *composite component output vector*. Similarly, $a = col(a_1, \ldots, a_n)$ is a *composite component input vector*, while the *composite* component transfer function is given by the matrix

$$
Z = diag(Z_1, \ldots, Z_n) = \begin{bmatrix} Z_1 & & & 0 \\ & Z_2 & & \\ & & \cdot & \\ & & & \cdot \\ 0 & & & & Z_n \end{bmatrix} \tag{6.5}
$$

Similarly, when using linear state models the *composite component state model* is

$$
\begin{aligned} \dot{x} &= Ax + Ba \\ b &= Cx + Da \end{aligned} \tag{6.6}
$$

where a and b are as above, the *composite component state vector*, x, is taken as $x = col(x_1, \ldots, x_n)$ and the four *composite component state matrices* are defined as $A = diag(A_1, \ldots, A_n)$, and similarly for B, C, and D. A straightforward matrix multiplication reveals[11] that 6.4 and 6.6 are simply concise expressions for the sets of simultaneous equations given in 6.1 and 6.2 respectively. As such, in developing our theory of interconnected dynamical systems, we will deal with these composite component models in lieu of the individual component models. However, this lumping together is done for notational convenience only. Normally in actual computer implementation of analysis and design algorithms component models are treated separately. In fact, it is this ability to deal separately with the individual system components which makes it possible to develop analysis and design algorithms for interconnected dynamical systems that are superior to the "classical" algorithms. The inadequacy of "classical" algorithms is that they deal with a single large system, thus lumping both *interconnection and component information* into a single set of equations.

Finally in the nonlinear case a *composite component nonlinear state model* takes the form

$$
\begin{aligned} \dot{x} &= f(x, a) \\ b &= g(x, a) \end{aligned} \tag{6.7}
$$

where the composite vectors, a, b, and x are as before and the composite functions are defined as $f = \text{col}(f_1, f, \ldots, f_n)$ and $g = \text{col}(g_1, g_2, \ldots, g_n)$.

EXAMPLE 6.8: Consider the network shown in figure 6.9.

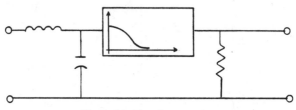

Figure 6.9. Electric network.

The network has four components: an *inductor, capacitor, resistor*, and *Butterworth filter*, characterized by the transfer functions

$$V_L(s) = [Ls]\, I_L(s) \tag{6.10}$$

$$I_c(s) = [Cs]\, V_c(s) \tag{6.11}$$

$$V_R(s) = [R]I_R(s) \tag{6.12}$$

and

$$V_{Out}(s) = \frac{1}{2s^2 + 2s + 1}\, V_{in}(s) \tag{6.13}$$

Note, the units for both the input and output variables used to characterize the network components differ from component to component. This is perfectly valid since the variables in our theory need not have a specific interpretation and may be hybrids of several types of variables. For this network, the composite component input vector is

$$a = \text{col}(I_L, V_C, I_R, V_{in}) \tag{6.14}$$

The composite component output vector is

$$b = \text{col}(V_L, I_C, V_R, V_{Out}) \tag{6.15}$$

Finally the composite transfer function matrix is

$$Z(s) = \text{diag}\left(Ls, Cs, R, \frac{1}{2s^2+2s+1} \right) \qquad (6.16)$$

Stacking everything together, the composite component equations become

$$\begin{bmatrix} V_L(s) \\ I_C(s) \\ V_R(s) \\ V_{out}(s) \end{bmatrix} = \begin{bmatrix} Ls & 0 & 0 & 0 \\ 0 & Cs & 0 & 0 \\ 0 & 0 & R & 0 \\ 0 & 0 & 0 & 1/2s+2s+1 \end{bmatrix} \begin{bmatrix} I_l(s) \\ V_C(s) \\ I_R(s) \\ V_{in}(s) \end{bmatrix} \qquad (6.17)$$

7. PROBLEMS

1. Show that the set of *admissible signals* form a *vector space* over the field of real numbers.
2. Show that the *derivative of an admissible signal* may fail to be admissible.
3. Let f be an admissible signal and let its translation f_T be defined by $f_T(t) = f(t-T)$. Show that the translation of an admissible signal is admissible.
4. Prove theorem 2.5 for *piecewise smooth "admissible" functions*.
5. Let $f(t)$ be an admissible function with *Laplace Transform* $f(s)$ and let $f^{(1)}(t)$ be its *derivative*. Show that the Laplace Transform of $f^{(1)}(t)$ is $sf(s)$. What is its *region of existence?*
6. Show that a *rational function*, f, is *strictly proper* if and only if $f(\infty)=0$.
7. Show that a rational function is *proper* if and only if $f(\infty)$ is finite and show that it is *improper* if and only if $f(\infty)$ is infinite.
8. Show that the *degree* of a rational function is equal to both the number of its *poles* and the number of its *zeros* (counting those at ∞ and multiple poles and zeros by their *multiplicity*).
9. Give a formula for computing the *residues* r_{ij} in equation 3.13.
10. Expand the rational function $s^2+s+1/s+1$ into the form required by equation 3.15.
11. Give a *time domain* description of the relationship between component input and output for the *transfer function* of problem 10.
12. Show that $1/s$ is the transfer function for an *integrator*.

13. Let (A,B,C,D) be a *state model* for a *system component* with *state vector* x and show that $(TAT^{-1}, TB, CT^{-1}D)$ is also a state model for the same component with state vector z = Tx where T may be any conformable non-singlular matrix.

14. Show that exp (A0) = I for any square matrix A.

15. Show that $_{exp}$ [diag($a_1, \cdots, a_{2)t}$] = diag($e^{a_1t}, e^{a_2t}, \cdots, e^{a_n^1t}$)

16. Compute

$$\exp\left(\begin{bmatrix} 1 & 0 \\ 1 & 0 \end{bmatrix} t\right)$$

17. Prove lemma 4.13.

18. Let B and C both be n by n matrices and t and s be real numbers. Show that $e^{Bt+Cs} = e^{Bt}e^{Cs}$ if and only if the matrices B and C commute.

19. Give an example to show that 7.18. fails with non-commuting matrices.

20. Verify that $e^{At} = T(e^{Bt})T^{-1}$ whenever $A = TBT^{-1}$.

21. Determine the *weighting matrix* $Ce^{A(t-q)}B$ associated with the state model

$$\left(\begin{bmatrix} 0 & 0 & 0 & 0 \\ 1 & 0 & 0 & 0 \\ 0 & -1 & 0 & 0 \\ 2 & 0 & 1 & 0 \end{bmatrix}, \begin{bmatrix} 1 \\ 2 \\ 3 \\ 4 \end{bmatrix}, \begin{bmatrix} 0 & 1 & 1 & 0 \\ 1 & 0 & 1 & 2 \end{bmatrix}, \begin{bmatrix} 0 \\ 0 \end{bmatrix}\right)$$

22. For any n by n matrix A show that $(sI-A)^{-1}$ is a *strictly proper rational matrix*.

23. Compute the *transfer function matrix* associated with the state model of problem 7.21.

24. Expand the transfer function $s+2/s^2+2s+2$ in powers of $1/s$.

25. Find a *recursive relationship* such as that of lemma 4.44 among the sequence of *Markov parameters* $(1,2,-1,1,2,-1,1,2,-1,\ldots)$.

26. Derive equation 4.51.

27. Determine a state model for the system component with transfer function matrix

$$Z(s) = \begin{bmatrix} \dfrac{s+1}{s+2} \\ \\ \dfrac{1}{s(s+1)} \end{bmatrix}$$

28. For multiple input multiple output transfer functions give an algorithm for *construction a state model* in which the dimension of x is rank(H_1).

29. Assuming the result of problem 28 show that the *degree* of a multiple-input – multiple-output component state model is equal to the *rank* of H_1.

30. Determine an improper state model for the system component with transfer function $s^4/s+1$.

31. Write state equations for a voltage controlled *nonlinear capacitor* and a flux controlled *nonlinear inductor*.

32. Show that the *hyperplanes* in R^3 are the set of all possible planes.

33. Show that the equation defining a *hyperplane* is non-unique. Can you characterize the degree of this non-uniqueness?

34. For the *piecewise linear function* of example 5.15 show that the rank of the difference of the A_i matrices for any two contiguous regions is less than or equal to one.

35. Show that the set of vectors, z, satisfying the equality $h^t z = 0$ where h is any non-zero n-vector is an $n-1$ dimensional *subspace* of R^n.'

36. Show that the n equations of 6.1 are equivalent to the single equation $B(s) = Z(s)a(s)$ where a is the *composite component input vector*, b is the *composite component output vector* and Z is the *composite component transfer function matrix*.

37. Write *composite component state equations* for the interconnected dynamical system with 3 components characterized by the state models

$$\left(\begin{bmatrix} 1 & -1 \\ 0 & -2 \end{bmatrix}, \begin{bmatrix} 0 \\ 2 \end{bmatrix}, \begin{bmatrix} 1 & 1 \end{bmatrix}, \begin{bmatrix} 1 \end{bmatrix} \right)$$

$$\left(\begin{bmatrix} 1 \end{bmatrix}, \begin{bmatrix} 1 & -3 \end{bmatrix}, \begin{bmatrix} 2 \end{bmatrix}, \begin{bmatrix} 0 \end{bmatrix} \right)$$

$$\left(\cdot , \cdot , \cdot , \cdot , \begin{bmatrix} 1 & 1 \\ 2 & 4 \end{bmatrix} \right)$$

where the dots in the state model for the third component denote the trivial zero dimensional matrices in the state model of a *memoryless* (zero degree) component.

8. REFERENCES

1. Brockett, R.W., *Finite Dimensional Linear Systems*, Wiley, New York, 1970.
2. Desoer, C.A., *Notes for a Second Course in Linear Systems*, Van Norstrand, New York, 1970.
3. Desoer, C.A., and M. Vidyasager, *Feedback Systems: Input Output Properties*, Academic Press, New York, 1975.
4. Gantmacher, F. R., *The Theory of Matrices*, Chelsea, New York, 1959.
5. Hale, J. K., *Ordinary Differential Equations*, Wiley, New York, 1969.
6. Lal, M., and M. E.Van Valkenburg, "Reduced-Order Modeling of Large-Scale Linear Systems", in *Large-Scale Dynamical Systems*, (ed. R. Saeks), Point Lobos Press, No. Hollywood, pp. 127-166, 1976.
7. Newcomb, R.W., *Linear Multivariable Synthesis*, McGraw-Hill, New York, 1966.
8. Rosenbrock, H.H., *State-Space and Multivariable Theory*, Wiley, New York, 1970.
9. Rudin, W., *Principles of Mathematical Analysis*, McGraw-Hill, New York, 1964.
10. Rudin, W., *Real and Complex Analysis*, McGraw-Hill, New York. 1967.
11. Saeks, R., *Generalized Networks*, Holt, Rinehart, and Winston, New York, 1972.
12. Sage, A.P., and J.L. Melsa, *System Identification*, Academic Press, New York, 1971.
13. Titchmarsh, E.C., *The Theory of Functions*, Oxford Univ. Press, London, 1932.

II. CONNECTIONS

1. INTRODUCTION

By intentionally ignoring the internal "physics" of a system component the previous chapter presented a policy of modeling a system component as a set of equations describing the component's input-output or input-state-output relationship independently of its actual structure. This begs the question of whether or not system theory is merely a subset of the theory of equations. Surely if one were dealing with a single component, this would be the case. However, a system is built up by interconnecting a (possibly very large) number of components. The investigation of the interrelationship between these various components is the heart of the theory.

The purpose of the present chapter is to formulate mathematical models of the system connectivity structure to complement the component models already developed. Such a mathematical model must satisfy several conflicting constraints. First, it must be *well defined* for most "real world" systems although, like the component model, our goal is to expose the fundamentals of the theory of interconnected dynamical systems and not to promulgate mathematical generalities. Secondly, if a viable theory is to be developed, the connection model should be amenable to *analytic manipulation*. Finally, since any "real world" system theory must ultimately be computer implemented, the *connection model* should lend itself to *efficient computerization*. This latter constraint is of fundamental importance since any system complex enough to require theoretical analysis will most certainly be too complex to permit a hand implemented theory.

Physically, the connections in a system usually manifest themselves in the form of some type of *conservation law* among the component input and output variables. For economic variables this is usually some type of *continuity equation* while in an electric network it is the *Kirchhoff laws*; and in a mechanical system the connection constraints are the *kinematic equations*. In practice, system geometry often determines the nature of these equations: the *linkage configuration* in a mechanical system, the *printed circuit board layout* in an electrical network, etc. As such, most connection models have, historically, been geometric in nature. These include the *schematic diagram* of an electric network, the *pert diagram* of an industrial process, the *blueprint* of a mechanical linkage or structure, and their abstractions such as the *linear graph* of an electric network, the *block diagram* and *signal flow graph* of an aerospace system, and the *bond graph* of a mechanical system.[2,9]

To date most research in the theory of interconnected dynamical systems has assumed some type of geometric connection model. In practice these models are always reduced to a set of linear algebraic equations when converting the theoretical results to a computer algorithm. Thus the central theme of the present chapter is the development of a *algebraic connection model* which absorbs the various *geometric connection models* while simultaneously satisfying the constraints of existence and amenability to both analytic manipulation and computer implementation. Such a connection model is described in the following section while its relationship to the common geometric connection models is discussed in section 3. Section 5 is devoted to a study of certain commonly encountered connectivity configurations, termed *hierarchical structures*. Section 6 develops a generalization of the algebraic connection model which proves useful in certain situations.

The key concept which makes a theory of interconnected dynamical systems possible is *sparsity*. Even though a system may have a large number of components, in practice it is found that each component is only connected directly to a small number of other components. Note however, that each component may affect the signals measured at the other components indirectly through a chain of components to which the particular component is connected. In the context of the algebraic connection model this sparsity manifests itself as *sparse connection matrices*, i.e., connection matrices in which most (often over 99%) of the entries are zero. Section 4, therefore, discusses the concept of *sparse matrices* and, in particular, techniques which allow the manipulation and storage of such matrices on a computer. It is often possible to develop algorithms whose complexity depends on the number of nonzero entries of a sparse matrix rather than on the total number of entries. Such techniques permit the analysis of extremely large systems on computers of reasonable size and speed. Thus they are the key to the practical implementation of our theory.

2. AN ALGEBRAIC CONNECTION MODEL

Intuitively, an interconnected dynamical system comprises a set of components, characterized by one of the models of Chapter I, interconnected by a set of linear algebraic constraints on their input-output variables. The assumption that the connection constraints are linear and algebraic is not universal. Fortunately, it is sufficient for many situations including Kirchhoff laws, block diagram equations, classical kinematics, continuity equations, etc. In addition this hypothesis is fundamental to the development of a viable theory.

Figures 2-a and 2-b symbolically depict a typical interconnected dynamical system. Here the system components have the *composite component model*, Z, with *composite component input vector*, a, and *composite component output vector*, b. The letter, Z, symbolically represents any of the *composite component models* described in Chapter I. The "donut shaped" region surrounding the composite component model in figure 2-a depicts the system connectivity structure. This connectivity structure characterizes such devices as adders, scalers, soldered wires, traffic intersections, pipes, circulators, etc. Finally, the *composite system input vector*, u, represents the external inputs to the *composite system* while y, the *composite system output vector*, characterizes the outputs from the composite system to the external world.

In general the only constraint on the system connectivity structure is that it satisfy a set of linear algebraic equations. The natural form for the algebraic connection model becomes apparent after redrawing figure 2-a as figure 2-b. This picture illustrates the decoupling of the overall system into into two distinct black boxes. The box denoted by Z is the composite component model with its corresponding composite component input vector, a, and composite component output vector, b.

$$b = Za \qquad\qquad (2.1)$$

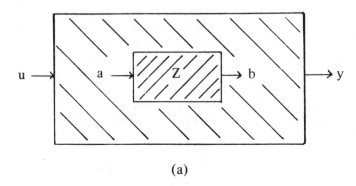

(a)

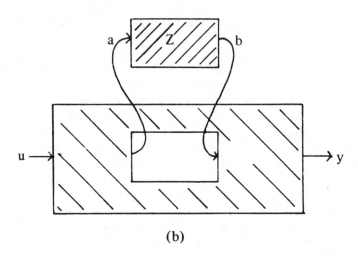

(b)

Figure 2.2. Symbolic representation of interconnected dynamical system.

The "donut shaped" box simulates the system connections. The inputs to the "donut" are the composite system input vector, u, and the composite component output vector, b, while the outputs from the connection subblock are the composite system output vector, y and the composite component input vector, a. Recall that by hypothesis, the connection subblock inputs and outputs satisfy a linear algebraic constraint. The mathematical expression for this phenomenon is thus

$$\left[\frac{a}{y}\right] = \left[\begin{array}{c|c} L_{11} & L_{12} \\ \hline L_{21} & L_{22} \end{array}\right] \left[\frac{b}{u}\right] \tag{2.3}$$

where the L_{ij} are real matrices of appropriate dimension. Again we emphasize that this construction is not always valid, yet it holds for numerous "real world" systems and is necessary for a viable theory.

The pair of equations, 2.2 and 2.3, comprise the *component connection model* for an interconnected dynamical system. This model is the foundation of the theory developed in the subsequent chapters.

Although more succinct formulations of an interconnected dynamical system exist (such as a *state model* which relates u and y) the component connection model effectively isolates the physics of the particular devices from their interconnections. The engineer can then readilly exploit this division of information both analytically and computationally. Essentially the model decouples the component information so that the behavior of one component does not mask the behavior of another component. This isolation is especially important for nonlinear systems in which the characteristics of a particular component are tractable whereas the overall system (input-output) nonlinearities are a hellish complexity. Futhermore the *connection matrices* are usually *sparse*. This provides an avenue for computationally efficient analysis and design procedures. Finally, the component connection model naturally divides the system into two sets of equations: *component equations* characterized by (block) decoupled differential equations, and the *connection equations* characterized by coupled linear algebraic equations.

Fortunately the notion of the component connection model is not a radical departure from traditional concepts but rather is a crystalization of both current and past ideas. The following "physical interpretation" indicates this point of view while simultaneously providing intuition and understanding to the model.

First consider the vector-matrix block diagram[5] of figure 2.4. By writing the equations characterizing the output of the two adders of the block diagram, it is clear that this diagram is equivalent to equations 2.1 and 2.3. Correlating this fact with the traditional *feedback control model*, one may view the component connection model as a *feedback system* where the components play the role of the *plant* and the connection matrices the role of the *compensators*. Explicitly, L_{11} represents a feedback matrix from the output of a component to the input of either another component or itself; L_{21} and L_{12} are post and precompensators while L_{22} is a feedforward gain. This is just one of several viewpoints which may prove beneficial in the sequal.

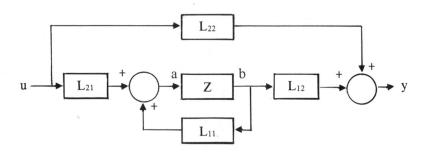

Figure 2.4. Vector-matrix block diagram.

A second point of view ensues by restricting all components to be *integrators*. All component equations then reduce to

$$a = \dot{b} \tag{2.5}$$

Substituting this into the connection equation (2.3) yields

$$\dot{b} = L_{11}b + L_{12}u$$

$$y = L_{21}b + L_{22}u \tag{2.6}$$

This is precisely the usual (A,B,C,D) *state model* in which $A = L_{11}$, $B = L_{12}$, $C = L_{21}$, and $D = L_{22}$. At first glance this conclusion may appear paradoxical since one commonly interpretes the state equations as modeling system dynamics but not system connections. Yet the implications of equations 2.5 and 2.6 indicate that the state model is a formula for interconnecting the components of a dynamical system whose components are all integrators. Surely·then, the proposed component connection model naturally extends the notion of a state model by permitting the interconnection of arbitrary components.

EXAMPLE 2.7:More than likely, the *block diagram* is the most commonly found model of system interconnections. Figure 2.8 is a typical example. The concept

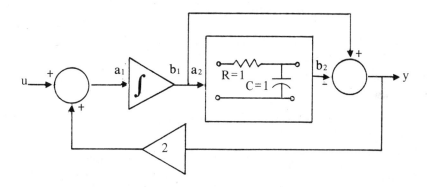

Figure 2.8. Block diagram example.

is self explanatory with the output of a component or the output of an adder either feeding the input of the same or another component or acting as a system output.

Clearly, the block diagram of figure 2.8 depicts an interconnected dynamical system with two components (the scalar in the feedback loop is lumped into the connectivity structure). One component is an integrator with transfer function $1/s$ while the other is an RC filter with transfer function $1/(s+1)$.

The composite component equations for this system are

$$
\begin{bmatrix} b_1 \\ b_2 \end{bmatrix} = \begin{bmatrix} 1/s & 0 \\ 0 & 1/(s+1) \end{bmatrix} \begin{bmatrix} a_1 \\ a_2 \end{bmatrix}
\tag{2.9}
$$

To construct the connection equations consider the output of each of the adders. Here

$$
\begin{aligned}
a_1 &= u + 2y \\
y &= b_1 - b_2
\end{aligned}
\tag{2.10}
$$

and considering that component 1 directly feeds component 2

$$
a_2 = b_1
\tag{2.11}
$$

Rearranging and putting into matrix form results in

$$
\begin{bmatrix} a_1 \\ a_2 \\ \hline y \end{bmatrix} = \left[\begin{array}{cc|cc} 2 & -2 & 1 & b_1 \\ 1 & 0 & 0 & b_2 \\ \hline 1 & -1 & 0 & u \end{array} \right]
\tag{2.12}
$$

The two sets of equations (equations 2.9 and 2.12) constitute the component connection model for figure 2.8.

EXAMPLE 2.13: Fortunately most "real world" systems admit a component connection model derivable from their block diagrams. Unfortunately, one may concoct *pathological systems* which fail to admit a component connection model. Figure 2.14 mirrors such a system. The

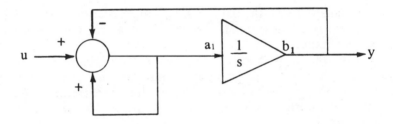

Figure 2.14. System with no component connection model.

block diagram of figure 2.14 displays a single component characterized by

$$
b_1 = (1/s)a_1
\tag{2.15}
$$

The pathology arises from the impossibility of writing a_1 explicitly as a linear combination of b_1 and u as the equation characterizing the adder clearly examplifies.

$$
a_1 = a_1 + u - y
\tag{2.16}
$$

Equation 2.16 is not solvable for a_1. Ostensibly systems not having a well defined solution reflect such pathological connections as figure 2.17 also shows. Here the only admissible

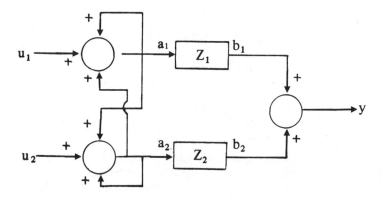

Figure 2.17. Block diagram of pathological connections.

input-output pair is $u_1 = u_2 = 0$ and $y = 0$. However, this is not always the case. The system depicted in figure 2.14 admits the solution $y = u$, $b_1 = u$, and $a_1 = u$ for all admissible signals u. Still this system lacks a well defined component connection model.

EXAMPLE 2.18: Another large class of interconnected dynamical systems is the ever present electrical network usually represented as a *schematic diagram*. Figure 2.19

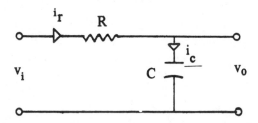

Figure 2.19. Schematic diagram of an electric network.

is a typical example. Unlike the components of a block diagram, the components of an electrical network are customarily *bilateral* with active devices serving as the major exceptions. This fact begets a degree of ambiguity in choosing a component's input or output.[6] A good rule of thumb is to choose the input and output of a component so as to simplify the formulation of the connection equations. For instance, given the RC filter of figure 2.19, if one models the resistor as an admittance and the capacitor as an impedance, the resultant composite component equations take the form

$$\begin{bmatrix} i_r \\ v_c \end{bmatrix} = \begin{bmatrix} 1/R & 0 \\ 0 & 1/Cs \end{bmatrix} \begin{bmatrix} v_r \\ i_c \end{bmatrix} \qquad (2.20)$$

while the connection matrix has the form

$$\begin{bmatrix} v_r \\ i_c \\ \overline{v_o} \end{bmatrix} = \left[\begin{array}{cc|c} 0 & -1 & 1 \\ 1 & 0 & 0 \\ \hline 0 & 1 & 0 \end{array} \right] \begin{bmatrix} i_r \\ v_c \\ \overline{v_i} \end{bmatrix} \qquad (2.21)$$

Here, the first and third equations in 2.21 represent the Kirchhoff voltage laws of the two loops of the circuit including source and load. The second equation is the Kirchhoff current law at the "tee" under the assumption that there is no load current.

Since the components of this circuit are bilateral, an equally valid component connection model is

$$\begin{bmatrix} v_r \\ i_c \end{bmatrix} = \begin{bmatrix} R & 0 \\ 0 & sC \end{bmatrix} \begin{bmatrix} i_r \\ v_c \end{bmatrix} \qquad (2.22)$$

and

$$\begin{bmatrix} i_r \\ v_c \\ \overline{v_o} \end{bmatrix} = \left[\begin{array}{cc|c} 0 & 1 & 0 \\ -1 & 0 & 1 \\ \hline -1 & 0 & 1 \end{array} \right] \begin{bmatrix} v_r \\ i_c \\ \overline{v_i} \end{bmatrix} \qquad (2.23)$$

In general, *passive electrical networks*, systems with bilateral components such as *traffic and economic systems* and *fluidic devices* reflect this type of behavior. Rarely is it encountered in *unilateral* systems.

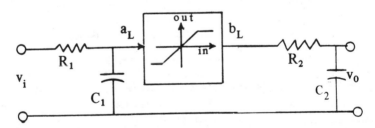

Figure 2.24. RC coupled limiter.

EXAMPLE 2.25: Although less amenable to analysis many interconnected dynamical systems contain a hybrid mixture of bilateral and unilateral devices. For instance, the *RC coupled limiter* of figure 2.24 has the following set of component connection equations

$$
\begin{bmatrix} i_{r_1} \\ v_{c_1} \\ i_{r_2} \\ v_{c_2} \\ b_L \end{bmatrix} = \begin{bmatrix} 1/R_1 & 0 & 0 & 0 & 0 \\ 0 & (1/C_1)\int & 0 & 0 & 0 \\ 0 & 0 & 1/R_2 & 0 & 0 \\ 0 & 0 & 0 & (1/C_2)\int & 0 \\ 0 & 0 & 0 & 0 & f \end{bmatrix} \begin{bmatrix} v_{r_1} \\ i_{c_1} \\ v_{r_2} \\ i_{c_2} \\ a_L \end{bmatrix} \tag{2.26}
$$

and

$$
\begin{bmatrix} v_{r_1} \\ i_{c_1} \\ v_{r_2} \\ i_{c_2} \\ a_L \\ \hline v_o \end{bmatrix} = \left[\begin{array}{ccccc|c} 0 & -1 & 0 & 0 & 0 & 1 \\ 1 & 0 & 0 & 0 & 0 & 0 \\ 0 & 0 & 0 & -1 & 1 & 0 \\ 0 & 0 & 1 & 0 & 0 & 0 \\ 0 & 1 & 0 & 0 & 0 & 0 \\ \hline 0 & 0 & 0 & 1 & 0 & 0 \end{array} \right] \begin{bmatrix} i_{r_1} \\ v_{c_1} \\ i_{r_2} \\ v_{c_2} \\ b_L \\ \hline v_i \end{bmatrix} \tag{2.27}
$$

Here the limiter is a unilateral component interpreted as a voltage transducer with limiting characteristic f, infinite input impedance and zero output impedance. Because of the presence of the nonlinearity, a time domain representation for all of the system components is more appropriate. Of course, the integrators can be equivalently represented by first order state models.

3. GEOMETRIC CONNECTION MODELS

The purpose of this section is to probe the kinship between the *algebraic connection model* and two of the more common graphical representations of a system. First we consider the signal flow graph as an abstraction of the block diagram. Basically the signal flow graph embodies most of the *unilateral connection models*. Secondly we will investigate the affinity between the algebraic connection model and the linear graph which basically represents *bilateral connection models*.

A *signal flow graph* is a directed graph composed of oriented line segments, termed *edges*, which interconnect a set of points termed *vertices*. Parallel edges, edges running in opposite directions between the same vertices and edges starting and ending on the same vertex, are all permitted.

Each has a physical significance in the signal flow graph. Figure 3.1a illustrates a typical signal flow graph. Physically an edge models a component, Z_i,

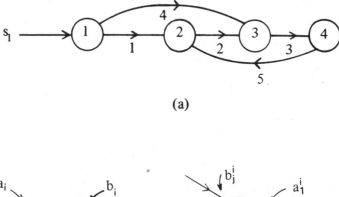

(a)

(b)   (c)

Figure 3.1. (a) Signal flow graph. (b) i-th edge as a component. (c) i-th vertex as an adder.

with the initial part of the edge representing the admissible component input, a_i, and the terminal part of the edge being the admissible output, b_i. Symbolically the individual component models take the form

$$b_i = Z_i a_i \qquad (3.2)$$

Where Z_i represents any of the possible component models developed in chapter 1. The *composite component model* for the interconnected dynamical system is

$$b = Za \qquad (3.3)$$

where $b = \text{col}(b_i)$, $a = \text{col}(a_i)$, and $Z = \text{diag}(Z_i)$.

The vertices of a signal flow graph are the abstractions of the physical notion of an *adder*. With this line of reasoning, we associate with the i-th vertex a *vertex signal* v_i, where v_i is the sum of all the component outputs entering the i-th vertex plus an *external input*, s_i. Of course, s_i may be taken as zero. Clearly then, any edge which leaves the i-th vertex has its component input equal to the i-th vertex signal v_i. Such an edge models nothing more than a component connected to the output of an adder. Finally the *composite input vector* to a system, modeled by the signal flow graph, may be taken as $s = \text{col}(s_i)$ or some subvector of s. Also the *composite system output vector* is $v = \text{col}(v_i)$ or some subvector of v.

EXAMPLE 3.4: For the signal flow graph of figure 3.1, the four summation equations are

$$
\begin{aligned}
v_1 &= s_1 \\
v_2 &= b_1 + b_5 \\
v_3 &= b_2 + b_4 \\
v_4 &= b_3
\end{aligned}
\qquad (3.5)
$$

whereas the equations characterizing the component inputs are $a_1 = v_1$, $a_2 = v_2$, $a_3 = v_3$, $a_4 = v_1$, and $a_5 = v_4$.

As with most diagrams which give rise to strings of similar equations, a vector matrix notation quite adequately consolidates them. In this vein we define the following *incidence matrices*.

DEFINITION 3.6: Let a signal flow graph have "p" edges and "n" vertices. Then we define its *incoming incidence matrix* A_+, to be the nxp matrix which has a "1" in the ij-th entry if edge j enters vertex i and is "0" elsewhere.

DEFINITION 3.7: Let a single flow graph have "p" edges and "n" vertices. The *outgoing incidence matrix*, A_-, is the nxp matrix having a "1" in the ij-th entry if edge j leaves vertex i and is "0" elsewhere.

Both incidence matrices are highly specialized with a single one in each column and all other entries zero.

EXAMPLE 3.8: The incoming and outgoing incidence matrices for the signal flow graph of figure 3.1 are

$$
A_+ = \begin{bmatrix}
0 & 0 & 0 & 0 & 0 \\
1 & 0 & 0 & 0 & 1 \\
0 & 1 & 0 & 1 & 0 \\
0 & 0 & 1 & 0 & 0
\end{bmatrix}
\qquad (3.9)
$$

and

$$A_- = \begin{bmatrix} 1 & 0 & 0 & 1 & 0 \\ 0 & 1 & 0 & 0 & 0 \\ 0 & 0 & 1 & 0 & 0 \\ 0 & 0 & 0 & 0 & 1 \end{bmatrix} \qquad (3.10)$$

These highly specialized incidence matrices completely characterize the algebraic constraints on a signal flow graph. The following lemma formalizes this fact.

LEMMA 3.11: Let a be the vector of component inputs, b be the vector of components outputs, v be the vector of vertex signals, and s be the vector of external inputs, all for a signal flow graph with incoming and outgoing incidence matrices, A_+ and A_-. Then

$$v = A_+b + s \qquad (3.12)$$

and

$$a = A_-^t v \qquad (3.13)$$

where "t" denotes matrix transpose.

PROOF: First consider equation 3.12. Here "s" is a composite input vector whose i-th entry is s_i which corresponds to an external input entering vertex i. Note further that the i-th row of A_+ has ones only in those columns which correspond to edges entering vertex i. Since each vertex signal v_i is nothing more than the sum of the signals entering vertex i, $v = A_+v + s$.

Now consider equation 3.13. A_- has a one in the ij-th entry if edge j leaves vertex i. Since an edge can leave at most one vertex, each column of A has a single one with the remaining entries of the column equal to zero. Clearly then a_j equals the transpose of the j-th column of A_- times v, thus picking out v_j. In matrix notation $a = A_-^t v$. The proof is now complete.

Suppose by chance we multiply equation 3.12 by A_-^t. This yields

$$a = A_-^t v = A_-^t A_+ b + A_-^t s \qquad (3.14)$$

which in combination with 3.12 is precisely the set of algebraic connection equations for a system defined by a signal flow graph. Adjoining equation 3.14 with lemma 3.11 yields the following theorem.

THEOREM 3.15: Let an interconnected dynamical system be characterized by a signal flow graph whose i-th component has the model

$$b_i = Z_i a_i \tag{3.16}$$

and whose j-th vertex is characterized by the vertex signal, v_j and has external input, s_j. then the system is characterized by the *component connection model*

$$b = Za \tag{3.17}$$

$$
\begin{bmatrix} a \\ \hline v \end{bmatrix} =
\begin{bmatrix} A_-^t A_+ & A_-^t \\ \hline A_+ & I \end{bmatrix}
\begin{bmatrix} b \\ \hline s \end{bmatrix} \tag{3.18}
$$

where $v = \text{col}(v_i)$, $s = \text{col}(s_i)$, A_+ and A_- are the incoming and outgoing incidence matrices, and Z, a, and b are the composite component matrices or vectors.

This theorem is rather general in that it assumes external inputs are applied to the system at all vertices and all vertex signals are taken as outputs. In practice inputs are applied to a small number of vertices and only a small number of vertex signals are taken as outputs. This implies that the connection matrix for an actual system is just a submatrix of the matrix of equation 3.18. In reducing equation 3.18, the L_{11} matrix remains unchanged. As we will see, L_{11} is by far the most important of the connection matrices.

EXAMPLE 3.19: The block diagram of figure 3.21a characterizes a nonlinear system equivalently represented as a signal flow graph of figure 3.21b. Since a wire has no direct representation in a signal flow graph, an artificial component with gain "-1" is used to model the feedback loop in figure 3.21a. Thus the composite component model for the system will have the form

$$
\begin{bmatrix} b_1 \\ b_2 \\ b_3 \end{bmatrix} =
\begin{bmatrix} f & & \\ & \int & \\ & & -1 \end{bmatrix}
\begin{bmatrix} a_1 \\ a_2 \\ a_3 \end{bmatrix} \tag{3.20}
$$

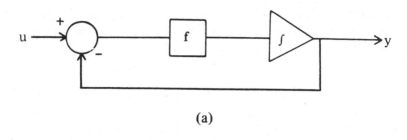

(a)

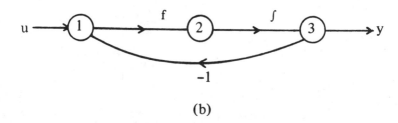

(b)

Figure 3.22. System characterized by a block diagram and its equivalent representation as a signal flow graph.

while the incidence matrices for the system are

$$A_+ = \begin{bmatrix} 0 & 0 & 1 \\ 1 & 0 & 0 \\ 0 & 1 & 0 \end{bmatrix} \tag{3.22}$$

and

$$A_- = \begin{bmatrix} 1 & 0 & 0 \\ 0 & 1 & 0 \\ 0 & 0 & 1 \end{bmatrix} \tag{3.23}$$

Consequently the algebraic connection model for the system has the form

$$\begin{bmatrix} a_1 \\ a_2 \\ a_3 \\ \overline{v_1} \\ v_2 \\ v_3 \end{bmatrix} = \left[\begin{array}{ccc|ccc} 0 & 0 & 1 & 1 & 0 & 0 \\ 1 & 0 & 0 & 0 & 1 & 0 \\ 0 & 1 & 0 & 0 & 0 & 1 \\ \hline 0 & 0 & 1 & 1 & 0 & 0 \\ 1 & 0 & 0 & 0 & 1 & 0 \\ 0 & 1 & 0 & 0 & 0 & 1 \end{array} \right] \begin{bmatrix} b_1 \\ b_2 \\ b_3 \\ \overline{s_1} \\ s_2 \\ s_3 \end{bmatrix} \tag{3.24}$$

where the composite system output is composed of all vertex signals and the composite system input includes sources at all vertices. However since $u = s_1$ and $y = v_3$, the system of figure 3.20 reduces to

$$
\begin{bmatrix} a_1 \\ a_2 \\ a_3 \\ \hline y \end{bmatrix}
=
\left[
\begin{array}{ccc|c|c}
0 & 0 & 1 & 1 & b_1 \\
1 & 0 & 0 & 0 & b_2 \\
0 & 1 & 1 & 0 & b_3 \\
\hline
0 & 1 & 0 & 0 & u
\end{array}
\right]
\tag{3.25}
$$

Note that for an interconnected dynamical system characterized by a signal flow graph, the component connection model is assured to exist. In particular, the type of pathology illustrated by the block diagram of figure 2.13 cannot occur. This follows because a signal flow graph models feedback paths through wires and scalars as components rather than connection constraints. As such these pathological situations, arising in block diagrams, can be characterized via a component connection model if they are first transformed into a signal flow graph model in which all wires and scalars are indentified as "artificial" components. This essentially hides the pathology. However, it will manifest itself again when the components and connections are combined.

The second commonly encountered geometric connection model is the *linear graph* used for *electric networks* and other *bilateral components.*[4] The linear graph, like the signal flow graph, is a directed graph composed of *vertices* and *edges*. Here each edge represents a single bilateral component or a part thereof. Associated with each edge is a pair of variables, v_i and i_i. For an electrical network v_i depicts the *port voltage* and i_i the *port current*. A completely analogous theory exists for other types of bilateral networks. In this theory v_i is often generically termed an *across variable* and i_i a *through variable*, for example pressure and flow in fluidic systems or cost and sales in economic systems.[3] The key parameters however for either the electric network or one of its analogs are the variables, v_i and i_i. They are not a-priori identified as the component input and output. Indeed, for most components either may serve as the input with the other taken as the output. The usual conservation laws constrain the variables v_i and i_i. In particular, the *Kirchhoff current law* or conservation of flow, etc. constrain i_i; the *Kirchhoff voltage law* or "sum of the pressure differences around a closed path equals zero" constrains v_i. With this intuition in mind let us develop some elementary *graph theory* as a preliminary to the formulation of the component connection model for the electric network and/or its analog based on these notions. To simplify the exposition we assume the concepts of *subgraph*, *path*, *connectivity*, and *circuit* of a graph are self explanatory.[4] These concepts are inutitive and the terminology aptly descriptive so we will not attempt to rigorize them. Moreover assume without loss of generality, that all graphs are connected. This is reasonable since in an electric network it is always possible, often desirable, to tie all components to a common ground.

A subgraph is said to be *minimal* with respect to a specific property if (i) the subgraph posesses the property and (ii) deletion of any edge of the subgraph implies the subgraph no longer posesses the property. Correspondingly a subgraph is *maximal* with respect to a particular property if (i) the subgraph has the property and (ii) the addition of any edge cancels the property. Observe that a *basis for a vector space* is a *maximal linearly independent set* and a *minimal spanning set*. Moreover either property completely characterizes the notion of basis for a vector space.

LEMMA 3.26: A subgraph is a minimal connected subgraph (minimal with respect to the property of being connected) if and only if it is a maximal subgraph containing no circuits (maximal with respect to the property of containing no circuits).

PROOF: If a subgraph is minimal with respect to the property of connectedness it cannot contain a circuit since the removal of an edge from a circuit does not disconnect the graph. Moreover, by adding any additional edge to the subgraph, the "new" subgraph would contain a circuit composed of the new edge and a path (in the original subgraph) connecting the vertices of the new edge. Such a path exists since the original subgraph is connected. As such a minimal connected subgraph is a maximal subgraph containing no circuits.

Conversely, a maximal subgraph containing no circuits must be connected. Otherwise it would contain at least two disconnected components. We could then add a path between these two components (we assume the original graph is connected) without introducing a circuit into the new subgraph. This, however, contradicts the assumption that the original subgraph was maximal with respect to the property of having no circuits. Moreover, by deleting any edge from the given maximal subgraph containing no circuits, the resultant subgraph will be disconnected. If this were false the given subgraph would have contained a circuit composed of the edge, just deleted, together with a path in the resultant subgraph connecting the two vertices of this edge (which exists if the resultant subgraph is connected). As such, a maximal subgraph containing no circuits is also a minimal connected subgraph. The proof is complete.

DEFINITION 3.27: A subgraph is a *tree* if it is characterized by either of the two equivalent properties of lemma 3.26.

Specifically a subgraph is a *tree* if it is a maximal subgraph containing no circuit or equivalently a minimal connected subgraph.

DEFINITION 3.28: A subgraph is a *co-tree* if it is the complement of a tree—i.e. it is composed of the edges remaining in a graph after deleting a tree. Basically any tree of a graph naturally decomposes the edges of the graph into two classes: edges in the tree and edges in the complementary co-tree.

EXAMPLE 3.29: Consider the graph of figure 3.30. Edges 1, 2, and 6 form a tree. To verity this, it is first necessary to show that this set of edges does not contain a circuit. This is obvious in this case. Secondly it is necessary to show that the addition of any new edges creates a circuit in the subgraph. Here the addition of edge 3 creates a circuit composed of edges 1, 2, and 3; the addition of edge 4 leads to a circuit composed of edges 2, 4, and 6; while the addition of edge 5 induces a circuit composed of edges 1, 5, and 6. Alternatively one may verify that edges 1, 2, and 6 form a tree by first verifying the obvious fact that they form a connected subgraph and then showing that the removal of any edge disconnects the subgraph. In particular, the removal of edge 1 isolates vertex b; the removal of edge 2 isolates vertex c; and the removal of edge 6 isolates vertex d. As such, we have verified by both of the equivalent characterizations that the set of edges, 1, 2, and 6, form a tree.

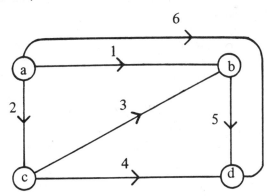

Figure 3.30. Linear graph.

Now in a given linear graph specify some tree. This specifies the complimentary co-tree. Each edge of the co-tree naturally associates itself with a circuit of the graph. The family of these circuits forms a set in which each element of the set must satisfy the Kirchhoff voltage law. In particular, if e_c is a co-tree edge, then the subgraph consisting of e_c and all of the tree edges must contain precisely one circuit. This follows since a tree is a maximal subgraph containing no circuits. Call $C(e_c)$ the specific circuit. Call the family of these circuits the *fundamental circuits* associated with the specified tree of the linear graph.

In a dual manner one may construct an appropriate family of subgraphs "satisfying" Kirchhoff's current law. Obviously the current law applies to a set of edges incident upon a given vertex. However, in a more general way, it is possible to "apply" it to any *minimal disconnecting set of edges*. A minimal disconnecting set of edges divides the graph into exactly two parts. Construe each "part" as a *generalized* vertex. Conservation of current (Kirchhoff's current law) requires that the sum of the currents entering the generalized vertex be zero, i.e., the sum of the currents crossing from one part to the other must be zero.

DEFINITION 3.31: A *cut-set* is a minimal disconnecting set.

Now for each edge of a specified tree in a given linear graph it is possible to naturally associate it with a particular cut-set. The family of these cut-sets determines a set constrained by the Kirchhoff current laws.

Since a tree is a minimal connected subgraph, the removal of a tree edge, say e_t, cuts the tree into precisely two sections, say t_a and t_b. As such, it is possible to define a set of edges, $S(e_t)$, which contains e_t together with all of the co-tree edges connecting t_a and t_b. Clearly $S(e_t)$ is a minimal disconnecting set since the deletion of any edge from $S(e_t)$ leaves an edge between a pair of vertices, one lying in t_a and the other in t_b. The set of cut-sets defined by the branches of a tree are the *fundamental cut-sets* associated with the specified tree of the graph. The concepts of fundamental circuits and fundamental cut-sets of a tree (and its complementary co-tree) although derived independently, are closely related. In fact each determines the other.

LEMMA 3.32: Specify a tree and thus its complementary co-tree in some linear graph. Let x be a co-tree edge and y a tree edge. then x is in S(y) if and only if y is in C(x).

PROOF: Recall that removal of edge y from the tree edge splits the tree into two parts, say t_a and t_b, because a tree is minimally connected. If x is in S(y) then x has one vertex in t_a and the other vertex in t_b. Clearly C(x) contains a path in the tree connecting the two vertices of x. This path must contain y since y is the only bridge between t_a and t_b.

Conversely, suppose y is in C(x). This implies there is a path in t_a

connecting one vertex of x and one of y. Also there is a path of t_b connecting the other vertex of x to the other one of y. The vertices of x are in two parts of the tree implying x connects the two parts of the tree. This in turn puts x in S(y).

EXAMPLE 3.33: For the graph of figure 3.30 using edges 1, 2, and 6 for the tree and 3, 4, and 5 for the co-tree, the fundamental circuits are:

$$C(3) = (1,2,1); \quad C(4) = (2,4,6); \tag{3.34}$$
$$\text{and } C(5) = (1,5,6)$$

while the fundamental cutsets are:

$$S(1) = (1,3,5); \quad S(2) = (2,3,4); \tag{3.35}$$
$$\text{and } S(6) = (4,5,6)$$

In this example all of the fundamental cutsets are sets of edges entering a vertex to which the Kirchhoff current law is classically applicable although this is not the case in general.

Lemma 3.32 demonstrates the equivalence of information carried by the fundamental circuits and fundamental cutsets. This suggests the possibility of writing both the Kirchhoff voltage and current law equations in terms of circuit information or alternatively in terms of cutset information. The possibility is real.

DEFINITION 3.36: Let a graph have a specified tree containing r edges and the complementary co-tree containing (p-r) edges. Here p is the number of edges in the graph. The number r always equals "n-1" where n is the number of vertices in the graph although this is not relavent to the present discussion.[4] The *D matrix* is defined to be an r by (p-r) matrix whose r rows are indexed by the tree edges and whose (p-r) columns are indexed by the co-tree edges in any fixed prespecified order. For a tree edge y and co-tree edge x the entry, D_{yx} in the y-th row and x-th column of D is either zero or plus/minus one as follows:

$$D_{yx} = \begin{cases} 0 & x \notin S(y) \\ 1 & x \in S(y) \text{ and its orientation} \\ & \quad \text{coincides with that of y.} \\ -1 & x \in S(y) \text{ and its orientation is} \\ & \quad \text{oposite to that of y.} \end{cases} \tag{3.37}$$

Again y splits the tree into exactly two parts connected by y and the elements of S(y). This fact suggests comparing the *orientation* of y with an edge in S(y) by checking whether they both leave and terminate on the same part ($D_{yx} = 1$) or whether one edge leaves one part terminating on the second part while the reverse holds true for the other edge, ($D_{yx} = -1$). Similarly the information inherent in C(x) also defines D. Here C(x) defines the entries in the x-th column in contrast to S(y) which defines the entries in the y-th row.

LEMMA 3.38: Given a linear graph and a matrix D as above, then

$$
D_{yx} \quad = \quad
\begin{cases}
0 & y \notin C(x) \\
1 & y \in C(x) \text{ and its orientation} \\
  & \text{is opposite to that of x.} \\
-1 & y \in C(x) \text{ and its orientation} \\
  & \text{coincides with that of x.}
\end{cases}
\qquad (3.39)
$$

where these *orientations* are relative to the circuit C(x).

To compute the relative orientation of the edges in a circuit consider following a path around the circuit in either direction. The edges are oppositely oriented ($D_{yx} = 1$) provided the path runs in the same direction as one edge and the opposite direction to the edge orientation for the other edge. The two edges are oriented in the same direction ($D_{yx} = -1$) if the path runs in the same direction or in the opposite direction as the edge orientation for *both* edges simultaneously. Clearly no *absolute edge orientation* exists for a circuit since one can choose a path which traverses the circuit in either direction. However *relative orientations* are well defined since a reversal of path direction reverses the effective orientation of both edges. Thus their relative orientation is unchanged.

PROOF OF LEMMA 3.38: Lemma 3.32 guarantees that D_{yx} of 3.39 is non-zero if and only if the D_{yx} of 3.37 is non-zero. It remains to verify that the signs are identical for the two cases. Suppose edges x and y have the same orientation relative to the two parts of the tree defined by deleting y from the tree. Then they have different orientations with respect to a path traversing the circuit C(x). This follows since a path must enter one part of the tree along one edge and leave it along the other provided the path tranverses the entire circuit.

Conversely suppose x and y have opposite orientations relative to the two parts of the tree. Here they have the same relative orientation with respect to a path traversing the entire circuit, C(x).

EXAMPLE 3.40: For the graph of figure 3.30, with tree (1,2,6) and co-tree (3,4,5), the D matrix is:

$$D = \begin{bmatrix} 1 & 0 & -1 \\ -1 & -1 & 0 \\ 0 & 1 & 1 \end{bmatrix} \qquad (3.41)$$

These somewhat tedious preliminaries complete, the next step is to algebraically fabricate the Kirchhoff voltage and current laws of an electrical network (or analog thereof) modeled by a linear graph. By assumption the network components are bilateral. Consequently the orientations of the edges of the linear graph do not necessarily denote input and output as in the case of the signal flow graph. They serve to define the *polarities* of the voltages (across variables) and currents (through variables) associated with the network. In particular the direction of positive current flow is the direction of the arrow while a positive potential is measured from the tail of the arrow to its head.

In a linear graph specify a given tree and allow external current sources to be placed in parallel with tree edges and external voltage sources to be placed in series with co-tree edges, both oriented in the same direction as the edge with which they are associated. Of course in practice, the network has fixed source locations forcing one to choose a tree consistent with the given sources. Since the number of sources is usually small this generally poses little difficulty. Let J_t denote the r-vector of current sources in parallel with tree edges and E_c the (p-r)-vector of voltage sources in series with co-tree edges. In both cases a zero entry is used for edges with no associated sources. Similarly, let v_t and v_c denote the voltages in the tree and co-tree edges respectively; let i_t and i_c denote the currents in the tree and co-tree edges respectively. Here, the ordering of the entries of J_t, v_t, and i_t must coincide with the ordering of the rows of D; and the ordering of the vectors, E_c, v_c, and i_c must coincide with the ordering of the columns of D.

THEOREM 3.42:

$$i_t = -Di_c - J_t \qquad (3.43)$$

and

$$v_c = D^t v_t - E_c \qquad (3.44)$$

PROOF: The theorem is merely a statement of the Kirchhoff current law for the fundamental cutsets defined by the tree edges and the voltage law for the fundamental circuits defined by the co-tree edges. Define D_y as the y-th row of D. Observe that $D_y i_c$ is the sum of the currents in co-tree edges of S(y) accounting for the polarity of the edges relative to edge y. Recall deletion of edge y splits the tree into two distinct parts. Upon adding the current in edge y, i_y, and the source current, J_y, which is in parallel with edge y, the Kirchhoff current law for S(y) becomes

$$i_y + J_y + D_y i_c = 0 \qquad (3.45)$$

Rearranging and transforming into matrix notation results in equations 3.43.

Similarly the Kirchhoff voltage law, accounting for source voltages, for the circuit C(x) is

$$v_x + E_x - D_x^t v_t = 0 \qquad (3.46)$$

This verifies 3.44

EXAMPLE 3.47: For the graph of figure 3.30 with tree (1,2,6) and co-tree (3,4,5) the Kirchhoff equations defined by 3.43 and 3.44 are

$$\begin{bmatrix} i_1 \\ i_2 \\ i_6 \end{bmatrix} = - \begin{bmatrix} 1 & 0 & -1 \\ -1 & -1 & 0 \\ 0 & 1 & 1 \end{bmatrix} \begin{bmatrix} i_3 \\ i_4 \\ i_5 \end{bmatrix} - \begin{bmatrix} J_1 \\ J_2 \\ J_6 \end{bmatrix} \qquad (3.48)$$

and

$$\begin{bmatrix} v_3 \\ v_4 \\ v_5 \end{bmatrix} = \begin{bmatrix} 1 & -1 & 0 \\ 0 & -1 & 1 \\ -1 & 0 & 1 \end{bmatrix} \begin{bmatrix} v_1 \\ v_2 \\ v_6 \end{bmatrix} - \begin{bmatrix} E_3 \\ E_4 \\ E_5 \end{bmatrix} \qquad (3.49)$$

The preceeding details have forged the tools needed to fabricate the *component connection model* for a given electric network. Specify a tree in a linear graph modeling the topology of an electric network. Let a = $col(i_t, v_c)$ and b = $col(v_t, i_c)$ be our *composite component input and output vectors*. Inherently tree edges correspond to components having impedence models and co-tree edges correspond to components with admittance models. Dealing with bilateral components forestalls any difficulty for the single-input single-output devices. For multiple-input multiple-output devices it is necessary to use the appropriate impedance, admittance or hybrid representation compatible with the tree or co-tree location of the edges corresponding to the different device ports. Usually the choice of a tree depends on the choice of model for the various components. Assuming that there is no intra-component coupling between components represented by tree edges and co-tree edges the *composite component model* for our network becomes

$$
\begin{bmatrix} v_t \\ \hline i_c \end{bmatrix} = \begin{bmatrix} Z_t & 0 \\ \hline 0 & Y_c \end{bmatrix} \begin{bmatrix} i_t \\ \hline v_c \end{bmatrix}
\tag{3.50}
$$

where Z_t is the composite component impedance matrix for components identified with tree edges and Y_c is the composite component admittance matrix for components identified with co-tree edges. Allowing intra-component coupling between tree and co-tree edges modifies equation 3.50. It becomes necessary to substitute a voltage transfer matrix for the "0" in the upper right hand corner of the matrix in 3.50; and correspondingly substitute a current transfer matrix for the lower left hand "0".

With our choice of "a" and "b" the connection matrices follow directly form theorem 3.42 taking the form

$$
\begin{bmatrix} i_t \\ \hline v_c \\ \hline v_t \\ \hline i_c \end{bmatrix} = \begin{bmatrix} 0 & -D & -1 & 0 \\ \hline D^t & 0 & 0 & -1 \\ \hline 1 & 0 & 0 & 0 \\ \hline 0 & 1 & 0 & 0 \end{bmatrix} \begin{bmatrix} v_t \\ \hline i_c \\ \hline J_t \\ \hline E_c \end{bmatrix}
\tag{3.51}
$$

Here we have deliberately chosen v_t and i_c as the composite system outputs since they are the "natural" responses to the source excitations, J_t and E_c. Moreover any other voltage of current output results from a linear combination of these variables. Of course, it is always possible to delete rows corresponding to variables not desired in the composite system output and/or columns corresponding tree and co-tree edges in which no source is located.

THEOREM 3.52: Let an interconnected dynamical system be characterized by a linear graph with specified tree. Suppose the tree edges correspond to components modeled by impedance matrices and co-tree edges to components modeled by admittance matrices, then equations 3.50 and 3.51 constitute a component connection model for the system.

EXAMPLE 3.53: The *resistive network* of figure 3.54 is characterized by the graph of figure 3.30. For this graph and tree (1,2,6) and co-tree (3,4,5) equation 3.41 defines the appropriate D matrix.

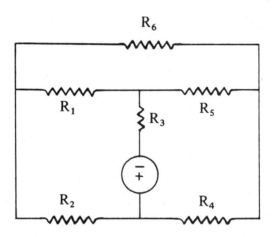

Figure 3.54. Resistive network.

The resistors represented by edges 1, 2, and 6 have a resistance matrix model since edges 1, 2, and 6 belong to the tree. On the other hand, the resistors corresponding to edges 3, 4, and 5 have a conductance model since these edges belong to the co-tree. This gives rise to the following composite component model.

$$
\begin{bmatrix} v_1 \\ v_2 \\ v_6 \\ \hline i_3 \\ i_4 \\ i_5 \end{bmatrix} = \left[\begin{array}{ccc|ccc} R_1 & 0 & 0 & 0 & 0 & 0 \\ 0 & R_2 & 0 & 0 & 0 & 0 \\ 0 & 0 & R_6 & 0 & 0 & 0 \\ \hline 0 & 0 & 0 & 1/R_3 & 0 & 0 \\ 0 & 0 & 0 & 0 & 1/R_4 & 0 \\ 0 & 0 & 0 & 0 & 0 & 1/R_5 \end{array} \right] \begin{bmatrix} i_1 \\ i_2 \\ i_6 \\ \hline v_3 \\ v_4 \\ v_5 \end{bmatrix} \qquad (3.55)
$$

and connection equations

$$
\begin{bmatrix} i_1 \\ i_2 \\ i_6 \\ v_3 \\ v_4 \\ v_5 \\ \hline v_1 \\ v_2 \\ v_6 \\ i_3 \\ i_4 \\ i_5 \end{bmatrix}
=
\left[\begin{array}{cccccc|c}
0 & 0 & 0 & -1 & 0 & 1 & 0 \\
0 & 0 & 0 & 1 & 1 & 0 & 0 \\
0 & 0 & 0 & 0 & -1 & -1 & 0 \\
1 & -1 & 0 & 0 & 0 & 0 & -1 \\
0 & -1 & 1 & 0 & 0 & 0 & 0 \\
-1 & 0 & 1 & 0 & 0 & 0 & 0 \\
\hline
1 & 0 & 0 & 0 & 0 & 0 & 0 \\
0 & 1 & 0 & 0 & 0 & 0 & 0 \\
0 & 0 & 1 & 0 & 0 & 0 & 0 \\
0 & 0 & 0 & 1 & 0 & 0 & 0 \\
0 & 0 & 0 & 0 & 1 & 0 & 0 \\
0 & 0 & 0 & 0 & 0 & 1 & 0
\end{array}\right]
\begin{bmatrix} v_1 \\ v_2 \\ v_6 \\ i_3 \\ i_4 \\ i_5 \\ \hline E_3 \end{bmatrix}
\qquad (3.56)
$$

Since the only source in the circuit is associated with edge three, we have deleted all columns associated with the *ficticious sources* in the other edges assumed in the derivation of theorem 3.52.

To recap this derivation, realize that there are certain assumptions necessary for the construction of the component connection model of an electric network. In particular we have assumed the existence of a tree such that:

(i)   All current sources are in parallel with tree edges.

(ii) All voltage sources are in series with co-tree edges.

(iii) All components whose ports are represented by tree edges have impedance models.

(iv) All components whose ports are represented by co-tree edges have admittance models.

(v)   All components whose ports are represented by some tree edges and some co-tree edges have an appropriate hybrid matrix model.

A component connection model as per theorem 3.52 always exists provided the appropriate tree exists. Of course the model is still independent of the kinship between the component and connection equations in which network degeneracies may also arise.

4. SPARSITY

The component connection model is founded on five matrices, the composite component matrix Z, and the four L_{ij} connection matrices. Experience dictates that the number of *composite system inputs and outputs* is generally small relative to the number of components of a system. For modern systems the number of components ranges from a half dozen or so to several thousand. Clearly, the computer memory required to store matrices describing such a large number of components and their interconnections may extend to several million words.

Fortunately for "real world" systems the *component connection matrices are sparse*, i.e., most of their entries are zero. By clever techniques for *computer storage* and manipulation of these matrices it is possible to reduce storage requirements from millions of words to perhaps several thousand words with a comparable reduction in *computational time*. Such techniques take account only of the location and value of the non-zero entries of the modeling matrices. In fact, such techniques are the key to the practical implementation of large scale system analysis. In many cases *sparse matrix techniques* diminish storage and CPU time below those required by classical techniques predicated on a much smaller, but non-sparse, composite system matrix.

A wide variety of possible techniques for *storing and manipulating sparse matrices* on a computer exists. The decision as to the appropriate technique is dependent on the special characteristics of the matrices involved and the desired manipulations of these matrices. For instance, the incoming and out going *incidence matrices* of a *signal flow graph* are characterized by the fact that each column contains exactly one non-zero entry, a single one. As such it suffices to store a single vector containing the row in which the non-zero entry appears.

EXAMPLE 4.1: The A_+ matrix of equation 3.9 is

$$A_+ \quad = \quad \begin{bmatrix} 0 & 0 & 0 & 0 & 0 \\ 1 & 0 & 0 & 0 & 1 \\ 0 & 1 & 0 & 1 & 0 \\ 0 & 0 & 1 & 0 & 0 \end{bmatrix} \tag{4.2}$$

Since each column contains a single "1," one may store A_+ in a computer memory as the vector

$$\hat{A}_+ \quad = \quad [2, 3, 4, 3, 2] \tag{4.3}$$

where (1) the position of the entry in the row vector corresponds to the column of A_+ and (2) the value of each position of the vector marks the row of the column containing the "1."

Clearly if a matrix is *diagonal* it may be stored as a single vector whereas if it is *tri-diagonal* it may be stored as a set of three vectors.

The unique characteristics of the above examples permit one to store the matrix in a vector whose word length exactly equals the number of nonzero entires. For the majority of sparse matrices no such scheme exists. Therefore the sparse matrix storage technique must allocate additional memory space to simultaneously mark the location and value of the nonzero entries. In many cases it is possible to do this with a number of words equal to two or three times the number of nonzero entires.

One storage technique requiring two words of memory for each nonzero matrix entry uses two row vectors for each row of the matrix—i.e. vectors say A_i and B_i store the i-th row of the matrix M. The first entry of A_i indicates the column in which the first nonzero entry in the i-th row of M appears; the first entry of B_i stores the value of that entry. The second entry of A_i marks the column for the second nonzero entry in the i-th row of M with the value of that entry being stored in the second entry of B_i; etc.

EXAMPLE 4.4: Using the above scheme the row vector

$$M_i = [0, 0, 0, 3, 0, 0, 0, -4, 0, 1, 0, 0, 0, 0, 0, 2, 0, 0, 0] \quad (4.5)$$

correspondingly the i-th row of M, say M_i, may be stored in the two vectors

$$A_i = [4, 8, 10, 16] \quad (4.6)$$

and

$$B_i = [3, -4, 1, 2] \quad (4.7)$$

This technique uses exactly two memory words for each nonzero entry. It is especially convenient for performing row operations on the matrix. Of course, a similar technique could store the columns of the matrix in preparation for column manipulations.

The primary difficulty with the above storage scheme arises when some matrix manipulation creates new nonzero entries. The problem is two-fold. First, a new nonzero entry croping up in the middle of a row necessitates shifting some of the entries of both A_i and B_i to the right, and then inserting a new entry in the middle of both A_i and B_i. Moreover this requires allocating additional blank memory space at the ends of the two vectors, over and above the two words per nonzero entry theoretically required for such a storage scheme.

The concept of a *linked list* kills all of the aforementioned difficulties. The trade off is three words of memory per nonzero matrix entry.

However, this *dynamic memory allocation* scheme permits easy storage of new non-zero entries. Moreover, it is unnecessary for the memory (allocated to store M) to be sequentially located in the computer memory. Rather each non-zero entry of the i-th row of M is stored in a separate cell comprising three sequentially located words of memory. A new entry to the i-th row of M is stored in any such cell regardless of its location in memory. Thus, rather than allocating specific memory locations for the storage of new entries in each row, one simply keeps track of a small number of empty cells for the storage of entries added anywhere in the matrix.

To implement a linked list scheme for the storage of M_i—i.e., the i-th row of M—fix a three word cell cataloging the information about the first non-zero entry of M_i. In the first word of the cell store the column containing this entry; in the second word store the value of the entry; and in the third word of the cell store the memory address for the cell storing the next non-zero entry of M_i. A cell of the form (7,0.23,500) indicates that the entry 0.23 appears in column seven of row M_i and the next non-zero entry of M_i is stored in the cell beginning at memory position 500.

EXAMPLE 4.8: For instance the M_i of equation 4.5 has the following as a possible linked list storage scheme.

$$(4, 3, 720), \quad (8, -4, 160), \quad (10, 1, 570), \quad (16, 2, 0) \qquad (4.9)$$

The first cell begins at some prespecified memory location (usually the cells for the first entry in each row are stored sequentially); the second cell begins at memory position 720; the third cell begins at position 160; and the fourth cell begins at position 570. The zero in the fourth cell indicates that it represents the last non-zero entry of the row.

A major advantage of the linked list is that the cells do not have to be sequentially located as with example 4.8. To add a new non-zero entry to M_i merely add a new cell and change the address in the preceding cell.

EXAMPLE 4.10: To add a "7" in the 12-th column of M_i change the third cell of 4.9 to (10,1,500) and add a new cell beginning at position 500. The new cell is (12,7,570).

Of course there are innumerable variations on the linked list theme.[6,7,10] Adopting a scheme depends on the manipulations the matrix will undergo. For instance, one may store the columns of M rather than the rows or instead by using larger cells, produce linkages simultaneously for both rows and columns. At worst this will require five or six words for each non-zero matrix entry. This is still tractable since storage allocation is linear in the number of non-zero entries.

Although the remainder of the text deemphasizes the specifics of manipulating sparse matrices on a computer the philosophy of the book acknowledges that such concepts underly the entire theory presented here. Most interconnected dynamical systems could not even be described on a computer, let alone be efficiently manipulated, without the aid of sparse matrix techniques. Clearly then, sparse matrix techniques are the foundation of the analysis of "real world" interconnected dynamical system both from the point of view of computational time and memory requirements and shear practicality of a functional model.

5. HIERARCHICAL SYSTEMS

Chapter IV will peruse a number of computational techniques based on the *component connection model* which calculate the response of a *composite system* given some input vector. Since large scale system theory is the thrust of the book, the matrices involved are extremely large. The analysis or simulation of a large scale system, economic, social or otherwise, is quite often performed numerous times. CPU-time is expensive. Sometimes it is possible to decompose a very large system into a hierarchy of subsystems. Many times such a decomposition will considerably reduce the cost of simulation. Prorating the cost of the decomposition over the instances of simulation may result in a net savings in computer costs. Unfortunately, most systems do not admit such a *hierarchical decomposition*. However, it is useful to do the decomposition whenever circumstances warrant.

Basically a hierarchical decomposition will simplify the connection equations by some sort of triangularization of the L_{11} matrix. Intuitively it is much easier to manipulate a triangular or block *triangular matrix* than an arbitrary one. The real worth of the decomposition will become evident after reading chapter IV, in particular the section on the *relaxation algorithm*.

The first type of hierarchical decomposition is the *series parallel system*, illustrated by figure 5.1. The trick is to break the composite system down into levels. The components on level one are driven *only* by external inputs. The components on level 2 are driven *only* by components on level one and/or external inputs. In general components on a given level are driven *only* by components on a lower level or by external inputs. Observe that components on a specific level may not interact. In particular the output of a component may not be fed back to its input.

Basically this decomposition requires a clever ordering or reordering of the components. Assuming the components are so ordered consider the L_{11} matrix. Here $L_{11}^{ij} = 0$ for $j \geqslant i$.

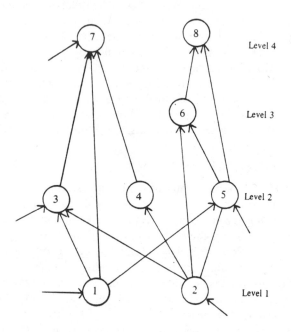

Figure 5.1. Series-parallel structure.

This discussion implies that L_{11} is *block strictly lower triangular* for a series parallel structure. For example consider the connection equation $a = L_{11}b$ as written in equation 5.2. Here there are n components numbered one through n with components on the first level having the lowest number etc.

$$
\begin{bmatrix} a_1 \\ \hline a_2 \\ \hline a_3 \\ \hline \cdot \\ \cdot \\ \cdot \\ \hline a_4 \end{bmatrix}
=
\begin{bmatrix}
0 & 0 & 0 & \dots & 0 \\
\hline
L_{11}^{21} & 0 & 0 & \dots & 0 \\
\hline
L_{11}^{31} & L_{11}^{32} & 0 & \dots & 0 \\
\hline
\cdot & & & & \cdot \\
\cdot & & & & \cdot \\
\cdot & & & & \cdot \\
\hline
L_{11}^{n1} & L_{11}^{n2} & L_{11}^{n3} & \dots & 0
\end{bmatrix}
\begin{bmatrix} b_1 \\ \hline b_2 \\ \hline b_3 \\ \hline \cdot \\ \cdot \\ \cdot \\ \hline b_n \end{bmatrix}
\qquad (5.2)
$$

Note that the submatrix L_{ij} represents the connection from the output of the j-th component to the input of the i-th component.

EXAMPLE 5.3: Typically a series-parallel system has a *feed-forward block diagram* characterization as per figure 5.4.

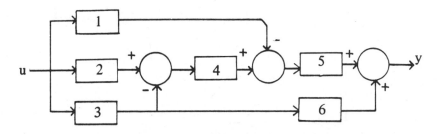

Figure 5.4. Feedforward block diagram.

Here interpret the hierarchical level as the position from left to right of the component groupings. Components one, two and three comprise the first level, component four the second level, components five and six the third level. On the other hand, observe that components one and six could legitimately be transferred to the second level without disrupting the hierarchical structure. Alternately, one could adopt a six level hierarchy characterized by the component numbering. Using this ordering, the system has the following L_{11} matrix.

$$
L_{11} = \begin{bmatrix}
0 & 0 & 0 & 0 & 0 & 0 \\
0 & 0 & 0 & 0 & 0 & 0 \\
0 & 0 & 0 & 0 & 0 & 0 \\
0 & 1 & -1 & 0 & 0 & 0 \\
-1 & 0 & 0 & 1 & 0 & 0 \\
0 & 0 & 1 & 0 & 0 & 0
\end{bmatrix}
\tag{5.5}
$$

This matrix has the aforementioned blocks strictly lower triangular form where all block are one by one.

The series-parallel structure permits one to assign a single component to each level. This in turn allows a level by level mode of analysis or simulation. First simulate level one; knowing the outputs of the components on level one, simulate level two, etc. Clearly the computational power of the decomposition rests in the sequential-type mode of analysis. Unfortunately few "real world" systems admit a series parallel structure since engineers intentionally design *feedback* into their systems. Fortunately a highly nontrivial and larger class of systems permits a *partial* decomposition. Such systems are termed *multilevel systems.*

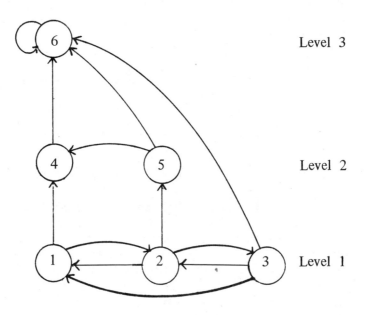

Figure 5.6. Example of multilevel structure.

Basically the multilevel structure accounts for component interaction, feedback, among the components on a given level. Specifically components on a given level are driven only by
   (i)   components on the same level,
   (ii)   components on lower levels, or
   (iii)   external inputs.
The analysis or simulation of this type of system again proceeds through a level by level process. However component interaction on a level increases the complexity of analysis of a level over that of the series parallel structure. The L_{11} connection matrix for multilevel structures has a *block lower triangular* form.

EXAMPLE 5.7: Figure 5.8 depicts a three level structure. Components one, two, and three form level one; component four forms level two; and components five and six make up level three.

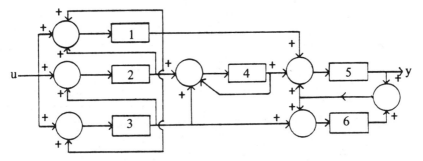

Figure 5.8. Diagram of multilevel structure.

The L_{11} matrix for this system is

$$\begin{bmatrix} a_1 \\ a_2 \\ a_3 \\ \hline a_4 \\ \hline a_5 \\ a_6 \end{bmatrix} = \left[\begin{array}{ccc|c|cc} 1 & 1 & 0 & 0 & 0 & 0 \\ 0 & 0 & 1 & 0 & 0 & 0 \\ 1 & 0 & 0 & 0 & 0 & 0 \\ \hline 0 & 1 & -1 & -1 & 0 & 0 \\ \hline -1 & 0 & 0 & 1 & -1 & -1 \\ 0 & 0 & 1 & 0 & 1 & 1 \end{array}\right] \begin{bmatrix} b_1 \\ b_2 \\ b_3 \\ \hline b_4 \\ \hline b_5 \\ b_6 \end{bmatrix} \qquad (5.9)$$

Clearly a level by level analysis offers the same type of computational savings as with the series parallel structure. Moreover the benefits increase as the number of levels increases. This suggests development of an algorithm for maximizing the number of levels of the multilevel structure; minimizing the number of components per level. There are two other motivational reasons for such an algorithm:

(i)   it is *not* obvious whether or not an interconnected dynamical system even has a hierarchical structure and

(ii)   even if it does, it will surely not exhibit a multilevel structure.

To determine an "optimal" hierarchical structure for a given interconnected dynamical system break the system down into a tier of levels with a minimum number of components per level. Each level becomes a minimal subsystem termed a *strongly connected subsystem.* Every component in a strongly connected subsystem affects the input of every other component in the subsystem either by direct connection or indirectly through other components. Basically there are two tasks: First it is necessary to identify the strongly connected subsystems and then one must place them on the appropriate level.

One approach to identifying the strongly connected subsystems uses the *system adjacency matrix,* A. Suppose there are n components in a system. The system adjacency matrix is an nxn matrix defined as follows:

$$A_{ij} = \begin{cases} 1 & \text{if the output of component j directly} \\ & \text{drives the input of component i and } i \neq j \\ 0 & \text{otherwise} \end{cases} \quad (5.10)$$

Clearly for $i \neq j$, $A_{ij} = 1$ if and only if $L_{ii}^{jj} \neq 0$ and $A_{ij} = 0$ if and only if $L_{ii}^{jj} = 0$ or $i = j$. Intuitively A describes direct coupling between distinct components. Now consider the product A^2 where multiplication and addition of entries is via *Boolean operations*, i.e., $1+1 = 1$ and $1 \times 1 = 1$, etc. The matrix accounts for component interaction taking place with exactly one intermediary component. For example component one may affect component three indirectly through component two. Similarly A^3 describes component interaction occurring with exactly two intermediaries, etc.

The problem is determining which components belong to the same strongly connected subsystem. To solve this problem define the *reachability matrix* as

$$R = (I + A)^n \quad (5.11)$$

where n is the number of components of the system; A is the system adjacency matrix; I is the identity matrix; and all manipulations are with respect to the *Boolean Algebra*.

THEOREM 5.12: Components i and j lie in the same strongly connected subsystem if and only if both R_{ij} and R_{ji} are non-zero.

PROOF: Applying the rules for Boolean operations and using the *binomial expansion*, the equation for R becomes

$$R = (I + A)^n = I + A + A^2 + \ldots + A^n \quad (5.13)$$

Now the identity matrix accounts for the possibility of individual component feedback. So in light of this expression for R, the components prior to the theorem indicate that R accounts for all affects either direct or indirect between components with up to a maximum of (n-1) intermediaries. Clearly this is sufficient since in a system of n components a single component will affect another through a maximum of (n-1) intermediaries.

This discussion implies that R_{ij} is non-zero if and only if component j directly or indirectly affects component i. Similarly R_{ji} is non-zero if and only if component i directly or indirectly affects component j. In a strongly connected subsystem, each component obviously affects the others either directly or indirectly. Therefore the conclusion of the theorem must follow.

This theorem provides the means of constructing the strongly connected subsystems of a composite system or equivalently minimizing the number of components per level of a multilevel structure. The remaining problem is to appropriately order these subsystems.

In this burden we define a *reduced adjacency* matrix $\hat{A}$. The reduced adjacency matrix $\hat{A}$ has a "1" in the ij-th entry if the j-th strongly connected subsystem directly drives the i-th strongly connected subsystem. Otherwise the entry is zero. A strongly connected subsystem does not drive itself.

THEOREM 5.14: $\hat{A}$ has at least one zero row.

PROOF: Implicit in the theorem is the understanding that we have broken the composite system down into more than one level. This forces A to have more than one row. The proof of the theorem is by contradiction. Assume that $\hat{A}$ has no zero row. We will show that if $\hat{A}$ has no zero row then the entire system is the only strongly connected subsystem. This means that there can be only one level and that $\hat{A}$ has only one row. This is the desired contradiction.

Consider the j-th row of $\hat{A}$. By assumption there is at least one non-zero entry, say A_{jk}. If $A_{jk} = 1$ then the k-th level drives the j-th level. By assumption the k-th row has at least one non-zero entry. Since there are only a finite number of strongly connected subsystems this process of backtracking eventually leads back to the j-th row. In particular by considering all rows of $\hat{A}$, the backtracking process will show that each strongly connected subsystem affects all the others either directly or indirectly. Thus there is at least one level which contradicts the initial assumption.

The procedure for numbering the levels is as follows: Choose for level one, a strongly connected subsystem which corresponds to one of the zero-rows in $\hat{A}$. Suppose this is the j-th row. Now delete the j-th row and j-th column from $\hat{A}$. This results in a new reduced adjacency matrix having at least one zero row by the same argument used to prove theorem 5.14. Choose for level two the strongly connected subsystem corresponding to this row. Delete the row and column associated with this subsystem and repeat the procedure until all strongly connected subsystems are numbered as levels.

EXAMPLE 5.15: Below are the connection equations for an interconnected dynamical system. The task is to decompose this system into a multilevel hierarchical structure. Primarily this entails some juggling of the entries of the L_{11} matrix resulting in a modified L_{11} matrix which is block lower triangular.

$$
\begin{bmatrix}
a_1^1 \\
a_2^2 \\
a_3^3 \\
a_4^3 \\
a_5^3 \\
a_6^4 \\
a_7^5 \\
a_8^5 \\
\hline
y
\end{bmatrix}
=
\left[
\begin{array}{cccccccc|c}
0 & 0 & 0 & 0 & 3 & 0 & 0 & 0 & 1 \\
0 & 0 & -1 & 0 & 0 & 0 & 0 & 0 & 0 \\
0 & 0 & 0 & 0 & 1 & 0 & 0 & 0 & 0 \\
1 & 0 & 0 & 3 & 7 & 0 & 0 & 0 & 1 \\
7 & 0 & 0 & 0 & 0 & 0 & 0 & 0 & 0 \\
0 & 0 & -2 & 0 & 0 & 0 & -2 & 0 & 0 \\
3 & 2 & 0 & 0 & 1 & 0 & 0 & 2 & 1 \\
0 & 4 & 1 & 1 & 0 & -1 & -2 & 0 & 0 \\
\hline
1 & 0 & 1 & 0 & 1 & 1 & 0 & 1 & 1
\end{array}
\right]
\begin{bmatrix}
b_1^1 \\
b_2^2 \\
b_3^3 \\
b_4^3 \\
b_5^3 \\
b_6^4 \\
b_7^5 \\
b_8^5 \\
\hline
u
\end{bmatrix}
\qquad (5.16)
$$

There are five components with the superscripts indicating the appropriate component. The first step is to place the components into strongly connected subsystems. By inspecting the L_{11} part of the connection matrix, one can write the 5x5 adjacency matrix

$$
A =
\begin{bmatrix}
0 & 0 & 1 & 0 & 0 \\
0 & 0 & 1 & 0 & 0 \\
1 & 0 & 0 & 0 & 0 \\
0 & 0 & 1 & 0 & 1 \\
1 & 1 & 1 & 1 & 0
\end{bmatrix}
\qquad (5.17)
$$

Applying the Boolean operations to the reachability matrix $R = I + A + \ldots + A^5$, we obtain

$$
R =
\begin{bmatrix}
1 & 0 & 1 & 0 & 0 \\
1 & 1 & 1 & 0 & 0 \\
1 & 0 & 1 & 0 & 0 \\
1 & 1 & 1 & 1 & 1 \\
1 & 1 & 1 & 1 & 1
\end{bmatrix}
\qquad (5.18)
$$

Now consider the *Boolean dot product* of R with R^t. This is nothing more than the entry by entry product of R and R^t as illustrated in equation 5.19. Inspecting the columns of $R \cdot R^t$ it is clear that components one and three form a strongly connected subsystem, component two another, and components four and five a third.

$$
\mathbf{R} \cdot \mathbf{R}^t = \begin{bmatrix} 1 & 0 & 1 & 0 & 0 \\ 0 & 1 & 0 & 0 & 0 \\ 1 & 0 & 1 & 0 & 0 \\ 0 & 0 & 0 & 1 & 1 \\ 0 & 0 & 0 & 1 & 1 \end{bmatrix} \tag{5.19}
$$

The next step is to determine the ordering of these subsystems into hierarchical levels. For this consider the reduced adjacency matrix

$$
\hat{\mathbf{A}} = \begin{bmatrix} 0 & 0 & 0 \\ 1 & 0 & 0 \\ 1 & 1 & 0 \end{bmatrix} \tag{5.20}
$$

Here the first row and column denote the strongly connected subsystem composed of components 1 and 3. This subsystem becomes level 1. Deleting row and col 1, we obtain

$$
\tilde{\mathbf{A}} = \begin{bmatrix} 0 & 0 \\ 1 & 0 \end{bmatrix} \tag{5.21}
$$

This implies that component 2 forms the second level. Clearly, components 4 and 5 make up the third level. Thus after appropriately modifying our connection equations, we obtain the hierarchical form as below.

$$
\begin{bmatrix} a_1^1 \\ a_3^3 \\ a_4^3 \\ a_5^3 \\ a_2^2 \\ a_6^4 \\ a_7^5 \\ a_8^5 \\ \hline y \end{bmatrix} = \left[\begin{array}{cccc|c|ccc||c} 0 & 0 & 0 & 3 & 0 & 0 & 0 & 0 & 1 \\ 0 & 0 & 0 & 1 & 0 & 0 & 0 & 0 & 0 \\ 1 & 0 & 3 & 7 & 0 & 0 & 0 & 0 & 1 \\ 7 & 0 & 0 & 0 & 0 & 0 & 0 & 0 & 0 \\ \hline 0 & -1 & 0 & 0 & 0 & 0 & 0 & 0 & 0 \\ \hline 0 & -2 & 0 & 0 & 0 & 0 & -2 & 0 & 0 \\ 3 & 0 & 0 & 1 & 2 & 0 & 0 & 2 & 1 \\ 0 & 1 & 1 & 0 & 4 & -1 & -2 & 0 & 0 \\ \hline 1 & 1 & 0 & 1 & 0 & 1 & 0 & 1 & 1 \end{array} \right] \begin{bmatrix} b_1^1 \\ b_3^3 \\ b_4^3 \\ b_5^3 \\ b_2^2 \\ b_6^4 \\ b_7^5 \\ b_8^5 \\ \hline u \end{bmatrix} \tag{5.22}
$$

The new L_{11} submatrix of these connections equations has the required block lower triangular form, where each "block" corresponds to a strongly connected subsystem.

Such a procedure as this would most likely find use (via sparse matrix techniques) in decomposing a large scale system into a multilevel hierarchy. However, so as to foster some intuitive insight, we present an alternate procedure. The procedure basically involves inspecting the *adjacency graph* of a (small) interconnected dynamical system. The adjacency graph has a *vertex* for each system component. An *edge* connects two vertices of the adjacency graph (representing components i and j) if the output of component j directly feeds the input of component i. Viewing the L_{11} matrix, this means that vertex j is connected to vertex i if and only if $L_{ij}^{11} \neq 0$.

Two components lie in the same strongly connected subsystem if and only if the two components lie in a common *directed circuit*. By a common directed circuit we mean a loop in which all edges are oriented in the same direction around the loop. The condition that the adjacency matrix have at least one zero row is equivalent to requiring that at least one subsystem have no edges entering it. By sequentially deleting such subsystems (common directed circuits and the edges leaving these circuits) one may sequentially construct the ordering of the hierarchy.

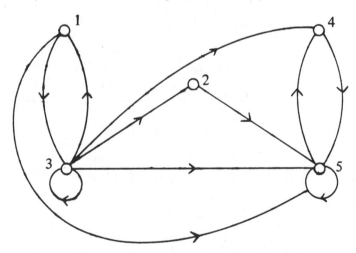

Figure 5.23. Adjacency graph of example 5.15.

EXAMPLE 5.24: Consider the L_{11} matrix of example 5.15 in which there are 5 components. The adjacency graph has five vertices. By inspecting the L_{11} matrix, one obtains the following adjacency graph for this system. The hierarchical structure can now be obtained by inspection: if not immediately, then after several minutes of contemplation.

For relatively small systems, inspection of the adjacency graph affords an easy way of determining the hierarchical structure. Large dynamical systems demand a computer for analysis. Thus an algorithm for a multilevel decomposition is necessarily based on some type of adjacency matrix. Yet even here, a graph theoretic point of view leads to highly efficient algorithms. In fact such agorithms exist and are far superior to the inspection techniques discussed above. Their advantage is that the number of computations increases only linearly with the system size.[9]

Determining a hierarchical decomposition of an interconnected dynamical system is commonly a toilsome process. In many cases, especially for large systems, the effort is well spent. One need perform a decomposition only once allowing its cost to be prorated over the number of times the system is analyzed. More importantly, however, hierarchical structures permit a "decoupled" analysis of the strongly connected subsystem. This effectively reduces computer time while providing intuition as to the behavior of the overall system. As an example of this philosophy, consider that the composition system is *stable* if and only if each strongly connected subsystem is stable.[1]

6. A GENERALIZATION OF THE ALGEBRAIC CONNECTION MODEL

As per example 2.13, not all interconnected dynamical systems admit a well defined component connection model. A circumvention of this difficulty appears if one embeds the *pathological connection* constraints into the components and therefore into the component equations. This course insures that the system has a well defined signal flow graph and thus a component connection model. However this suppression of the inherent pathology of the system postpones dealing with it until the component and connection equations are combined. (We discuss this idea in chapter III.) Consequently we define a *generalization of the component connection model* which openly displays the pathology.

The generalized connection equations take the form

$$\left[\frac{a}{y}\right] = \left[\begin{array}{c|c} L_{11} & L_{12} \\ \hline L_{21} & L_{22} \end{array}\right] \left[\frac{b}{u}\right] + \left[\begin{array}{c|c} R_{11} & R_{12} \\ \hline R_{21} & R_{22} \end{array}\right] \left[\frac{a}{y}\right] \quad (6.1)$$

The vector-matrix block diagram of figure 6.2 depicts this model. Clearly (vaguely at first) if the matrix $(I-R)$ is invertible, equation (6.1) is equivalent to the regular connection equations. The degree of the linear dependence of the rows (or columns) of (I-R) serves as a measure of the pathology inherent in the system interconnections.

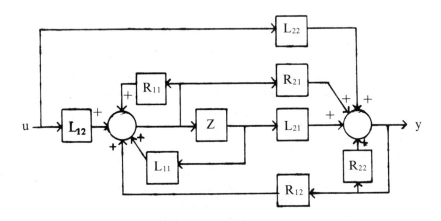

Figure 6.2. Vector-matrix block diagram of generalized algebraic connection equations.

EXAMPLE 6.3: Consider the pathological system of figure 6.4.

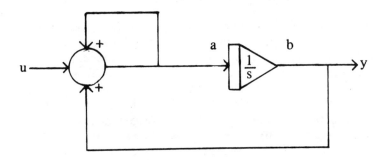

Figure 6.4. Block diagram of pathological system.

The set of connection equations are

$$a_1 = a_1 + u - y$$
$$y = b_1$$

(6.5)

Representing this set of equations in the generalized algebraic connection model we obtain

$$\begin{bmatrix} a_1 \\ \overline{} \\ y \end{bmatrix} = \begin{bmatrix} 0 & | & 1 \\ \overline{} & | & \overline{} \\ 1 & | & 0 \end{bmatrix} \begin{bmatrix} b \\ \overline{} \\ u \end{bmatrix} + \begin{bmatrix} 1 & | & -1 \\ \overline{} & | & \overline{} \\ 0 & | & 0 \end{bmatrix} \begin{bmatrix} a_1 \\ \overline{} \\ y \end{bmatrix}$$ (6.6)

Inspecting $(I-R)$ one observes that $(I-R)$

$$(I-R) = \begin{bmatrix} 0 & 1 \\ 0 & 1 \end{bmatrix}$$ (6.7)

is singular which implies that equation 6.6 cannot be written in the usual form. This agrees with the result of example 2.13.

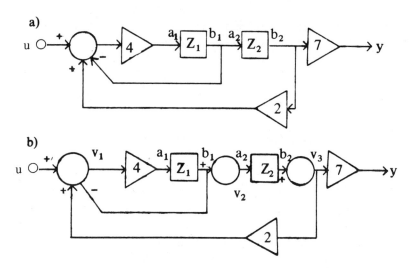

Figure 6.8. Block diagram and its equivalent with "artificial adders".

The pathology of this particular example arises from the fact that the output of an adder is fed back as an input to the adder with the right amount of gain to cause $(I-R)$ to be singular. It turns out that whenever a system displays a connection pathology, the above phenomena is always the crux of the problem. As such, a system is said to have *improperly connected adders* whenever $(I-R)$ is singular.[8]

At any rate, the generalized algebraic connection model may characterize any interconnected dynamical system described by a *block diagram* irregardless of whether the adders are properly connected. The process of writing the connection equations by inspecting the block diagram is relatively straight forward. However by assuming that all composite system outputs feed an adder directly and all composite system inputs are fed from an adder (possibly through a scalar), one can find a more direct route. This statement is valid if one inserts an "artificial adder" between a component output or system input which directly feeds a component input or system output. Such adders have only one input and their output equals their input. As an example, consider figure 6.8-b. Here two such "artificial adders" have been inserted into the block diagram of figure 6.8-a to achieve the desired form.

With this transformation complete, it is a simple chore to mathematically construct the connection equations. Let v denote the vector of adder outputs with the vectors a,b,u, and y having their usual interpretation. The adder outputs are sums of component outputs and/or outputs of other adders all of which may be passed through scalars. Specifically v is a linear combination of b,u, and v, which in matrix notation takes the form

$$v\Big] = \Big[P_1 \mid P_2\Big]\dfrac{b}{u}\Big] + \Big[Q\Big]v\Big] \qquad (6.9)$$

By assumption each entry of a and y is equal to an appropriately scaled entry of v. Thus we also have

$$\dfrac{a}{y}\Big] = \Big[\dfrac{S_1}{S_2}\Big]v\Big] \qquad (6.10)$$

where each row of S has exactly one nonzero entry and each column of S is nonzero. If a column of S had all its entries equal to zero, then an adder would not have its output connected to anything. These properties imply that the matrix S is left invertible. Interestingly, examples exist for which neither S_1 nor S_2 are left invertible. Thus multiplying equation 6.10 on the left by the *left inverse* of S we obtain

$$v\Big] = \Big[\dfrac{S_1}{S_2}\Big]^{-L}\dfrac{a}{y}\Big] \qquad (6.11)$$

By first substituting equation 6.11 into equation 6.9 and then substituting this result into equation 6.10, we have the desired generalized algebraic connection equations for an arbitrarily given block diagram of an interconnected dynamical system.

$$
\begin{bmatrix} a \\ \hline y \end{bmatrix} = \begin{bmatrix} S_1 \\ \hline S_2 \end{bmatrix} \begin{bmatrix} P_1 & \vdots & P_2 \end{bmatrix} \begin{bmatrix} b \\ \hline u \end{bmatrix} + \begin{bmatrix} S_1 \\ \hline S_2 \end{bmatrix} Q \begin{bmatrix} S_1 \\ \hline S_2 \end{bmatrix}^{-L} \begin{bmatrix} a \\ \hline y \end{bmatrix} \tag{6.12}
$$

EXAMPLE 6.13: For the modified block diagram of figure 6.8-b equations 6.9, 6.10, and 6.11 are as follows.

$$
\begin{bmatrix} v_1 \\ v_2 \\ v_3 \end{bmatrix} = \begin{bmatrix} -1 & 0 & \vdots & 1 \\ 1 & 0 & \vdots & 0 \\ 0 & 1 & \vdots & 0 \end{bmatrix} \begin{bmatrix} b_1 \\ b_2 \\ \hline u \end{bmatrix} + \begin{bmatrix} 0 & 0 & 2 \\ 0 & 0 & 0 \\ 0 & 0 & 0 \end{bmatrix} \begin{bmatrix} v_1 \\ v_2 \\ v_3 \end{bmatrix} \tag{6.14}
$$

$$
\begin{bmatrix} a_1 \\ a_2 \\ \hline y \end{bmatrix} = \begin{bmatrix} 4 & 0 & 0 \\ 0 & 1 & 0 \\ \hline 0 & 0 & 7 \end{bmatrix} \begin{bmatrix} v_1 \\ v_2 \\ v_3 \end{bmatrix} \tag{6.15}
$$

and

$$
\begin{bmatrix} v_1 \\ v_2 \\ v_3 \end{bmatrix} = \begin{bmatrix} 1/4 & 0 & \vdots & 0 \\ 0 & 1 & \vdots & 0 \\ 0 & 0 & \vdots & 1/7 \end{bmatrix} \begin{bmatrix} a_1 \\ a_2 \\ \hline y \end{bmatrix} \tag{6.16}
$$

Combining these equations as in 6.12, yields the algebraic connection equations

$$
\begin{bmatrix} a_1 \\ a_2 \\ \hline y \end{bmatrix} = \begin{bmatrix} -4 & 0 & \vdots & 4 \\ 1 & 0 & \vdots & 0 \\ \hline 0 & 7 & \vdots & 0 \end{bmatrix} \begin{bmatrix} b_1 \\ b_2 \\ \hline u \end{bmatrix} + \begin{bmatrix} 0 & 0 & \vdots & 8/7 \\ 0 & 0 & \vdots & 0 \\ \hline 0 & 0 & \vdots & 0 \end{bmatrix} \begin{bmatrix} a_1 \\ a_2 \\ \hline y \end{bmatrix} \tag{6.17}
$$

Finally observe that

$$
(I-R) = \begin{bmatrix} 1 & 0 & \vdots & -8/7 \\ 0 & 1 & \vdots & 0 \\ \hline 0 & 0 & \vdots & 1 \end{bmatrix} \tag{6.18}
$$

which is invertible. Consequently the adders are properly connected which allows the transformation of 6.17 into the regular connection model as below.

$$
\begin{bmatrix} a_1 \\ a_2 \\ \hline y \end{bmatrix} = \begin{bmatrix} 1 & 0 & -8/7 \\ 0 & 1 & 0 \\ \hline 0 & 0 & 1 \end{bmatrix}^{-1} \begin{bmatrix} -4 & 0 & 4 & b_1 \\ 1 & 0 & 0 & b_2 \\ \hline 0 & 7 & 0 & u \end{bmatrix}
\tag{6.19}
$$

$$
= \begin{bmatrix} -4 & -8 & 4 & b_1 \\ 1 & 0 & 0 & b_2 \\ \hline 1 & +7 & 0 & u \end{bmatrix}
$$

7. PROBLEMS

1. Verify that the system constraints implied by the *vector-matrix block diagram* of figure 2.4 are equivalent to the constraints defined by equations 2.2 and 2.3.
2. Formulate a *component connection model* for a series RLC circuit with a series voltage generator. Repeat for a parallel RLC circuit with a parallel current source.
3. Formulate a component connection model for the circuit of figure I.6.9.
4. Formulate a component connection model for a four stage RC ladder.
5. Give an example of a system which fails to admit a component connection model in which the output of an adder is not directly fed back to its own input.
6. Write incoming and outgoing *incidence matrices* for the directed graph shown below.

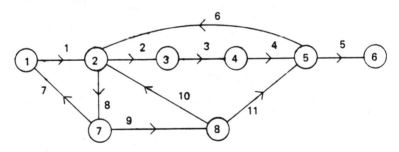

7. Write an *algebraic connection equation* for the system characterized by the graph.

8. Show that the connection matrix of equation 3.18 is always singular providing $e > n$.

9. Show that the L matrix characterizing the connection in a *signal flow graph* as derived in equation 3.18 is composed wholly of ones and zeros.

10. Formulate a pair of variables with which to model a *traffic system* whose properties are analogous to those of voltage and current in an electric network.

11. Give formal definitions for the concepts of *graph*, *sub-graph*, *path*, *connectivity*, and *circuit*.

12. For a subset of the vectors in a *vector space* show that the following are equivalent:
    (i) It is a *basis*.
    (ii) It is a minimal spanning set.
    (iii) It is a maximal linearly independent set.

13. List all *trees* for the graph of figure 3.30.

14. A graph is said to be *complete* if it has exactly one edge connecting every vertex to every other vertex. How many edges does a complete graph on n vertices have? How many circuits? How many trees?

15. For the graph of figure 3.30 list the set of *fundamental circuits and cut-sets* associated with the tree composed of edges 2, 3, and 6.

16. Show that a *co-tree* is a maximal subgraph containing no cut-sets.

17. Show that every tree of a graph has $r = n-1$ edges.

18. Exhibit a cut-set of the graph shown in figure 3.30 which is not composed of the edges entering a vertex.

19. Write the *D matrix* for the graph of figure 3.30 using the tree composed of edges 2, 3, and 6.

20. *Conservation of energy* implies that the power entering a system from its external sources equals the power dissipated in its components (one has to be careful to include all sources including bias, etc. for this to hold). In an electric network, however, a stronger result holds wherein one measures the voltages in the network at time t_1 and the currents at time t_2 and still gets a "conservation of energy like" equality among the products of the voltages and currents. This result is known as *Tellegen's theorem* and in our notation takes the form

$$i^t(t_1)v(t_2) = -J_t^t(t_1)v_t(t_2) - i_c^t(t_i)E_c(t_2)$$

where the minus signs are due to our choice of polarities. Verify this equation.

21. Write the *Kirchhoff voltage and current law* equations for the graph of figure 3.30 in matrix form using the tree composed of edges 2,3, and 6.

22. Show that the connection matrix of equation 3.51 is skew-hermitian (i.e., $L = -L^t$).

23. Formulate a method for storing a large *sparse matrix*, all of whose entries are ones or zeros, on a computer using exactly one word of memory per non-zero matrix entry. If you were given two words of memory per non-zero matrix entry could you develop a superior technique?

24. Assume that you have three word memory cells starting at locations 1,4,7,10,13,16, ... , 3k+1, ... Using these cells indicate how the matrix of equation 5.16 might be stored as a *linked list* of rows.

25. Develop a technique for *storing a matrix* using linked lists simultaneously for both its rows and columns. How do you go about inserting a new non-zero entry into the matrix when using this scheme?

26. Interpret the graph of figure 7.6 as the *adjacency* graph of an interconnected dynamical system and write its *adjacency and reachability matrices*.

27. Interpret the graph of figure 7.6 as the adjacency graph of an interconnected dynamical system and formulate its *multilevel structure*.

28. Let R be the reachability matrix of an interconnected dynamical system and define R • R^t to be the entry by entry Boolean product of R and its transpose. Show that the rows (or columns) of R • R^t represent the strongly connected subsystems of the given interconnected dynamical system.

29. Verify that the *block diagram* of figure 6.2 is equivalent to equation 6.1.

30. Prove that the matrix col(S_1,S_2) defined in equation 6.10 has a *left inverse*.

31. Write a *generalized component connection model* for the block diagram of figure 2.17 using the matrix algebraic formula given in equation 6.12. Verify analytically that this block diagram has *improperly connected adders*.

8. REFERENCES

1. Collier, F. M., Chan, W. S., and C. A. Desoer, Stability Theory of Interconnected Systems", Meno. ERL-MSCF, University of California at Berkeley, 1975.
2. Frank, H., and I. Frisch, *Communication, Transmission, and Transportation Networks*, Addison-Wesley, Reading, 1971.
3. Koenig, H., Tokas, Y., and H. K. Kesavan, *Analysis of Discrete Physical Systems*, McGraw-Hill, New York, 1966.

4. Mayeda, W., *Graph Theory*, Wiley, New York, 1972.
5. Kuh, E.S., and R. A. Rorher, *Theory of Linear Active Networks*, Holden-Day, San Francisco, 1967.
6. Reid, J. K., *Large Sparse Sets of Linear Equations*, Academic Press, New York, 1971.
7. Rose, D. J., and R. A. Willoughby, *Sparse Matrices and their Applications*, Plenum Press, New York, 1972.
8. Singh, S. P., and R. W. Liu, "Existence of State Equation Representation of Linear Large-Scale Dynamical Systems", IEEE Trans. on Circuit Theory, Vol. CT-20, pp. 239-246, (1973).
9. Tarjan, R., "Depth-first Search and Linear Graph Algorithms", SIAM Jour. on Comp., Vol. 1, pp. 146-160, (1972).
10. Tewerson, R. P., *Sparse Matrices*, Academic Press, New York, 1972.

III. SYSTEMS

1. INTRODUCTION

The *component connection model* for an interconnected dynamical system carries a complete description of the system. The information rests in two separate equations: a decoupled dynamical equation characterizing the components and a coupled algebraic equation characterizing the connections. An alternate approach directly relates the composite system input vector, u, to the composite system output vector, y, via an equality of the form

$$y = Su \qquad (1.1)$$

interpreting the sympoblic operator, S, as either a *transfer function matrix*, or a set of linear or nonlinear state equations.

Dimensionally, the *composite system model* of 1.1 is usually much smaller than the component connection model. In practice, however, the matrices representing S are usually non-sparse whereas those of the component connection model are sparse. Thus computer storage requirements (a truer measure of system dimensionality) for the component connection model are often far less than those of the composite system model. Indeed, this advantage usually grows with system complexity. Moreover, the component connection model displays the separate attributes of the system distinctly and independently. Each component is modeled independently of the other components and the connections, whereas the effect of the connectivity structure is displayed explicitly. No attribute of the system is masked. Finally, for *nonlinear systems*, the component connection model eliminates the necessity of confronting the rather complex composite system nonlinearities which result from combining several component nonlinearities through the connectivity structure. The structure allows one to deal with the more facile component nonlinearities.

For such reasons, in using the theory of interconnected dynamical systems, it is almost always advantageous to work directly with the system's component connection model rather than its composite system model. This reasoning applies equally well to the problems of *system simulation, analysis, and design*. Such arguments not withstanding, the primary goal of the present chapter is to investigate the properties of the composite system model for an interconnected dynamical system while simultaneously outlining the deep relationship between the composite system model and the component connection model.

The composite system model is denoted symbolically by S. Again S may represent any of the standard models previously formulated for individual components. For example one may use a transfer function matrix S(s) satisfying the equality

$$y(s) = S(s)u(s) \tag{1.2}$$

for all s such that $Re(s) > c$. Here, u(s) and y(s) are the Laplace transforms of the *composite system input and output vectors*, respectively, while S(s) is a matrix of rational functions in the complex variable, s. The *composite system transfer function matrix*, S(s), has precisely the same properties as the component transfer function matrix, Z(s), defined in I.3.1. All of the concepts appearing in section I.3 similarly apply to the composite system transfer function matrix. For instance, S(s) is *proper* if all of its entries are proper rational functions.

Likewise, the composite system has a state model characterized by

$$\begin{aligned} \dot{x} &= Fx + Gu \\ y &= Hx + Ju \end{aligned} \tag{1.3}$$

or

$$\begin{aligned} \dot{x} &= F(x,u) \\ y &= G(x,u) \end{aligned} \tag{1.4}$$

in the linear and nonlinear cases respectively. Here, u is the composite system input vector; y is the composite system output vector; and, as we will show later, the *composite system state vector*, x, can usually (but not always) be taken as the *composite component state vector*. Again, the obvious properties and terminology carry over from chapter I.

Sections 2 and 3 of this chapter study the composite system transfer function and its relationship to the composite component model where the components will be assumed to admit a transfer function model. The key to this relationship is a matrix valued function of a matrix valued variable symbolically represented as

$$S(s) \;\; = \;\; f(Z(s)) \tag{1.5}$$

where $Z(s)$ is the composite component transfer function matrix. The function, f, is the *connection function* and is entirely determined by the L_{ij} connection matrices. It completely describes the relationship between $Z(s)$ and $S(s)$. Thus section 3 investigates the decomposition, differentiation, and inversion of f.

Section 4 studies the *inversion of sparse matrices*, a process required for the computation of $S(s)$ from the component connection model. Moreover, sparse matrix inversion is the key numerical process required throughout our entire theory. The main difficulty with such inversion is that the inverse of a sparse matrix is not sparse. Thus one must invoke special techniques to permit the inverse to be computed and stored sparsely.

Sections 5, 6, and 7 describe the composite system state models and techniques for their computation in the linear and nonlinear cases. in the linear case our interest focuses on the determination of an appropriate composite system state vector for those cases where it cannot be taken as the composite component state vector. In the nonlinear case the computation of F and G requires the *global inversion* of appropriate nonlinear vector valued functions, hence sections 6 and 7 study this subject. Finally, section 8 discusses several extensions of the previous results; to systems with *improper models*, *Rosenbrock models*, etc.

2. COMPOSITE SYSTEM TRANSFER FUNCTIONS

What is the relationship between the *component connection model* and the *composite system transfer function*? This section presents a smattering of the answer to this question. For the past two chapters we have built and discussed the component connection model; now it is time to use these newly forged tools to analyze composite systems. The first step is to write the composite system transfer function in terms of the *component connection equations.* Recall that

$$b = Za \qquad (2.1)$$

and

$$a = L_{11}b + L_{12}u$$
$$y = L_{21}b + L_{22}u \qquad (2.2)$$

where all symbols have their usual denotations.

The task is to construct a rational matrix S such that

$$y = Su \qquad (2.3)$$

from equations 2.1 and 2.2. By substituting the equation for "a" into equation 2.1 we have

$$b = ZL_{11}b + ZL_{12}u \qquad (2.4)$$

Assuming for the moment that $(I - ZL_{11})^{-1}$ exists and solving for b yields

$$b = (I - ZL_{11})^{-1}ZL_{12}u \qquad (2.5)$$

Finally substituting equation 2.5 into the second equation of 2.2 yields the desired relationship

$$y = [L_{22} + L_{21}(I - ZL_{11})^{-1}ZL_{12}]u \qquad (2.6)$$
$$= Su$$

To date, the most convenient point of view is to consider S as a matrix valued function of the composite component transfer function, Z.

$$S = f(Z) = L_{22} + L_{21}(I - ZL_{11})^{-1}ZL_{12} \qquad (2.7)$$

Here f is a nonlinear function entirely determined by the L_{ij} connection matrices. Thus we call f the *connection function.* The point of view is consistent with the previous philosophy of separating component and connection information. The matrix Z represents the component information whereas the nonlinear function f represents the connection information. This separation is, in fact, the power behind the viewpoint.

Before developing the properties of f and S, it is necessary to investigate the existent questions surrounding the matrix inverse $(I-ZL_{11})^{-1}$. To begin, consider the classical but not well known *lemma of matrix algebra* which describes the relationship between the matrices $(I-ZL_{11})^{-1}$ and $(I-L_{11}Z)^{-1}$

THEOREM 2.8: Let X and Y be nxm and mxn matrices respectively. Then $(I+XY)^{-1}$ exists if and only if $(I+YX)^{-1}$ exists. Moreover when they exist $X(I+YX)^{-1} = (I+XY)^{-1}X$.

PROOF:

Step 1: Assume $(I+YX)^{-1}$ exists. Consider the product $(I - X(I+YX)^{-1}Y)$ $(I+XY)$. If this product is I, then $(I - X(I+YX^{-1}Y)$ is the inverse of $(I+XY)$ which implies that $(I+XY)^{-1}$ exists. Consider now the following string of equalities:

$$(I - X(I+YX)^{-1}Y) (I+XY) \tag{2.9}$$

$$= I - X(I+YX)^{-1}Y + XY - X(I+YX)^{-1}YXY$$

$$= I + X(I - (I+YX)^{-1} - (I+YX)^{-1}YX)Y$$

$$= I + X(I+YX)^{-1}(I + YX - I - YX)Y$$

$$= I$$

Therefore $(I+XY)^{-1}$ exists.

Step 2: A symmetric argument shows that if $(I+XY)^{-1}$ exists then so does $(I+YX)^{-1}$.

Step 3.: Finally we must show that

$$X(I+YX)^{-1} = (I+XY)^{-1}X \tag{2.10}$$

From step 1

$$(I+XY)^{-1}X = (I - X(I+YX)^{-1}Y)X \tag{2.11}$$

$$= X - X(I+YX)^{-1}YX$$

$$= X(I - (I+YX)^{-1}YX)$$

$$= X(I+YX)^{-1}(I + YX - YX)$$

$$= X(I + YX)^{-1}$$

The proof is now complete.

Applying this result to equation 2.6, we have two alternate forms of the connection function.

$$S = f(Z) = L_{22} + L_{21}(I - ZL_{11})^{-1}ZL_{12} \qquad (2.12)$$
$$= L_{22} + L_{21}Z(I - L_{11}Z)^{-1}L_{12}$$

Since Z and L_{11} are not necessarily square matrices, the inversion part of the connection function may be easier to compute in one form than in the other.

EXAMPLE 2.13: Consider the *Butterworth Filter* as shown in figure 2.14.

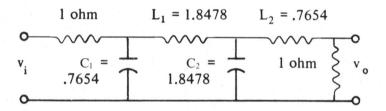

Figure 2.14. Butterworth Filter.

The component connection model is

$$
\begin{bmatrix} v_{C_1} \\ i_{L_1} \\ v_{C_2} \\ i_{L_2} \end{bmatrix}
=
\begin{bmatrix} 1/sC_1 & & & \\ & 1/sL_1 & & \\ & & 1/sC_2 & \\ & & & 1/sL_2 \end{bmatrix}
\begin{bmatrix} i_{C_1} \\ v_{L_1} \\ i_{C_2} \\ v_{L_2} \end{bmatrix}
\qquad (2.15)
$$

and

$$
\begin{bmatrix} i_{C_1} \\ v_{L_1} \\ i_{C_2} \\ v_{L_2} \\ \hline v_0 \end{bmatrix}
=
\left[\begin{array}{cccc|c} -1 & -1 & 0 & 0 & 1 \\ 1 & 0 & -1 & 0 & 0 \\ 0 & 1 & 0 & -1 & 0 \\ 0 & 0 & 1 & -1 & 0 \\ \hline 0 & 0 & 0 & 1 & 0 \end{array}\right]
\begin{bmatrix} v_{C_1} \\ i_{L_1} \\ v_{C_2} \\ i_{L_2} \\ \hline v_i \end{bmatrix}
\qquad (2.16)
$$

Here, we have chosen to embed the source and load resistors into the connection structure since they represent inaccessible physical entities and are not components in the usual sense of the term. Mathematically, however, it was equally possible to define a component connection model in which the source and load resistors were components.

Upon invoking either equation 2.7 or equation 2.12, and plugging in the component values

$$C_1 = L_2 = .7654 \tag{2.17}$$

and

$$C_2 = L_1 = 1.8478 \tag{2.18}$$

we obtain the normalized **4th order Butterworth filter function**

$$S(s) = \frac{1}{s^4 + 2.61\ s^3 + 3.41\ s^2 + 2.61\ s + 1} \tag{2.19}$$

The problem remaining is whether or not the inverse matrix $(I - ZL_{11})^{-1}$ exists for a sufficiently large class of interconnected dynamical systems. A yes answer justifies the use of equations 2.7 and 2.12 since S will then be a well defined transfer function matrix. It is important to realize that $(I-Z(s)L_{11})^{-1}$ will not exist for every value of s since $(I-Z(s)L_{11})^{-1}$ will be a rational matrix having poles at some finite number of points. Mathematically it is required that $\det(I-Z(s)L_{11})$ *not* be identically equal to zero for each $Z(s)$ and "almost all connections." In particular if $\det(I-Z(s)L_{11}) \not\equiv 0$, then by *Cramers Rule*

$$(I - Z(s)L_{11})^{-1} = \frac{Adj(I - Z(s)L_{11})}{\det(I - Z(s)L_{11})} \tag{2.20}$$

The obvious existence condition is that $(I - Z(s)L_{11})^{-1}$ exists if and only if $\det(I - Z(s)L_{11}) \not\equiv 0$ with appropriate conditions on the set of components and their interconnections. The intended goal is to show for a given set of components and a specified *connection structure* (L_{11} structure) that $\det(I - Z(s)L_{11}) \not\equiv 0$ for "almost all connections."

First, for a given set of components, define a connection structure to be an L_{11} matrix in which the zero entries are specified while the possible non-zero entries are considered variables. Physically this implies that we have specified a given connection structure (interconnection of components) in which the connection gains are left arbitrary. The notion allows for comments on sparse matrix connection structures. To rigorize these notions we take a *dimension theoretic* avenue. The vehicle will be the concepts of an algebraic hypersurface and an algebraic variety. Suppose

$f(x_1,...,x_n)$ is a *polynomial in n real variables* with possibly complex coefficients. The *zero set* of the polynomial is the set of points $(x_1,...,x_n)$ in $\mathbf{R}^n$ such that $f(x_1,...,x_n) = 0$.

EXAMPLE 2.21: Suppose $f(x_1,x_2) = (x_1+1)(x_2-1)$. The zero set of this polynomial is plotted in figure 2.22.

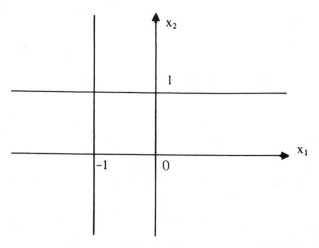

Figure 2.22. Zero set of $f(x_1,x_2) = (x_1+1)(x_2-1)$.

If f is a polynomial is n real variables, then the zero set of f has real *dimension* (n-1). On the other hand, if f is a polynomial in n-complex variables, then the zero set has complex dimension (n-1) which is equivalent to a real dimension of 2n-2. Since our approach is dimension oriented it is essential to be careful about such dimensional details.

Suppose a set V is a subset of $\mathbf{R}^n$, then V is an *algebraic hypersurface* if and only if V is the zero set of a non constant polynomial.

EXAMPLE 2.23: Consider the connection structure specified by

$$L_{11} = \begin{bmatrix} x & 0 \\ y & z \end{bmatrix} \tag{2.24}$$

Such an L_{11} connection structure is a three dimensional subspace of $\mathbf{R}^4$. Suppose

$$Z = \begin{bmatrix} s & 0 \\ 0 & 1 \end{bmatrix} \tag{2.25}$$

A set of hypersurfaces, parameterized by s, may be constructed as follows.

First consider the $\det(I+Z(s)L_{11})$.

$$\det(I-Z(s)L_{11}) = (1-sx)(1-z) \qquad (2.26)$$

For each value of s in **C**, one obtains the hypersurface

$$V_s = \left\{ (\frac{1}{s},y,z) \mid y \text{ and } z \text{ arbitrary;} \qquad (2.27) \right.$$

$$\left. (x,y,1) \mid x \text{ and } y \text{ arbitrary} \right\} .$$

For each non-zero value of s in **C**, one obtains the hypersurface with V_s being two dimensional in each case.

With these notions of hypersurface, define an *algebraic variety* to be the intersection of some finite set of hypersurfaces.

EXAMPLE 2.28: From example 2.23

$$\det(I-ZL_{11}) = (1-sx)(1-z) \qquad \textbf{(2.29)}$$

The algebraic variety associated with the set of hypersurfaces $\{V_s \mid s = s_1, s_2, s_3, \ldots , s_n\}$ is

$$V = \bigcap_{i=1}^{n} V_s = \{(x,y,1) \mid x \text{ and } y \text{ arbitrary} \} \qquad (2.30)$$

Finally we remark that a given connection structure may be viewed as an algebraic variety. As an example consider the connection structure

$$L_{11} = \begin{bmatrix} x & 0 \\ y & z \end{bmatrix} = \begin{bmatrix} x & w \\ y & z \end{bmatrix} \qquad (2.31)$$

with $w = 0$. This structure may be identified with the algebraic variety (hypersurface) defined by the zero set of the polynomial $p(x,y,z,w) = w$. Clearly such an algebraic variety is three dimensional. In the following assume that a given connection structure is identified with an r-dimensional algebraic variety.

For a given set of components and a given connection structure, (some r dimensional algebraic variety) a *property* is said to hold for *almost all connections* if it holds for all L_{11} matrices in the given structure except (possibly) those lying in an r–1 dimensional (or less) algebraic variety. Relative to the r dimensional algebraic variety, an r–1 dimensional (or less) algebraic variety is considered "small." In measure theoretic terms, the

Lebesque measure [18] of an r–1 dimensional variety is zero, relative to the r dimensional variety.

THEOREM 2.32: Let an interconnected dynamical system have a given connection structure and a given set of components with the connection model

$$b \;=\; Z(s)a$$

$$\left[\dfrac{a}{y}\right] \;=\; \left[\begin{array}{c|c} L_{11} & L_{12} \\ \hline L_{21} & L_{22} \end{array}\right]\left[\dfrac{b}{u}\right] \tag{2.33}$$

Then the composite system transfer function

$$S(s) \;=\; L_{22} + L_{21}(I{-}ZL_{11})^{-1}ZL_{12} \tag{2.34}$$
$$\;=\; L_{22} + L_{21}Z(I{-}L_{11}Z)^{-1}L_{12}$$

exists for almost all connections within the given connection structure.

PROOF: For S(s) to exist for almost all connections in a given connection structure it is sufficient to verify that $\det(I{-}Z(s)L_{11}) \not\equiv 0$ for almost all connection matrices in the given L_{11} structure.

First it is necessary to address the question as to whether or not there exists an L_{11} structure for which $\det(I{-}Z(s)L_{11}) \not\equiv 0$ for all L_{11} matrices in the structure. Fortunately the answer is no, since $L_{11} = [0]$ is an admissible connection matrix for any structure. Clearly $\det(I) = 1$. Therefore no connection structure, sparse or otherwise, is in general "bad."

Now suppose the given L_{11} structure is identified with an r-dimensional algebraic variety (L_{11} has r possible non-zero entries). It is sufficient to show that the L_{11} matrices for which $\det(I{+}Z(s)L_{11}) \equiv 0$ belongs to an algebraic variety of dimension r–1 or less.

For this, fix $s = s_0$ in $\bar{\mathbf{C}}$ and consider the zero set, V_{s_0}, of the polynomial $\det(I{-}Z(s_0)L_{11})$ where this determinant is a polynomial in the r variable entries of the L_{11} structure. For each s, the zero set of this polynomial determines a hypersurface Vs of dimension r–1 or less. Therefore the set of connections for which $\det(I{-}Z(s)L_{11} \equiv 0$, i.e., the determinant is equal to zero for all s, is

$$V \;=\; \bigcap_{s\in\mathbf{C}} V_s \subset V_{s_0} \tag{2.35}$$

Clearly any matrix for which $\det(I{-}Z(s)L_{11}) = 0$ for all s in $\mathbf{C}$ must necessarily belong to V_{s_0}. Thus all the "bad" connection matrices must belong to the at most $(r-1)$ dimensional algebraic variety V_{s_0}. This completes the proof.

EXAMPLE 2.36: Let

$$Z = \begin{bmatrix} 1 & 0 & 0 \\ 0 & \dfrac{1}{s} & 0 \\ 0 & 0 & s \end{bmatrix} \tag{2.37}$$

and let the connection structure be given by

$$L_{11} = \begin{bmatrix} 0 & x & 0 \\ 0 & 0 & y \\ z & 0 & 0 \end{bmatrix} \tag{2.38}$$

Clearly such a connection structure belongs to a three dimensional algbebraic variety. Consider that

$$\det(I-ZL_{11}) = 1 - xyz \tag{2.39}$$

Now $\det(I-Z(\underline{s})L_{11}) \not\equiv 0$ for any L_{11} matrix whose entries lie in the algebraic variety $V = \{(x,y,1/xy)\,|\,x \text{ and } y \text{ arbitrary}\}$. Observe that V is two dimensional and thus small relative to the given three dimensional connection structure.

Notions of the kind developed in this section are related to the theory of *Algebraic Geometry*.[5] Within this theory there are some interesting physical consequences of the above theorem. First, for any L_{11} matrix with $\det(I-Z(s)L_{11}) \not\equiv 0$ then "for almost all" arbitrarily small perturbations of the nonzero entries of L_{11}, resulting in say $\hat{L}_{11}$, then $\det(I+Z(s)\hat{L}_{11}) \not\equiv 0$.

A second view is probabilistic. For any absolutely continuous probability distribution[8] characterizing the nonzero entries of a given connect on structure, the probability that $\det(I-Z(s)L_{11}) \equiv 0$ for any L_{11} matrix in the connection structure is zero. All of these viewpoints indicate that $f(Z)$ or $S(s)$ is a well defined entity for *most* real world connections.

If the system is a *series parallel system* then the connection function is always well defined. This follows since series parallel systems always have a *block strictly lower triangular* L_{11} matrix implying that $\det(I-ZL_{11}) = 1$ for all s.

For a general *hierarchical system* $(I-Z(s)L_{11})$ os *block lower triangular* with diagonal blocks $(I-Z(s)_i L_{11}^{ii})$. Here $Z(s)_i$ denotes the composite system transfer function for the i-th level and L_{11}^{ii} denotes the corresponding diagonal block of L_{11}. Invoking the formula for inverting a block triangular matrix, the matrix $(I-Z(s)L_{11})$ is invertible if and only if each of the diagonal blocks is invertible. Moreover the inverse of $(I-Z(s)L_{11}^{ii})$ may be explicitly written in terms of the inverses of $(I-Z(s)_iL_{11}^{ii})$.[19]

In any case whether the system is hierarchical or not the connection function will be well defined for most real world systems. With this important question settled it is appropriate to begin investigating the properties of the connection function.

3. THE CONNECTION FUNCTION

The *connection function*, f, is a mapping from the space of *composite component transfer function matrices*, Z, to the space of *composite system transfer function matrices*, S.

$$S = f(Z) = L_{22} + L_{21}(I-ZL_{11})^{-1}ZL_{12} \qquad (3.1)$$

The properties of the connection function completely describe the relationship between Z and S. Basically, then, this section delineates some of the properties of f, illustrating the power of this point of view.

First recall the interpretation of the *component connection model* being a *generalized state model*. By letting $Z=1/s$, $L_{11}=A$, $L_{12}=B$, $L_{21}=C$, and $L_{22}=D$, then the formula I.4.39 for computing the transfer function matrix of a state model is consistent with the connection function formula, 3.1. This analogy permits the application of state space theory to the study of the connection function, however, we defer this to a later section.

Clearly the connection function is a nonlinear matrix valued function of a matrix variable. Fortunately it is surprisingly tractable. The key to this tractability lies in the *decomposition of f* into two functions, h and g, as $f = h \circ g$. The function g is nonlinear with an analytically expressible global inverse and h is an *affine function* (linear plus a constant). Much of the ambiguity in the system synthesis problem lies with the failure of h to admit an inverse so that point of view will shortly offer a quick harvest of results.

Since we will encounter numerous nonlinear functions in the remainder of the text before proceeding to our analysis of the connection function it will be helpful to formalize some of the notation and terminology required for the analysis of such functions. Formally, a *function*, f, is a "relation" between the elements of two sets, A and B, termed the *domain* and *range* of the function, respectively. In particular, the function f associates with each element, $a \epsilon A$, a unique element $f(a) \epsilon B$. In general, a given element $b \epsilon B$ may be the image of several elements in A, say b = f(a) = f(a'), or it may not lie in the image of any a ∈ A. The set of elements, a ∈ A, such that f(a) = b is termed the *inverse image* of b and is denoted by f⁻¹(b). As indicated above f⁻¹(b) may contain several elements, a unique element or no elements of A at all. The concept of an inverse image can be extended from a single point b ∈ B to a subset $\hat{B} \subset B$ via the equality

$$f^{-1}(\hat{B}) = \{a \in \hat{A} \mid f(a) \in \hat{B}\} \qquad (3.2)$$

while the image of a set $\hat{A} \subset A$ may be defined via

$$f(\hat{A}) = b \in B \mid b = f(a) \text{ for some } a \in \hat{A} \qquad (3.3)$$

These ideas are illustrated pictorially in figure 3.4.

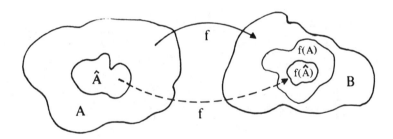

Figure 3.4. The image and inverse image of f.

If the function f maps A to B then an *inverse function* is a function mapping B to A which in some sense reverses the action of f. In particular, a *left inverse*, f⁻ᴸ, for f is a function mapping B to A such that f⁻ᴸ of = I_A. That is, the application of f followed by the application of f⁻ᴸ is the identity on A (I_A(a) = a for all a ∈ A). Similarly, a *right inverse*, f⁻ᴿ, for f is a function mapping B to A such that fof⁻ᴿ = I_B where I_B is the *identity function* on B. Finally, if a function f⁻¹ is simultaneously both a right and left inverse of f we call it the inverse of f. This terminology is justified since f⁻¹ is unique if it exists though, in general, neither f⁻ᴸ nor f⁻ᴿ need be unique. Indeed, a little manipulation of these concepts will reveal that a left inverse exists if and only if f is *one-to-one* while f^{-R} exists if and only if f is *onto*. I.e., f is one-to-one if at most one element in A is mapped to each element in $B(f^{-1}(b)$ contains exactly one element or it is void) whereas it is onto if every element of B is the image of at least one element of A $(f^{-1}(b)$ contains of least one element in A or equivalently f(A) = B). Using these criteria the reader may now readily verify the above stated *existence and uniqueness conditions* simply by constructing the required inverse functions. Of course it follows that an inverse exists if and only if f is both one-to-one and onto.

While the basic concept of a function is defined in terms of an abstract set most of the functions which we will encounter map R^n to R^m or, as in the case of the connection function, they may map n by m matrices to p by q matrices. In either case the additional structure inherent in these *topological vector spaces* allows us to formalize the concepts of linearity, continuity, and differentiation for our functions. Recall from classical *linear algebra* that a function defined on a vector space is *linear* if f(cx + dy) = cf(x) + df(y) for any two vectors (or matrices) x and y and scalars c and d. Moreover, in the case where f maps R^n to R^m f may be represented by a matrix, F, defined with respect to the natural *bases* for these vector spaces such that f(x) = Fx. Of course, in this case f^{-1} is represented by the inverse matrix F^{-1} and similarly for f^{-t} and f^{-R}.

In the case where the range and domain for our functions are spaces of matrices this simple matrix representation for f is subsumed by a more complex *tensor product* representation. Alternatively, one may transform the matrices into vectors in terms of which a matrix representation for f may be formulated. Here, we simply take a matrix M = row(M_i) and identify it with the vector vec(M) = col(M_i) which transforms an n by m matrix into an mn-vector. For instance

$$\text{vec} \begin{bmatrix} 1 & 3 \\ 0 & 7 \end{bmatrix} = \begin{bmatrix} 1 & 0 & 3 & 7 \end{bmatrix}^t \tag{3.5}$$

Clearly, the *"vec operation"* is simply a change of notation which preserves all of the properties of the given matrix while displaying it in a new pattern. It can, however, be employed to permit us to transform a matrix valued function of a matrix valued variable into the more classical vector valued function of a vector valued variable. For instance, the connection function S = f(Z) may be transformed into the equivalent function vec(S) = f(vec(Z)) where convenient. Now, if this function were linear (verify that this is true if and only if $L_{11} = 0$) a matrix representation for f could be formulated.

Intuitively, one can define *continuity* for a vector valued function of a vector valued variable or a matrix valued function of a matrix valued variable by defining an appropriate *norm* on the range and domain spaces for the function and using it in lieu of the absolute value operation in a classical epsilon-delta definition. For our purposes, however, the intuitive concept of continuity to the effect that "small" changes in the input are reflected in "small" changes in the output suffices. Clearly, if a function f: $\mathbf{R}^n \rightarrow \mathbf{R}^m$ is broken up into m *coordinate functions*; $f(x) = \text{col}(f_i(x))$, where f_i: $\mathbf{R}^n \rightarrow \mathbf{R}$ is the scalar valued function of n variables representing the response of f in the ith coordinate to x; then f is continuous if and only if each f_i is continuous. Finally, we call a continuous function which admits a continuous inverse a *homeomorphism*.

As with continuity the *derivative* of a nonlinear function can be defined abstractly in terms of a norm on the spaces in which f is defined. For our purposes, however, the derivative of such a function will always take the form of an appropriate matrix of partial derivatives which we may thus adopt as a definition for df/dx. In particular, if f: $\mathbf{R}^n \rightarrow \mathbf{R}^m$ is a vector valued function of a vector valued variable its derivative reduces to the *Jacobian matrix*

$$J_f = \frac{df}{dx} = \begin{bmatrix} \dfrac{df_1}{dx_1} & \dfrac{df_1}{dx_2} & \cdots & \dfrac{df_1}{dx_n} \\[2mm] \dfrac{df_2}{dx_1} & \dfrac{df_2}{dx_2} & & \dfrac{df_2}{dx_n} \\[2mm] \cdot & \cdot & & \cdot \\ \cdot & \cdot & & \cdot \\ \cdot & \cdot & & \cdot \\[1mm] \dfrac{df_m}{dx_1} & \dfrac{df_m}{dx_2} & \cdots & \dfrac{df_m}{dx_n} \end{bmatrix} \tag{3.6}$$

where df_i/dx_j is the *partial derivative* of f_i with respect to x_j. Note, however, that *total derivative* notation is used exclusively throughout the text whether we are dealing with functions of one or several variables. Furthermore, the reader should not confuse the Jacobian matrix with the *Jacobian* which is the determinant of the Jacobian matrix for the case where m = n. Of course, if n = m = 1 the Jacobian matrix is just the classical derivative whereas if m = 1 it reduces to the classical *gradient vector* (which we express as a row matrix).

In the case where one is dealing with a matrix valued function of a matrix valued variable f(x), one would like the derivative to take the form of an array which displays all of the partial derivatives of the coordinate functions f_{ij} with respect to each entry x_{rs} in the matrix X. Since the text is written on two dimensional paper the representation of this four dimensional array may present a problem. Fortunately, we often deal with matrix valued functions of a scalar variable for which we may write

$$
J_f = \frac{df}{dx} = \begin{bmatrix} \dfrac{df_{11}}{dx} & \dfrac{df_{12}}{dx} & \cdots & \dfrac{df_{1n}}{dx} \\[2ex] \dfrac{df_{21}}{dx} & \dfrac{df_{22}}{dx} & & \dfrac{df_{2n}}{dx} \\[1ex] \cdot & \cdot & & \cdot \\ \cdot & \cdot & & \cdot \\ \dfrac{df_{m1}}{dx} & \dfrac{df_{m2}}{dx} & \cdots & \dfrac{df_{mn}}{dx} \end{bmatrix} \tag{3.7}
$$

In the general case where one desires to work with a true matrix valued function of a matrix valued variable we generally transform it into a vector valued function of a vector valued variable via the *"vec operation"* and then work with the Jacobian matrix for the resultant function, Finally, we note that most of the classical rules of the calculus also extend to these arrays of partial derivatives so long as one takes care to assure that the array employed is compatible with the matrix algebra.[1] The required details will be discussed as required in the sequel.

Returning to our investigation of the connection function we have the following theorem.

THEOREM 3.8: The connection function given by the equality $S = f(Z) = L_{22} + L_{21}(I - ZL_{11})^{-1}ZL_{12}$ has the decomposition

$$
f = hog \tag{3.9}
$$

where

$$
Q = g(Z) = (I{-}ZL_{11})^{-1}Z \tag{3.10}
$$

and

$$
S = h(Q) = L_{22} + L_{21}QL_{12} \tag{3.11}
$$

Moreover, h is affine and g is globally invertible as

$$
Z = g^{-1}(Q) = (I{+}QL_{11})^{-1}Q. \tag{3.12}
$$

PROOF: Showing $f = h \circ g$ is an **exercise** in the art of substitution. Clearly h is affine since $L_{21}QL_{12}$ is linear in Q. The only troublesome step of the proof is showing that g is globally invertible with inversion formula 3.8. For this we show that g has a left and right inverse. Consider that $g(Z) = Q$ implies $g^{-1}(Q) = (I+QL_{11})^{-1}Q$ as per 3.8. By assumption $Q = (I-ZL_{11})^{-1}Z$ which implies

$$\begin{aligned} g^{-1}(Q) &= (I+(I-ZL_{11})^{-1}ZL_{11})^{-1}(I-ZL_{11})^{-1}Z \qquad (3.13)\\ &= [(I-ZL_{11}) + (I-ZL_{11})(I-ZL_{11})^{-1}ZL_{11}]^{-1}Z\\ &= Z \end{aligned}$$

This shows that $g^{-1}(g(Z)) = Z$ or that g is left invertible. Similarly one may verify that $g(g^{-1}(Q)) = Q$. The conclusion of the theorem now follows.

The matrix Q plays a fundamental role in the theory thus taking the name, the *internal composite system transfer function matrix*. The terminology follows because if the external system variables coincided with the internal system variables (i.e. $L_{12} = L_{21} = I$ and $L_{22} = 0$) then Q would be the composite system transfer function matrix.

EXAMPLE 3.14: Consider the system described by the block diagram of figure 3.16. Here, both components are *integrators* hence

$$Z(s) \quad = \quad \begin{bmatrix} 1/s & 0 \\ 0 & 1/s \end{bmatrix} \qquad (3.15)$$

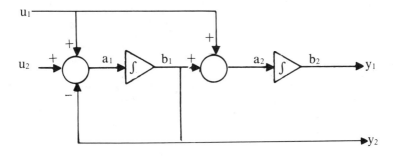

Figure 3.16. System composed of two integrators.

The connection equations are

$$
\begin{bmatrix} a_1 \\ a_2 \\ -- \\ y_1 \\ y_2 \end{bmatrix}
=
\left[
\begin{array}{cc:cc}
-1 & 0 & 1 & 1 \\
1 & 0 & 1 & 0 \\
\hdashline
0 & 1 & 0 & 0 \\
1 & 0 & 0 & 0
\end{array}
\right]
\begin{bmatrix} b_1 \\ b_2 \\ -- \\ u_1 \\ u_2 \end{bmatrix}
\tag{3.17}
$$

For this system

$$
(I-ZL_{11}) =
\begin{bmatrix} 1 & 0 \\ 0 & 1 \end{bmatrix}
-
\begin{bmatrix} 1/s & 0 \\ 0 & 1/s \end{bmatrix}
\begin{bmatrix} -1 & 0 \\ 1 & 0 \end{bmatrix}
\tag{3.18}
$$

$$
=
\begin{bmatrix} \dfrac{s+1}{s} & 0 \\ \dfrac{-1}{s} & 1 \end{bmatrix}
$$

and

$$
(I-ZL_{11})^{-1} =
\begin{bmatrix} \dfrac{s}{s+1} & 0 \\ \dfrac{1}{s+1} & 1 \end{bmatrix}
\tag{3.19}
$$

Thus the internal composite system transfer function matrix is given by

$$
Q = (I-ZL_{11})^{-1}Z =
\begin{bmatrix} \dfrac{1}{s+1} & 0 \\ \dfrac{1}{s(s+1)} & \dfrac{1}{s} \end{bmatrix}
\tag{3.20}
$$

while the external composite system transfer function matrix is given by

$$S \;=\; L_{22} + L_{21}QL_{12}$$

$$= \begin{bmatrix} 0 & 0 \\ \\ 0 & 0 \end{bmatrix} + \begin{bmatrix} 0 & 1 \\ \\ 1 & 0 \end{bmatrix} \begin{bmatrix} \dfrac{1}{s+1} & 0 \\ \\ \dfrac{1}{s(s+1)} & \dfrac{1}{s} \end{bmatrix} \begin{bmatrix} 1 & 1 \\ \\ 1 & 0 \end{bmatrix}$$

$$\hspace{11cm}(3.21)$$

$$= \begin{bmatrix} \dfrac{s+2}{s(s+1)} & \cdot & \dfrac{1}{s(s+1)} \\ \\ \dfrac{1}{s+1} & & \dfrac{1}{s+1} \end{bmatrix}$$

The *decomposition theorem* splits the apparently complex connection function into two parts, each being amenable to analysis; herein lies the power and value of the theorem.[12]

Next consider the question of *inverting the connection function*: physically given a composite system transfer function matrix find a set of connections and/or components which *synthesize* it. The immediate inversion formula for f is

$$Z \;=\; f^{-1}(S) \;=\; (g^{-1} \circ h^{-1})(S) \hspace{3cm}(3.22)$$

Theorem 3.8 assures that g^{-1} exists globally and gives a specific formula for its computation. Consequently the inversion of f reduces to computing the inverse of the affine function h. After adopting appropriate notation, this again simplifies to a matrix inversion problem, nicely lending itself to computational techniques.

DEFINITION 3.23: Let X be an mxn matrix and Y be pxq. The *matrix tensor product* of X with Y, denoted $X \otimes Y$, is an mpxnq matrix

$$X \otimes Y \;=\; \begin{bmatrix} x_{11}Y & x_{12}Y & \cdots & x_{1n}Y \\ x_{21}Y & x_{22}Y & \cdots & x_{2n}Y \\ \cdot & \cdot & & \cdot \\ \cdot & & \cdot & \cdot \\ & & \cdot & \\ x_{m1}Y & x_{m2}Y & \cdots & x_{mn}Y \end{bmatrix} \hspace{2cm}(3.24)$$

where x_{ij} is the ij-th entry of the matrix X.

3.25. PROPERTIES OF THE MATRIX TENSOR PRODUCT:

(a) $(X \otimes Y)^t = X^t \otimes Y^t$

(b) $(X+Y) \otimes (W+Z) = X \otimes W + X \otimes Z + Y \otimes W + Y \otimes Z$

(c) $(X \otimes Y)(W \otimes Z) = XW \otimes YZ$

(d) $vec(XYZ) = [Z^t \otimes X] vec(Y)$

(e) $(X \otimes Y)^{-1} = X^{-1} \otimes Y^{-1}$

(f) $(X \otimes Y)^{-L} = X^{-L} \otimes Y^{-L}$

(g) $(X \otimes Y)^{-R} = X^{-R} \otimes Y^{-R}$

Since g is globally invertible, our analysis of the connection function centers on inverting $S = h(Q) = L_{22} + L_{21}QL_{12}$. Taking "vec" of both sides yields

$$vec(S) = vec(L_{22}) + vec(L_{21}QL_{12}) \qquad (3.26)$$

which reduces to

$$vec(S-L_{22}) = \left[L_{12}^t \otimes L_{21} \right] vec(Q) \qquad (3.27)$$

Via 3.25d. Equation 3.27 indicates the dependence of h^{-1} on the inversion of the matrix $L_{12}^t \otimes L_{21}$. This matrix is seldom invertible in the classical sense forcing a study of its *left and right invertibility*. Consider the following derivation if $L_{12}^t \otimes L_{21}$ is assumed right invertible. This implies

$$\begin{aligned} vec(Q) &= (L_{12}^t \otimes L_{21})^{-R} vec(S-L_{22}) \\ &= (L_{12}^{-1})^t \otimes L_{21}^{-R} vec(S-L_{22}) \end{aligned} \qquad (3.28)$$

Applying the *inverse vec operator*[13] we obtain

$$Q = L_{21}^{-R}(S-L_{22})L_{12}^{-L} \qquad (3.29)$$

A similar manipulation arises if $(L_{12}^t \otimes L_{21})$ is left invertible. However in this case Q would be unique when it exists whereas in the above derivation Q is not unique but is assured to exist. Summarizing the above and recalling that g is always invertible we have the following theorem.

THEOREM 3.30: The connection function is invertible if and only if L_{12} and L_{21} are invertible matrices. It has a left inverse if and only if L_{12} has a right inverse and L_{21} has a left inverse; it has a right inverse if and only if L_{12} has a left inverse and L_{21} has a right inverse.

For many "real world" systems, the dimension of the composite system input and output vectors, p and q respectively, are much smaller than the dimensions of the composite component input and output vectors, m and n respectively. In other words, L_{12} has more rows than columns while L_{21} has more columns than rows. Thus one expects L_{12} to have a left inverse, provided its columns are *linearly independent*,[13] but not a right inverse; similarly one suspects that L_{21} will have a right inverse, provided its rows are linearly independent, but not a left inverse. Therefore it is highly likely that the connection function will admit a right but not a left inverse.

Physically this implies it is reasonably likely to find a set of components to satisfy the equality $S = f(Z)$. However knowing S and given the system connections it would be difficult to determine which of the several possible Z's satisfying this equation we had.

The former problem is the essence of *system synthesis*. Here given S, one desires a set of components realizing the system transfer function matrix. Unfortunately, in practice there are usually additional constraints on Z such as Z *positive definite* for *passive network synthesis* and $Z = 1/s$ for *state space realizability theory*. This transforms the system synthesis problem into a much more delicate and complex task. On the other hand the left inversion of the connection function is the essence of the *system diagnosability* problem which is insoluble without the addition of extra composite system inputs and outputs (*test points*) and/or making an a-priori assumption about Z to aid in the solution of the equation $S = f(Z)$.

EXAMPLE 3.31: Consider an interconnected dynamical system with connection equations

$$\begin{bmatrix} a_1 \\ a_2 \\ \hline y \end{bmatrix} = \begin{bmatrix} 0 & 0 & | & 1 \\ 2 & 0 & | & 2 \\ \hline 3 & 0 & | & 1 \end{bmatrix} \begin{bmatrix} b_1 \\ b_2 \\ \hline u \end{bmatrix} \tag{3.32}$$

Here L_{12} and L_{21} have the following left and right inverses respectively

$$L_{21}^{-L} = \begin{bmatrix} 1 & 0 \end{bmatrix}, \quad L_{21}^{-R} = \begin{bmatrix} 1/3 \\ 1 \end{bmatrix} \tag{3.33}$$

Recalling that $Z = g^{-1}(Q) = (I+QL_{11})^{-1}Q$ and substituting equation 3.29 for

Q symbolically yields

$$Z = (I+L_{21}^{-R}(S-L_{22})L_{12}^{-L}L_{11})^{-1}L_{21}^{-R}(S-L_{22})L_{12}^{-L} \quad (3.34)$$

Substituting the matrices of equation 3.33 into 3.34 yields

$$Z = \begin{bmatrix} \frac{S-1}{3} & 0 \\ S-1 & 0 \end{bmatrix} \quad (3.35)$$

Suppose it is required that S be the *all pass function*

$$S = \frac{s-1}{s+1} \quad (3.36)$$

then one composite component connection matrix realizing S with the specified connection structure is

$$Z(s) = \begin{bmatrix} \frac{-2}{3(s+1)} & 0 \\ \frac{-2}{s+1} & 0 \end{bmatrix} \quad (3.37)$$

Of course, since this system is not diagnosable, 3.37 is not the only, and most likely, not the best $Z(s)$. All one need do is change the appropriate left and right inverses in 3.33 to obtain a different $Z(s)$ also realizing the all pass function.

One final aspect of the connection function having important ramifications is its *differentiation* with respect to a parameter affecting Z but not the L_{ij} connection matrices. Suppose the matrix Z varies with respect to a parameter, r. Our goal is to develop an expression for the variation in the composite system transfer function with respect to r, dS/dr, in terms of the connection matrices. The derivation assumes dZ/dr is known a priori. This investigation will serve as the foundation of the sensitivity analysis discussed in chapter V.

Recall that $S = h(Q) = L_{21}QL_{12} + L_{22}$. This implies

$$\frac{dS}{dr} = \frac{d}{dr}(L_{21}QL_{12} + L_{22}) \quad (3.38)$$

$$= L_{21}\frac{dQ}{dr}L_{12}$$

since the L_{ij} matrices are constant with respect to r. Thus our task reduces to developing an expression for dQ/dr. The following identity[1] stated without proof facilitates this endeavor.

$$\frac{dX^{-1}}{dr} = -X^{-1} \frac{dX}{dr} X^{-1} \tag{3.39}$$

Recall that $Q = g(Z) = (I-ZL_{11})^{-1}Z$. Taking the derivative of Q with respect to r yields

$$\frac{dQ}{dr} = (I-ZL_{11})^{-1} \frac{dZ}{dr} + [\frac{d}{dr}(I-ZL_{11})^{-1}]Z \tag{3.40}$$

Applying the formula of **3.39** to this expression yields

$$\frac{dQ}{dr} = [(I-ZL_{11})^{-1}]\frac{dZ}{dr}[I-L_{11}(I+ZL_{11})^{-1}Z] \tag{3.41}$$

Combining **3.38** and **3.41**, the required formula is

$$\frac{dS}{dr} = [L_{21}(I-ZL_{11})^{-1}]\frac{dZ}{dr}[I+L_{11}(I-ZL_{11})^{-1}Z]L_{12} \tag{3.42}$$

We formalize this discussion in the following theorem.

THEOREM 3.43: Let an interconnected dynamical system have the usual algebraic connection matrices. Suppose Z varies with respect to r and the L_{ij} matrices are independent of r. Then

$$\frac{dS}{dr} = [L_{21}(I-ZL_{11})^{-1}]\frac{dZ}{dr}[I+L_{11}(I-ZL_{11})^{-1}Z]L_{12} \tag{3.44}$$

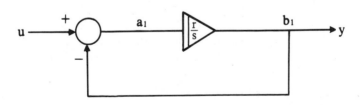

Figure 3.45. Feedback loop containing integrator.

EXAMPLE 3.46: Applying equation 3.44 to the feedback loop of figure 3.45 and differentiating with respect to the integrator gain r we obtain

$$\frac{dS}{dr} = \frac{s}{(s+r)^2} \tag{3.47}$$

which is a function of the complex variable s and the parameter r.

4. SPARSE MATRIX INVERSION

The difficulty in *computing the composite system transfer function matrix*, S, from the component connection model lies in the inversion of the matrix $(I-ZL_{11})$. Similarly, in analyzing the connection function, the inversion of $(I-ZL_{11})$ again is essential. The nxn matrix $(I-ZL_{11})$ typically has extremely large dimensions. Therefore, its efficient inversion is fundamental to developing viable algorithms for the analysis and design of interconnected dynamical systems. Fortunately, $(I-ZL_{11})$ is generally *sparse*. Hence it is amenable to specialized *inversion techniques* exploiting this sparseness.

The first impediment underlying the design of inversion algorithms for sparse matrices is that their inverse is habitually non-sparse.

EXAMPLE 4.1: Below is a 10x10 matrix M. It has twenty non-zero entries.

$$M = \begin{bmatrix} 1 & 0 & 1 & 0 & 0 & 0 & 0 & 0 & 0 & 0 \\ 0 & 1 & 0 & 1 & 0 & 0 & 0 & 0 & 0 & 0 \\ 1 & 0 & 0 & 0 & 0 & 0 & 0 & 0 & 1 & 0 \\ 0 & 0 & 0 & 0 & 0 & 1 & 1 & 0 & 0 & 0 \\ 0 & 1 & 0 & 0 & 0 & 0 & 0 & 1 & 0 & 0 \\ 0 & 0 & 0 & 0 & 1 & 1 & 0 & 0 & 0 & 0 \\ 0 & 0 & 0 & 0 & 0 & 0 & 1 & 0 & 0 & 1 \\ 0 & 0 & -1 & 1 & 0 & 0 & 0 & 0 & 0 & 0 \\ 0 & 0 & 0 & 0 & 1 & 0 & 0 & 0 & 1 & 0 \\ 0 & 0 & 0 & 0 & 0 & 0 & 0 & 1 & 0 & 1 \end{bmatrix} \tag{4.2}$$

A little algebra reveals that M^{-1} has all entries non-zero as equation 4.3 elucidates. This is not an artificial example. The *non-sparseness of the inverse of a sparse matrix* is the norm rather than the exception. Another example of the phenomena is the *tri-diagonal matrix* which always has a full inverse unless the entries are meticulously chosen to force cancellation of the terms in the cofactors of the matrix.[19]

$$M^{-1} = .5 \begin{bmatrix} 1 & -1 & 1 & -1 & 1 & 1 & 1 & 1 & -1 & -1 \\ -1 & 1 & 1 & -1 & 1 & 1 & 1 & -1 & -1 & -1 \\ 1 & 1 & -1 & 1 & -1 & -1 & -1 & -1 & 1 & 1 \\ 1 & 1 & -1 & 1 & -1 & -1 & -1 & 1 & 1 & 1 \\ 1 & -1 & -1 & -1 & 1 & 1 & 1 & 1 & 1 & -1 \\ -1 & 1 & 1 & 1 & -1 & 1 & -1 & -1 & -1 & 1 \\ 1 & -1 & -1 & 1 & 1 & -1 & 1 & 1 & 1 & -1 \\ 1 & -1 & -1 & 1 & 1 & -1 & -1 & 1 & 1 & 1 \\ -1 & 1 & 1 & 1 & -1 & -1 & -1 & -1 & 1 & 1 \\ -1 & 1 & 1 & -1 & -1 & 1 & 1 & -1 & -1 & 1 \end{bmatrix} \qquad (4.3)$$

Given this reality, two problems arise in formulating *sparse matrix inversion algorithms*. First it is necessary to blueprint inversion algorithms whose complexity grows with the number of non-zero matrix entries, not the dimension of the matrix. Secondly, it is necessary to resolve the problem of storing the inverse matrix which typically takes far more memory than the original matrix. Both problems unravel by observing that *the inverse of a sparse triangular matrix is sparse.*

Suppose an nxn $M = [m_{ij}]$ is *lower triangular*—i.e., $M_{ij} = 0$ for $i < j$. Then M is invertible if and only if all diagonal entries are non-zero and M^{-1} is also lower triangular. As such, one can compute M^{-1} entry by entry simply by solving the n^2 scalar equations of the matrix equality

in the proper sequence. The name of the process is *back substitution* and it is outlined as follows. The k-k entry of the product matrix satisfies the equality

$$\sum_{i=1}^{n} m_{ki}^{-1} m_{ik} = 1 \qquad (4.5)$$

implying

$$m_{kk}^{-1} = \frac{1}{m_{kk}} \qquad (4.6)$$

This follows because the triangularity of both M and M^{-1} guarantees that $m_{ij} = 0$ for $i < k$ and $m_{ki}^{-1} = 0$ for $k < i$. Clearly then the diagonal entries of M^{-1} are the scalar inverses of the diagonal entries of M. To compute the off-diagonal entries of M^{-1}, say the k-j-th entry, $j < k$, consider the formula

$$\sum_{i=1}^{n} m_{ki}^{-1} m_{ij} = 0 \tag{4.7}$$

The next step is to sequentially solve equation 4.7 for all off diagonal entries in the k-th row of M^{-1}. First solve for $m_{k,k-1}^{-1}$ which requires knowledge only of m_{kk}^{-1}. Once $m_{k,k-1}^{-1}$ is known we then use this value to compute $m_{k,k-2}^{-1}$ which is used in computing $m_{k,k-3}^{-1}$. Essentially this process sequentially solves for all entries in the k-th row of M^{-1} by "back substitution." Repeating the process for each row, one fabricates the entire inverse matrix. Of course a similar algorithm applies to the inversion of an upper triangular matrix.

More important than the capacity to readily invert a triangular matrix is the fact that the number of non-zero terms in the summation of equation 4.7 is extremely small for a sparse M. This follows since a term is non-zero if and only if both m_{ij} and m_{ki}^{-1} are non-zero. Thus the computation of the summation in equation 4.7 is relatively easy and has good numerical accuracy. Moreover, if M is sparse the probability of either m_{ij} or m_{ki}^{-1} being zero for all values of i in the interval between $j+1$ and k is sufficiently high to assure that M^{-1} is also sparse.

EXAMPLE 4.8: Below is a 10x10 lower triangular matrix, M.

$$M = \begin{bmatrix}
1 & 0 & 0 & 0 & 0 & 0 & 0 & 0 & 0 & 0 \\
1 & -1 & 0 & 0 & 0 & 0 & 0 & 0 & 0 & 0 \\
0 & 0 & 1 & 0 & 0 & 0 & 0 & 0 & 0 & 0 \\
0 & 0 & 1 & 2 & 0 & 0 & 0 & 0 & 0 & 0 \\
1 & 0 & 0 & 0 & 1 & 0 & 0 & 0 & 0 & 0 \\
0 & 1 & 0 & 0 & 0 & -1 & 0 & 0 & 0 & 0 \\
1 & 0 & 0 & 0 & 1 & 0 & 1 & 0 & 0 & 0 \\
0 & 0 & 0 & 0 & 0 & 0 & 0 & 1 & 0 & 0 \\
0 & 0 & 0 & 0 & 1 & 0 & 0 & 0 & 1 & 0 \\
2 & 0 & 0 & 0 & 0 & 0 & 0 & 0 & 0 & 1
\end{bmatrix} \tag{4.9}$$

One computes the inverse of M one row at a time. For example to compute the fourth row of M^{-1} first computs $m_{44}^{-1} = .5$ via equation 4.6. Computing m_{43}^{-1} via equation 4.7 yields

$$m_{43}^{-1} = -.5 \tag{4.10}$$

Using this value in equation 4.7, we may now compute m_{42}^{-1} via

$$m_{42}^{-1} = \frac{-1}{-1} [.5(0) - .5(0)] = 0 \qquad (4.11)$$

Repeating the process again we have

$$m_{41}^{-1} = \frac{-1}{-1} [.5(0) - .5(0) + 0(1)] = 0 \qquad (4.12)$$

continuing this procedure for each row yields the inverse matrix.

$$M^{-1} = \begin{bmatrix}
1 & 0 & 0 & 0 & 0 & 0 & 0 & 0 & 0 & 0 \\
1 & -1 & 0 & 0 & 0 & 0 & 0 & 0 & 0 & 0 \\
0 & 0 & 1 & 0 & 0 & 0 & 0 & 0 & 0 & 0 \\
0 & 0 & -.5 & .5 & 0 & 0 & 0 & 0 & 0 & 0 \\
-1 & 0 & 0 & 0 & 1 & 0 & 0 & 0 & 0 & 0 \\
1 & -1 & 0 & 0 & 0 & 1 & 0 & 0 & 0 & 0 \\
0 & 0 & 0 & 0 & -1 & 0 & 1 & 0 & 0 & 0 \\
0 & 0 & 0 & 0 & 0 & 0 & 0 & 1 & 0 & 0 \\
1 & 0 & 0 & 0 & -1 & 0 & 0 & 0 & 1 & 0 \\
-2 & 0 & 0 & 0 & 0 & 0 & 0 & 0 & 0 & 1
\end{bmatrix} \qquad (4.13)$$

Observe that M has eighteen non-zero entries while M^{-1} has nineteen. This example manifests our theoretical prediction that the degree of sparsity for a triangular matrix and its inverse are relatively equal.

The key to the solution of the general sparse matrix inversion problem is the above phenomena— the inverse of a sparse triangular matrix is sparse. In the general problem one factors an arbitrary sparse matrix as the product of a sparse upper and a sparse lower triangular matrix. After inverting each by back substitution one stores the resultant sparse triangular inverses separately without taking their product to explicitly express the inverse. This construction is called an *LU-factorization*. It completely dissolves both fundamental problems connected with sparse matrix inversion: the actual inversion process and the storage of the inverses as a product of triangular sparse matrices.

Needless to say the viability of the LU factorization is dependent on the existence of an efficient algorithm for the upper and lower triangular decomposition. As a simple illustration of a factorization algorithm, consider the *Crout algorithm*. It is one of numerous such algorithms.[19] The specific project is to factor a matrix M as

$$M = LU \qquad (4.14)$$

where L is lower and U upper triangular. Since the factorization is non-unique (scale one factor up and the other down) we may without loss of generality assume that the diagonal entries of U are all ones—i.e.,

$$u_{ii} = 1 \qquad (4.15)$$

With this assumption the first column of U is of the form $U_1 = \text{col}(1,0,0,...,0)$. Upon explicitly writing out the formula for the entries in the first column of M we have

$$m_{i1} = \sum_{j=1}^{n} \ell_{ij} u_{j1} \qquad (4.16)$$

Observing that $u_{j1} = 0$ for $j > 1$ while $u_{11} = 1$ we obtain

$$\ell_{i1} = m_{i1} \qquad (4.17)$$

Equation 4.17 specifies the first column of L. Similarly, by using the fact that the first row of L is of the form $\text{row}(m_{11},0,...,0)$ we have the formula

$$u_{1j} = \frac{m_{1j}}{m_{11}} = \frac{m_{1j}}{\ell_{11}} \qquad (4.18)$$

which readily computes the first row of U.

Now assume that the first $k-1$ columns of L and the first $k-1$ rows of U have been computed. The chore is to formulate an inductive algorithm for computing the k-th column of L and the k-th row of U. Observe that the summation form of matrix multiplication for computing m_{ij} ($i \geq k$) of $M = LU$ requires

$$m_{ik} = \sum_{j=1}^{n} \ell_{ij} u_{jk} = \ell_{ik} + \sum_{j=1}^{k-1} \ell_{ij} u_{jk} \qquad (4.19)$$

where the reduction in the range of the summation follows since U is upper triangular and $u_{kk} = 1$. Rearranging equation 4.19 as

$$\ell_{ik} = m_{ik} - \sum_{j=1}^{k-1} \ell_{ij} u_{jk} \qquad i \geq k \qquad (4.20)$$

gives us a formula for the entries in the k-th column of L in terms of m_{ik} and those entries of L and U already computed. Similarly, the formula for computing the entry m_{ki}, $i > k_j$ of $M = LU$ yields

$$u_{ki} = \frac{m_{ki} - \sum_{j=1}^{k-1} \ell_{kj} u_{ji}}{\ell_{kk}} \qquad i > k \qquad (4.21)$$

Again this computes the entries in the k-th row of U in terms of m_{ki} and entries of L and U already computed. Thus by computing the first column of L and first row of U via equations 4.17 and 4.18, then equations 4.20 and 4.21 will sequentially generate the remaining rows and columns of L and U.

As with the back substitution method, the Crout algorithm permits one to exploit the sparseness of M by deleting terms from the summations of 4.20 and 4.21 for which either ℓ_{kj} or u_{ji} are zero. This both accelerates the speed of the algorithm and improves its accuracy. Similarly, the sparseness of M is transferred to both L and U. Finally, we note that in terms of *storage requirements* the same memory words originally used to store M may simultaneously store L, U and M during the computation of L and U. Observe that once one computes the first (k−1) columns of L and the first (k−1) rows of U, the corresponding columns and rows of M are unneeded. Thus the memory originally allocated to the first (k−1) columns of M may store the first (k−1) columns of L. Likewise the memory allocated to the first (k−1) rows of M may store the first (k−1) rows of U. Here, since L is lower triangular and U is upper triangular with a diagonal of ones (which need not be stored) the first (k−1) columns of L and rows of U will not overlap when stored in the same matrix.

EXAMPLE 4.22: Consider the 3x3 matrix

$$M = \begin{bmatrix} 1 & 0 & 2 \\ 0 & 1 & 0 \\ -2 & 0 & 1 \end{bmatrix} \qquad (4.23)$$

By equation 4.17 the first column of L is col(1,0,-2) and by equations 4.15 and 4.18 the first row of U is row(1,0,2). This completed, row one and column one of M are now unneeded. This permits storage of col(1,0,-2) in the first column of M and row(0,2) in the first row and last two columns of M. Note that it is unnecessary to store the first entry of row(1,0,2) since equation 4.15 fixes this entry as a one. At this point in the algorithm the computer would store the following matrix in memory.

$$M_1 = \begin{bmatrix} 1 & 0 & 2 \\ 0 & 1 & 0 \\ -2 & 0 & 1 \end{bmatrix} \tag{4.24}$$

M_1 contains all information needed to continue the Crout algorithm. Now invoke equation 4.20 to compute $\ell_{22} = 1$ and $\ell_{32} = 0$. Equation 4.21 requires $u_{23} = 0$. Storing this information in the second column and row of M we have

$$M_2 = \begin{bmatrix} 1 & 0 & 2 \\ 0 & 1 & 0 \\ -2 & 0 & 1 \end{bmatrix} \tag{4.25}$$

Again equation 4.20 requires $\ell_{33} = 5$. Realizing that the diagonal entries of U are all one, the necessary information for the LU factorization is stored as

$$M_3 = \begin{bmatrix} 1 & 0 & 2 \\ 0 & 1 & 0 \\ -2 & 0 & 5 \end{bmatrix} \tag{4.26}$$

In other words the explicit LU factorization is

$$M = \begin{bmatrix} 1 & 0 & 0 \\ 0 & 1 & 0 \\ -2 & 0 & 5 \end{bmatrix} \begin{bmatrix} 1 & 0 & 2 \\ 0 & 1 & 0 \\ 0 & 0 & 1 \end{bmatrix} \tag{4.27}$$

as required.

Each step of the Crout algorithm requires that ℓ_{kk} be non-zero. This is also a requirement for the inversion of L. This restriction is satisfied if one properly permutes the last $(n-k+1)$ rows of M_{k-1} to place a row for which $\ell_{kk} \neq 0$ in the k-th position. Such permutations may also help minimize *numerical error* in the algorithm and/or minimize the generation of non-zero entries in the factorization process. These aspects of the algorithm find ample discussion in the sparse matrix literature,[19] and will not be repeated here.

Again, the inversion of M proceeds by first decomposing M as the LU product via the Crout algorithm. By applying back substitution to the inversion of L and U, the resultant inverse of M becomes $M^{-1} = U^{-1}L^{-1}$. Obviously each step of the algorithm both exploits and preserves sparsity as the example above illustrates. In the final step, storage of M^{-1} is via separate storage of the sparse matrices L^{-1} and U^{-1} instead of the usually non sparse M^{-1}.

An alternative to the Crout algorithm and back substitution is the *Gaussian elimination* method which computes L^{-1} and U^{-1} directly from M without first executing the LU factorization. A careful inspection of the Gaussian elimination process[19] shows that it actually generates the sparse factors L^{-1} and U^{-1} in computing M^{-1}. This algorithm provides another means of inverting M and expoiting sparsity.

These techniques pertain to inverting a single sparse matrix. In our application it is necessary to invert $(I-Z(s)L_{11})$—i.e. a family of sparse matrices parameterized by the complex variable s. Fortunately, it usually suffices to consider only the frequency response, s restricted to the imaginary axis. In many cases one assumes that the *sparsity pattern* of $(I-Z(s)L_{11})$ is independent of s. One then designs an inversion algorithm for the specific sparsity pattern. The increased overhead for the specific design is prorated over the entire family of inversions, a cost-effective power in applications wherein repeated inversion of sparse matrices with the same sparsity pattern is required. For instance power system stability analysis or circuit optimization requires repeated analysis of the same system with different parameter values. Therefore it pays to write a special purpose routine.

Some general purpose *sparse matrix inversion packages* include a special purpose compiler designed to write routines for the inversion of sparse matrices with a specified pattern. If the entire family of matrices with the same sparsity pattern is large the overhead cost of the special purpose program, prorated over many inversions, is negligible. In fact the increased efficiency may offset the entire cost since the overhead is paid once while the benefits are repeated each time a matrix is inverted.

An alternative approach utilizes a *continuation algorithm*. A continuation algorithm solves a family of algebraic problems (in this case matrix inversion) via a differential equation whose underlying parameter characterizes the elements of the family of algebraic problems (s in our case). The trajectories of this differential equation are the solutions to the family of algebraic problems as a function of the underlying parameter. Basically one solves a single case via a standard technique and uses this solution as an initial condition for the differential equation. For example set $s = s_0$, invert $(I-Z(s_0)L_{11})$, and use the answer as an initial condition. Integrating the equation yields the solutions for the remaining problems in the family.

Specifically the task at hand is to invert the family of sparse matrices $M(s) = (I-Z(s)L_{11})$ via a continuations algorithm. It is first necessary to formulate a differential equation in the form

$$M(\dot{s})^{-1} = f(M(s)^{-1}, s) \qquad (4.28)$$

Then one inverts $M(s)$ at some fixed value s_0 by a standard routine to obtain the initial condition $M(s_0)^{-1}$. This becomes the starting point for the computation of $M(s)^{-1}$ as a function of s through the integration of equation 4.28.[6] Differentiating the equality

$$MM^{-1} = I \qquad (4.29)$$

sets up the required differential equation.

$$0 = \dot{I} = \dot{\overline{MM^{-1}}} = \dot{M} M^{-1} + M \dot{M^{-1}} \qquad (4.30)$$

Rearranging yields the more tractable form

$$\dot{M}^{-1} = -M^{-1} \dot{M} M^{-1} \qquad (4.31)$$

in which "·" denotes differentiation with respect to s. The explicit dependence of the various matrices on s has been suppressed.

The immediate difficulty with equation 4.31 is the presence of non-sparse inverses. However, an LU factorization of M alleviates this problem. This entails deriving a differential equation whose solution permits updating the individual sparse matrices L^{-1} and U^{-1}. The theorem below constructs the necessary differential equation. For this purpose we adopt the notation $^{\ell}X$ for the *lower triangular part* of a matrix X. Specifically

$$(^{\ell}X)_{ij} = \begin{cases} X_{ij} & i \geq j \\ 0 & i < j \end{cases} \qquad (4.32)$$

The notation ^{u}X for the *strictly upper triangular* part of X is defined as

$$(^{u}X)_{ij} = \begin{cases} 0 & i \geq j \\ X_{ij} & i < j \end{cases} \qquad (4.33)$$

Clearly, $X = {}^{\ell}X + {}^{u}X$.

THEOREM 4.34: Let $M(r)$ be a parameterized family of matrices and let

$M(r_0)^{-1}$ be computed via an LU factorization as

$$M(r_0)^{-1} = U(r_0)^{-1}L(r_0)^{-1} \tag{4.35}$$

If $U(r)^{-1}$ and $L(r)^{-1}$ satisfy the differential equations

$$\dot{U}^{-1} = -U^{-1} \ ^u(L^{-1}\dot{M} U^{-1})$$

$$\dot{L}^{-1} = -{}^\ell(L^{-1}\dot{M} U^{-1})L^{-1} \tag{4.36}$$

then the equality

$$M(r)^{-1} = U(r)^{-1}L(r)^{-1} \tag{4.37}$$

is valid where $U(r)$ is upper triangular with ones along the diagonal (provided $U(r_0)$ has ones along the diagonal) and $L(r)$ is lower triangular. The proof of the theorem is relatively straightforward. So as to foster intuition, the following is a derivation which underlies the statement of the theorem.

Proof:

$$\dot{M}^{-1} = -M^{-1}\dot{M} M^{-1} \tag{4.38}$$

The matrices M^{-1} and $\dot{M}^{-1}$ are not sparse. Using the sparse factorization $M^{-1} = U^{-1}L^{-1}$ and $\dot{M}^{-1} = \dot{U}^{-1}L^{-1} + U^{-1}\dot{L}^{-1}$ equation 4.38 becomes

$$\dot{U}^{-1}L^{-1} + U^{-1}\dot{L}^{-1} = -U^{-1}L^{-1}\dot{M}U^{-1}L^{-1} \tag{4.39}$$

Every matrix in equation 4.39 is sparse. The goal is to construct a differential equation whose solution will determine U^{-1} and L^{-1} via the continuations process. First multiply 4.39 on the right by L and then take the strictly upper triangular part. This yields

$$\dot{U}^{-1} = -U^{-1} \ ^u(L^{-1}\dot{M}U^{-1}) \tag{4.40}$$

(Does $\dot{U}^{-1}$ have diagonal entries?) Now multiply equation 4.39 on the left by U and take the lower triangular part to obtain

$$\dot{L}^{-1} = -{}^\ell(L^{-1}\dot{M} U^{-1})L^{-1} \tag{4.41}$$

Both of these equations are subject to the initial conditions $U(r_0)^{-1}$ and $L(r_0)^{-1}$ where $M(r_0)^{-1} = U(r_0)^{-1}L(r_0)^{-1}$. Clearly then the continuations algorithm provides a utile means for constructing $M(r)^{-1} = U(r)^{-1}L(r)^{-1}$ via sparse matrix techniques.

EXAMPLE 4.42: Consider the family of matrices

$$M(r) = \begin{bmatrix} 1 & r \\ -1 & 1 \end{bmatrix} \tag{4.43}$$

Here $M(0)$ is lower triangular and hence has the following trivial LU factorization

$$\begin{bmatrix} 1 & 0 \\ -1 & 1 \end{bmatrix} = M(0) = L(0)U(0) = \begin{bmatrix} 1 & 0 \\ -1 & 1 \end{bmatrix}\begin{bmatrix} 1 & 0 \\ 0 & 1 \end{bmatrix} \tag{4.44}$$

whereas

$$\dot{M}(r) = \begin{bmatrix} 0 & 1 \\ 0 & 0 \end{bmatrix} \tag{4.45}$$

Calculating $L(0)$ inverse and $U(0)$ inverse we have

$$L(0)^{-1} = \begin{bmatrix} 1 & 0 \\ 1 & 1 \end{bmatrix} \tag{4.46}$$

and

$$U(0)^{-1} = \begin{bmatrix} 1 & 0 \\ 0 & 1 \end{bmatrix} \tag{4.47}$$

Using an *Euler integration formula*, $X(h) \approx X(0) + h\dot{X}(0)$, one may estimate $U(.1)^{-1}$ and $L(.1)^{-1}$ via the following string of equalities

$$\begin{aligned} U(.1)^{-1} &\approx U(0)^{-1} + .1\,\dot{U}(0)^{-1} \\ &= U(0)^{-1} - .1\,U(0)^{-1}\,^{u}(L(0)^{-1}\dot{M}(0)U(0)^{-1}) \\ &= \begin{bmatrix} 1 & 0 \\ 0 & 1 \end{bmatrix} - \begin{bmatrix} 0 & .1 \\ 0 & 0 \end{bmatrix} = \begin{bmatrix} 1 & -.1 \\ 0 & 1 \end{bmatrix} \end{aligned} \tag{4.48}$$

and

$$L(.1)^{-1} \approx L(0)^{-1} + .1 \, \dot{L}(0)^{-1}$$

$$= L(0)^{-1} - \ell \, (L(0)^{-1} \dot{M}(0) U(0)^{-1}) L(0)^{-1}$$

$$\begin{bmatrix} 1 & 0 \\ 0 & 1 \end{bmatrix} - \begin{bmatrix} 0 & 0 \\ .1 & .1 \end{bmatrix} = \begin{bmatrix} 1 & 0 \\ .9 & .9 \end{bmatrix} \qquad (4.49)$$

Multiplying these estimates out then yields

$$\dot{M}(.1)^{-1} = U(.1)^{-1} L(.1)^{-1} = \begin{bmatrix} .91 & -.09 \\ .9 & .9 \end{bmatrix} \qquad (4.50)$$

This compares favorably with the exact inverse

$$M(.1)^{-1} = \begin{bmatrix} .90909 & -.090909 \\ .90909 & .90909 \end{bmatrix} \qquad (4.51)$$

The error here is due to the approximation inherent in the numerical integration process and can be reduced by use of a more accurate integration procedure. Of course, the result of theorem 4.34 is exact and the computed value for $M(r)^{-1}$ will be as accurate as the integration routine employed.

Theorem 4.34 ideally exhibits the merit of sparse matrix techniques in the study of interconnected dynamical systems. Even though our primary interest is in the analytical aspects of the theory and not the details of the numerical methods, our analytical constructions always demand a formulation amenable to efficient numerical execution. In particular, LU factorizations should underlie and shape the analytical results necessitating the inversion of large matrices as in theorem 4.34. Consistent with this view future analytical results will build around the concept of the LU factorization where needed making the eventual implementation of the theory numerically feasible.

As a final point consider the case of a *hierarchical system*. Here Z(s) is *block diagonal* and L_{11} block lower triangular making $(I-Z(s)L_{11})$ block lower triangular. Hence one may implement a *"block back substitution"* process analogous to that described in equations 4.6 and 4.7 so that one may invert the entire matrix in terms of the inverses of the individual diagonal blocks. These blocks being inverted by standard sparse matrix techniques such as those already described. In particular one may apply the continuations algorithm to each block separately. Note, in the case of a *series parallel system* the diagonal blocks reduce to identity matrices thus trivializing the inversion procedure.

5. COMPOSITE SYSTEM STATE MODELS

Question: What is the affinity between a *composite component state model* and a *composite system state model?* Consider an interconnected dynamical system with composite component state model

$$
\begin{aligned}
\dot{x} &= Ax + Ba \\
b &= Cx + Da
\end{aligned}
\tag{5.1}
$$

and the usual algebraic connection equations

$$
\begin{aligned}
a &= L_{11}b + L_{12}u \\
y &= L_{21}b + L_{22}u
\end{aligned}
\tag{5.2}
$$

With this information one can build a state equation description of the composite system in the form

$$
\begin{aligned}
\dot{x} &= Fx + Gu \\
y &= Hx + Ju
\end{aligned}
\tag{5.3}
$$

Clearly the matrices F, G, H, and J will depend on the composite component connection model. Moreover the matrices of the formulation will require the *state vector for the composite system state model* to coincide with the composite component state vector. This prerequisite constitutes a *non-trivial* restriction. However it is valid for *almost all* component connection models. The remainder of this section presupposes this assumption. Section 8 discusses the general case which allows and in some cases demands an arbitrary state vector for the composite system state model.

To derive F, G, H, and J from the connection model matrices solve the second equation of 5.1 and the first equation of 5.2 simultaneously to obtain

$$
a = L_{11}Cx + L_{11}Da + L_{12}u
\tag{5.4}
$$

or equivalently assuming $(I-L_{11}D)^{-1}$ exists

$$a = (I-L_{11}D)^{-1}L_{11}Cx + (I-L_{11}D)^{-1}L_{12}u \qquad (5.5)$$

Substituting equation 5.5 into equations 5.1 and 5.2 yields the desired set of equations

$$\begin{aligned}
\dot{x} &= [A+B(I-L_{11}D)^{-1}L_{11}C]x + [B(I-L_{11}D)^{-1}L_{12}]u \\
y &= [L_{21}C+L_{21}D(I-L_{11}D)^{-1}L_{11}C]x + [L_{21}D(I-L_{11}D)^{-1}L_{12}+L_{22}]u
\end{aligned} \qquad (5.6)$$

THEOREM 5.7: Let an interconnected dynamical system have a component connection model in the form of equations 5.1 and 5.2. Then it has a state model in the form of equations 5.3 if and only if $(I-L_{11}D)^{-1}$ exists. Moreover if the inverse exists

$$F = A + B(I-L_{11}D)^{-1}L_{11}C \qquad (5.8)$$

$$G = B(I-L_{11}D)^{-1}L_{12} \qquad (5.9)$$

$$H = L_{21}C + L_{21}D(I-L_{11}D)^{-1}L_{11}C = L_{21}(I-DL_{11})^{-1}C \qquad (5.10)$$

and

$$J = L_{21}D(I-L_{11}D)^{-1}L_{12} + L_{22} \qquad (5.11)$$

PROOF: If the inverse of $(I-L_{11}D)$ exists, the state model is as derived above. On the other hand, suppose the inverse fails to exist, then there exists a composite component input vector $a \neq 0$ satisfying equation 5.4 when x and u are both zero. This in turn forces the system to have a non zero output with zero input and zero state. Although this is quite possible for an interconnected system (see II.2.13) it is clearly impossible for a system with a composite state model in the form of equation 5.3. One concludes that the state model of equation 5.3 fails to exist if the inverse of $(I-L_{11}D)$ fails to exist. Finally, the second equality in 5.10 follows upon an application of theorem 2.8 thereby completing the proof.

COROLLARY 5.12: Let an interconnected dynamical system be characterized by the component connection model of equations 5.1 and 5.2. Then for a given composite component state model the system admits a composite system state model in the form of equations 5.3 for *almost all connections* within any given L_{11} *connection structure.*

PROOF: The proof is identical to that of theorem 2.32 and will not be repeated here.

EXAMPLE 5.13: Consider the feedback system shown in figure 5.14.

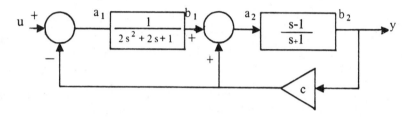

Figure 5.14. Feedback system composed of a Butterworth filter and an all pass function.

The system is composed of a *Butterworth filter* and *all pass function* in a feedback loop. The state model for the Butterworth filter was derived in example I.4.57. The model is

$$\left(\begin{bmatrix} 0 & 1 \\ -.5 & -1 \end{bmatrix}, \begin{bmatrix} 0 \\ .5 \end{bmatrix}, [1 \quad 0], [0] \right) \tag{5.15}$$

A state model for the all pass function is

$$([-1], [1], [-2], [1]) \tag{5.16}$$

Therefore the D matrix for the composite component state model is

$$\begin{bmatrix} 0 & 0 \\ 0 & 1 \end{bmatrix} \tag{5.17}$$

and the algebraic connection equations for the system are

$$\begin{bmatrix} a_1 \\ \underline{a_2} \\ y \end{bmatrix} = \begin{bmatrix} 0 & -c & 1 \\ 1 & c & 0 \\ \hline 0 & 1 & 0 \end{bmatrix} \begin{bmatrix} b_1 \\ b_2 \\ u \end{bmatrix} \tag{5.18}$$

This results in

$$(I - L_{11}D) = \begin{bmatrix} 1 & c \\ 0 & 1-c \end{bmatrix} \tag{5.19}$$

This matrix is invertible whenever the scale factor "c" is not equal to 1.

For $c \neq 1$ we have

$$(I - L_{11}D)^{-1} = \begin{bmatrix} 1 & \dfrac{-c}{1-c} \\ 0 & \dfrac{1}{1-c} \end{bmatrix} \qquad (5.20)$$

For the sake of example, set $c = -1$. Then the composite system state model is

$$\left(\begin{bmatrix} 0 & 1 & 0 \\ -.25 & -1 & -.5 \\ .5 & 0 & 0 \end{bmatrix}, \begin{bmatrix} 0 \\ .5 \\ 0 \end{bmatrix}, [.5 \ \ 0 \ \ -1], [0] \right) \qquad (5.21)$$

Theorem 5.7 bases the existence of the composite system state equations only on knowledge of the relationship between L_{11} and D regardless of the properties of A, B, and C. This implies that the existence of the composite system state model as per equation 5.3 is independent of the particular state model chosen to characterize the individual system components. This is intuitively reasonable since $D = Z(\infty)$ is the same for any of the possible state models for a given component.

In a similar vein, it is possible to exploit the dependence of the composite system state model on $(I - L_{11}D)^{-1}$ for numerical reasons. In many cases a simple *reordering* of the components produces a *lower triangular* product, $L_{11}D$ where $L_{11}D$ corresponds to a reordered composite component state model. This concept is akin to the notion of a *hierarchical structure*. The best method for illustrating the idea is by example.

EXAMPLE 5.22: Suppose we have an interconnected dynamical system whose adjacency graph is illustrated in figure 5.23. Using the hierarchical system jargon, the system forms a single *strongly connected component* (sub-system) since every pair of vertices shares at least one *directed loop*. Suppose vertex 3 represents a component for which D=0, say a Butterworth filter or an integrator. Deleting vertex 3 breaks all directed loops. The remaining vertices form a reduced system which admits a "*series parallel structure.*" Specifically the system has no *algebraic loops* and hence by appropriately reordering the components, $L_{11}D$ changes to a lower triangular matrix.

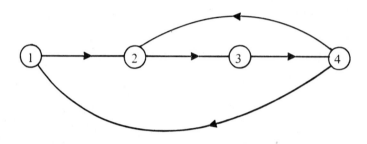

Figure 5.23. Adjacency graph for interconnected dynamical systems.

For example the *adjacency graph* of figure 5.23 is representative of the matrix

$$L_{11} = \begin{bmatrix} 0 & 0 & 0 & 1 \\ 2 & 0 & 0 & 10 \\ 0 & -1 & 0 & 0 \\ 0 & 0 & 1 & 0 \end{bmatrix} \qquad (5.24)$$

A possible composite component D matrix for which component three has no algebraic term is

$$D = \begin{bmatrix} 1 & 0 & 0 & 0 \\ 0 & 1 & 0 & 0 \\ 0 & 0 & 0 & 0 \\ 0 & 0 & 0 & -1 \end{bmatrix} \qquad (5.25)$$

Now upon neglecting vertex 3 in figure 5.23 a series parallel structure results in the *reduced system* by the component reordering 4,1,2. Now placing component three last should result in a strictly lower triangular $L_{11}D$ matrix. After the reordering 5.24 becomes

$$L_{11} = \begin{bmatrix} 0 & 0 & 0 & 1 \\ 1 & 0 & 0 & 0 \\ 10 & 2 & 0 & 0 \\ 0 & 0 & -1 & 0 \end{bmatrix} \qquad (5.26)$$

and equation 5.25 becomes

$$D = \begin{bmatrix} -1 & 0 & 0 & 0 \\ 0 & 1 & 0 & 0 \\ 0 & 0 & 1 & 0 \\ 0 & 0 & 0 & 0 \end{bmatrix} \qquad (5.27)$$

Inspecting the $L_{11}D$ product shows that

$$L_{11}D = \begin{bmatrix} 0 & 0 & 0 & 0 \\ -1 & 0 & 0 & 0 \\ -10 & 2 & 0 & 0 \\ 0 & 0 & -1 & 0 \end{bmatrix} \qquad (5.28)$$

which is as predicted, lower triangular.

All of the above notions focus on state modeling of *linear* systems. Much of the theory generalizes to nonlinear interconnected systems. For these systems the composite component state model takes the symbolic form

$$\dot{x} = f(x,a)$$
$$b = g(x,a) \qquad (5.29)$$

Combining 5.29 with the connection equations of 5.2 forms the component connection model for an arbitrary nonlinear system. As with the linear case, we assume that the composite system state vector coincides with the composite component state vector. The symbolic functional notation of the composite system state model is

$$\dot{x} = F(x,u)$$
$$y = G(x,u) \qquad (5.30)$$

Substituting $b = g(x,a)$ into $a = L_{11}b + L_{12}u$ and rearranging yields

$$a - L_{11}g(x,a) - L_{12}u = h(a,x,u) = 0 \qquad (5.31)$$

This is the nonlinear analog of equation 5.4. Solving equation 5.31 involves much more than the inversion of $(I-L_{11}D)$ as with equation 5.4. Here one confronts the more despairing task of computing an *implicit function*,

$k(x,u) = a$, such that

$$h(k(x,u),x,u) = 0 \qquad (5.32)$$

Assume for the moment that such an implicit function exists. (The next section picks up the burdening question of existence.) In particular assume

$$a = k(x,u) \qquad (5.33)$$

The composite system state model then becomes

$$\dot{x} = f(x,k(x,u))$$
$$y = L_{21}g(x,k(x,u)) + L_{22}u \qquad (5.34)$$

The formalization of this derivation is given by the following theorem.

THEOREM 5.35: Let an interconnected dynamical system have a component connection model in the form of equations 5.29 and 5.2. Then it has a state model in the form of equation 5.30 provided there exists an implicit function $k(\cdot,\cdot)$ satisfying $h(k(x,u),x,u) = 0$. Moreover if $k(,)$ exists

$$F(x,u) = f(x,k(x,u)) \qquad (5.36)$$

and

$$G(x,u) = L_{21}k(x,u) + L_{22}u \qquad (5.37)$$

EXAMPLE 5.38: Figure 5.39 depicts a nonlinear feedback system. The system comprises an *all pass network* cascaded with a *diode-like function* both within a negative feedback loop.

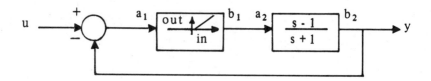

Figure 5.39. Nonlinear feedback system.

The composite component state equations for the system are

$$\dot{x} \;=\; -x + a_2 \;=\; f(x,a)$$

$$\left.\begin{matrix} b_1 \\ b_2 \end{matrix}\right] \;=\; \left.\begin{matrix} \lambda(a_1) \\ -2x+a_2 \end{matrix}\right] \;=\; g(x,a) \tag{5.40}$$

Here $\lambda(\cdot)$ denotes the diode function sketched in figure 5.41.

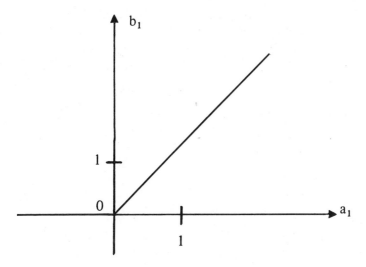

Figure 5.41. Sketch of nonlinear diode function.

The connection equations for this system are

$$\left.\begin{matrix} a_1 \\ a_2 \\ \hline y \end{matrix}\right] = \left[\begin{array}{cc|c} 0 & -1 & 1 \\ 1 & 0 & 0 \\ \hline 0 & 1 & 0 \end{array}\right] \left.\begin{matrix} b_1 \\ b_2 \\ \hline u \end{matrix}\right] \tag{5.42}$$

Combining 5.40 and 5.42 the function $h(\cdot,\cdot,\cdot)$ defined in equation 5.31 takes the form

$$h(x,a,u) \;=\; \left.\begin{matrix} a_1-2x+a_2-u \\ a_2-\lambda(a_1) \end{matrix}\right] = \left.\begin{matrix} 0 \\ 0 \end{matrix}\right] \tag{5.43}$$

A little arithmetic reveals that the required implicit function takes the form

$$a_1 = k_1(x,u) = \begin{cases} 2x+u & 2x+u \leq 0 \\ x+\dfrac{u}{2} & 2x+u > 0 \end{cases}$$

$$a_2 = k_2(x,u) = \begin{cases} 0 & 2x+u \leq 0 \\ x+\dfrac{u}{2} & 2x+u > 0 \end{cases} \tag{5.44}$$

These are both *piecewise linear functions* and are sketched in figure 5.44 a and b. Finally, substituting 5.44 into 5.40 yields the *piecewise linear composite system state model below*

$$\begin{aligned} \dot{x} &= -x \\ y &= -2x \end{aligned} \quad ; 2x + u \leq 0$$

$$\begin{aligned} \dot{x} &= \dfrac{u}{2} \\ y &= x + \dfrac{u}{2} \end{aligned} \quad ; 2x + u > 0 \tag{5.45}$$

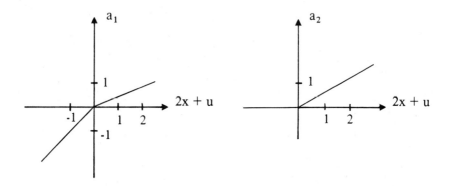

Figure 5.46. Nonlinear functions arising in example 5.38.

Needless to say the key to the practical formulation of a nonlinear state model for a composite system is the computation of the implicit function. As with the inversion of $(I-L_{11}D)$ this process may be simplified by a *component reordering*. Basically one deletes components in which y is not a function of a (i.e. components with *strictly proper state models*) to form a *reduced system*. When the system has no *algebraic loops* the implicit function can be computed by a *back substitution* process similar to that outlined in the preceding section.[6] As such the *optimal reordering*[6] of the components in a nonlinear system is of primary importance since it can lead to the complete trivialization of the potentially tedious computation of k(x,u). Indeed it suffices to avoid algebraic loops containing nonlinearities since linear algebraic loops can be eliminated by matrix inversion.[6]

6.   PIECEWISE LINEAR IMPLICIT FUNCTIONS

The last part of the previous section briefly discussed the details of building a *composite nonlinear state model* from the nonlinear component connection equations. During the discussion the following equation arose.

$$a - L_{11}g(x,a) - L_{12}u = h(a,x,u) = 0 \qquad (6.1)$$

The solution to this equation required postulation of the existence of a globally defined *implicit function*, k(x,u) = a, satisfying equation 6.2.

$$h(k(x,u),x,u) = 0 \qquad (6.2)$$

As noted earlier this section undertakes a discussion of the existence of such objects for the *piecewise linear* case. Before proceeding note that "h(a,x,u) = 0" implicitly defines the desired function $k(\cdot,\cdot) = a$. The terminology is unusually descriptive.

Unfortunately, the general theory surrounding the existence of globally defined implicit functions either uses intuitive but very sophisticated mathematics or long tedious elementary topological arguments. Moreover, the theory notwithstanding, the actual computation of globally defined implicit functions is a highly onerous undertaking. The major exception is the piecewise linear case as with example 5.36 wherein g(x,a) is a piecewise linear function of x and a. Hence h(a,x,u) = a − $L_{11}g(x,a) - L_{12}u$ is a piecewise linear function.

Before wading through the unavoidable abstractions, let us discuss a technique for constructing the implicit function algorithmically, for the piecewise linear case along while motivating the general results by considering this special case. The technique utilizes the *Householder formula.*[4]

What the above discussion hints at, is that a composite system state model will exist, given the component connection model, only if it is possible to uniquely determine the composite component input vector, a, in terms of the composite system input vector, u, and the composite component state vector, x. Recall that we tacitly assume x also serves as the *composite system state vector.* Further consideration of this assumption is taken up in section 8.

Clearly then if there exists a globally defined implicit function, k(x,u) = a, satisfying

$$h(k(x,u),x,u) = 0 \tag{6.3}$$

then the composite system state model becomes

$$x = f(x,k(x,u)) \overset{\Delta}{=} F(x,u)$$
$$y = L_{21}g(x,k(x,u)) + L_{22}u \overset{\Delta}{=} G(x,u) \tag{6.4}$$

Let us now pin down the notion of piecewise linearity for the equation, $b = g(x,a)$. For each instant of time suppose a is in $\mathbf{R}^m$, x is in $\mathbf{R}^k$, and b is in $\mathbf{R}^n$. As such $g(\cdot,\cdot)$: $\mathbf{R}^k x \mathbf{R}^m \rightarrow \mathbf{R}^n$. For $g(\cdot,\cdot)$ to be piecewise linear it must satisfy the following definition

DEFINITION 6.5: The function g(x,a) as described above is *piecewise linear* in both x and a if and only if

$$g(x,a) = P_i x + p_i + Q_j a + q_j \tag{6.6}$$

where (i) P_i and Q_j arc real matrices
      (ii) p_i and q_j are vectors in $\mathbf{R}^n$
      (iii) P_i: $\mathbf{R}^k \rightarrow \mathbf{R}^n$ and Q_j: $\mathbf{R}^m \rightarrow \mathbf{R}^n$
      (iv) $1 \leqslant i \leqslant kk$ for some integer kk
      (v) $1 \leqslant j \leqslant mm$ for some integer mm.

Basically the subscripts i and j index the regions of $\mathbf{R}^k$ and $\mathbf{R}^m$ over which the appropriate "piece" of the piecewise linear map, $g(\cdot,\cdot)$ is defined. Also for this business to make sense and fit the desired framework it is required that $g(\cdot)$ be continuous. Not unexpectedly, however, continuity of

$g(\cdot,\cdot)$ is not enough to insure the existence of $k(\cdot,\cdot)$, as will shortly become evident.

To construct the appropriate piecewise linear function, recall that

$$h(a,x,u) \;=\; a - L_{11}g(x,a) - L_{12}u \;=\; 0 \tag{6.7}$$

Plugging 6.6 into 6.7 yields

$$a - L_{11}(P_ix + p_i) - L_{11}(Q_ja + q_j) - L_{12}u \;=\; 0 \tag{6.8}$$

Rearranging and simplifying, we have

$$(I - L_{11}Q_j)a - L_{11}q_j \;=\; L_{11}(P_ix + p_i) + L_{12}u \tag{6.9}$$

For each fixed time, x and u become fixed vectors in a suitable Euclidean space. Under these circumstances, the right side of equation 6.9 becomes a fixed vector in $\mathbf{R}^m$. This then allows us to rewrite 6.9 more succinctly as

$$D_ja + d_j \;=\; y \;\overset{\Delta}{=}\; f(a) \tag{6.10}$$

where D_j, d_j, and y are related to the matrices of 6.9 in the obvious way.

Intuitively, for each fixed y, determination of a unique vector, a, for each y requires that each D_j be invertible. Moreover the mapping, $f(\cdot)$, (all pieces glued together) must be one-to-one. Equivalently the regions R_j, defined as the domains of the particular D_j, must both partition $\mathbf{R}^m$ and satisfy $f(R_i) \cap f(R_j) = \phi$ for all pairs (i,j) with $i \neq j$. Finally $f(\cdot)$ must be onto $\mathbf{R}^n$ for otherwise it would be possible (in general) to find a vector y in $\mathbf{R}^n$ for which equation 6.10 is never satisfied. Lumping all this together yields the following theorem.

THEOREM 6.11: The continuous map $f(\cdot)$ as defined in 6.10 is globally invertible if and only if

  (i) $f(\cdot)$ is onto,
  (ii) $f(\cdot)$ is locally invertible in each region R_j, and
  (iii) $f(R_i) \cap f(R_j) = \phi$ for each pair (i,j) with $i \neq j$.

First we point out that the map $f(\cdot)$ as constructed from $g(\cdot,\cdot)$ is necessarily continuous, so the added requirement is superfluous. Secondly, condition (ii) is equivalent to each of the D_j's being invertible.

The point of the theorem is simple. The sought after implicit function $k(x,u) = a$, becomes identified with the inverse of the map $f(\cdot)$ in the piecewise linear case. In other words

$$k(x,u) = f^{-1}(y) = a \qquad (6.12)$$

The computation of this inverse in numerically facilitated by use of the Householder formula.

In essence the Householder formula describes the effect of a rank "r" perturbation on the inverse of a matrix. It turns out that the set of matrices D_j of equation 6.10 differ by a rank one perturbation provided the respective domains are contiguous. In particular if R_j and R_i are adjacent regions whose common border is characterized by the *hyperplane*

$$c^t a = k \qquad (6.13)$$

where k is a constant, a is the vector of specific variables, and c an appropriate column vector, then there exists a vector r in $\mathbf{R}^{\prime\prime\prime}$ such that

$$D_j = D_i + rc^t \qquad (6.14)$$

This relationship is nothing more than a restatement of theorem I.5.19. This fact together with the Householder formula offers the numerical avenue toward the calculation of $k(x,u)$.

LEMMA 6.15: *Householder's formula:* Let D be an nxn matrix. Define $\underline{D} = D + REC$ where R is mxr, E is rxr, and C is rxn. Then

$$\underline{D}^{-1} = [I - D^{-1}RE(I + CD^{-1}RE)^{-1}C]D^{-1} \qquad (6.16)$$

PROOF: Consider the product $\underline{D}\,\underline{D}^{-1}$ as follows

$$
\begin{aligned}
\underline{D}\,\underline{D}^{-1} &= (D + REC)[I - D^{-1}RE(I + CD^{-1}RE)^{-1}C]D^{-1} \\
&= I - RE(CD^{-1}RE + I)^{-1}CD^{-1} + RECD^{-1} \\
&\quad - RECD^{-1}RE(CD^{-1}RE + I)^{-1}CD^{-1} \\
&= I - RE[I - (CD^{-1}RE + I) + CD^{-1}RE](CD^{-1}RE + I)^{-1}CD^{-1} \\
&= I
\end{aligned}
\qquad (6.17)
$$

Observe that to initialize the algorithm, it is necessary to compute D^{-1} by standard means. All subsequent inverses are then computable by the formula. As mentioned above

$$D_j \;\; = \;\; D_i + rc^t \tag{6.18}$$

where the common boundary of the regions R_j and R_i are given by the hyperplane $c^t a = k$. With this fact and the equalities $C = c^t$, $R = r$, and $E = 1$ we form the following simplification of the Householder formula

$$D_j^{-1} \;\; = \;\; [I - D_i^{-1} \frac{rc^t}{(1 + c^t D_i^{-1} r)}]D_i^{-1} \tag{6.19}$$

whenever R_i and R_j are adjacent regions. The obvious numerical advantage is that $(I + CD^{-1}RE)^{-1} = (1 + c^t D_i^{-1} r)^{-1}$ is a scalar inverse.

THEOREM 6.20: Let a continuous piecewise linear function

$$y \;\; = \;\; f(a) \;\; = \;\; D_j a + d_j \tag{6.21}$$

(for a in R_j) have a global inverse. Then $f^{-1}(\ )$ is defined by

$$f^{-1}(y) \;\; = \;\; D_j^{-1} y - D_j^{-1} d_j \tag{6.22}$$

where (i) y is in $f(R_j)$, and (ii) the matrix inverses may be computed sequentially via equation 6.19 whenever R_j and R_i have a common border. Note that beyond the computation of the inverse matrices the only additional computation required for the solution of 6.21 is the determination of the region, $f(R_j)$, in which y lies.

This completes the pertinent discussion of the tools necessary to construct the desired implicit function. We cement these ideas together with the following example.

EXAMPLE 6.23: In this example we will calculate the implicit function, $k(x,u) = a$, for a given nonlinear circuit. Figure 6.24 depicts the circuit.

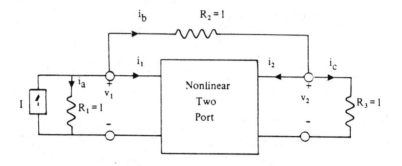

Figure 6.24. Circuit containing nonlinear two port.

The circuit consists of three resistors which we embed in the connection equations; the nonlinear two port; one circuit input, the current source; and one output which we will take as v_2. Suppose the nonlinear two port has a piecewise linear state model symbolically written as

$$\begin{aligned} \dot{x} &= f(x,v) \\ i &= g(x,v) \end{aligned} \qquad (6.25)$$

where

$$x = \begin{bmatrix} x_1 \\ x_2 \end{bmatrix}; \quad v = \begin{bmatrix} v_1 \\ v_2 \end{bmatrix}; \quad i = \begin{bmatrix} i_1 \\ i_2 \end{bmatrix} \qquad (6.26)$$

In particular suppose $i = g(x,v)$ is given by the following piecewise linear relationship.

$$\begin{bmatrix} i_1 \\ i_2 \end{bmatrix} = \begin{bmatrix} 1 & 0 \\ 0 & 1 \end{bmatrix}\begin{bmatrix} x_1 \\ x_2 \end{bmatrix} + \frac{1}{3}\begin{bmatrix} 3 & 0 \\ -1 & 5 \end{bmatrix}\begin{bmatrix} v_1 \\ v_2 \end{bmatrix} + \frac{1}{3}\begin{bmatrix} 0 \\ 1 \end{bmatrix} \qquad (6.27)$$

for v in R_1

$$\begin{bmatrix} i_1 \\ i_2 \end{bmatrix} = \begin{bmatrix} 1 & 0 \\ 0 & 1 \end{bmatrix}\begin{bmatrix} x_1 \\ x_2 \end{bmatrix} + \frac{1}{3}\begin{bmatrix} 3 & 0 \\ 0 & 4 \end{bmatrix}\begin{bmatrix} v_1 \\ v_2 \end{bmatrix} \qquad (6.28)$$

for v in R_2

$$\begin{bmatrix} i_1 \\ i_2 \end{bmatrix} = \begin{bmatrix} 1 & 0 \\ 0 & 1 \end{bmatrix}\begin{bmatrix} x_1 \\ x_2 \end{bmatrix} + \frac{1}{3}\begin{bmatrix} 3 & 0 \\ 1 & 5 \end{bmatrix}\begin{bmatrix} v_1 \\ v_2 \end{bmatrix} + \frac{1}{3}\begin{bmatrix} 0 \\ 1 \end{bmatrix} \qquad (6.29)$$

for v in R_3, and finally

$$\begin{bmatrix} \ddot{i}_1 \\ i_2 \end{bmatrix} = \begin{bmatrix} 1 & 0 \\ 0 & 1 \end{bmatrix}\begin{bmatrix} x_1 \\ x_2 \end{bmatrix} + \begin{bmatrix} 1 & 0 \\ 0 & 2 \end{bmatrix}\begin{bmatrix} v_1 \\ v_2 \end{bmatrix} \qquad (6.30)$$

for v in R_4, where the regions R_1, R_2, R_3, and R_4 are depicted on figure 6.31.

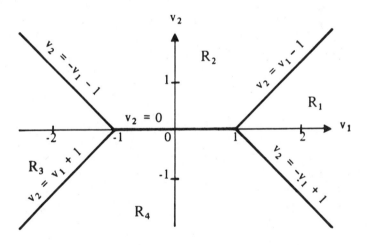

Figure 6.31. Definitions of the regions R_1, R_2, R_3, and R_4.

By embedding the resistors into the connection equations we arrive at the following connection model.

$$\begin{bmatrix} v_1 \\ v_2 \\ \hline v_2 \end{bmatrix} = \frac{1}{3}\left[\begin{array}{cc|c} -2 & -1 & 2 \\ -1 & -2 & 1 \\ \hline -1 & -2 & 1 \end{array}\right]\begin{bmatrix} i_1 \\ i_2 \\ \hline I \end{bmatrix} \qquad (6.32)$$

Note that it is seldom advantageous to embed the resistors into the connection equations. In this case it was done only to reduce the dimension of the matrices involved. From equation 6.10 recall that

$$f(a) = y \triangleq L_{11}P_i x + L_{11}p_i + L_{12}u$$

$$\triangleq \frac{1}{3}\begin{bmatrix} -2 & -1 \\ -1 & -2 \end{bmatrix} + \begin{bmatrix} 2 \\ 1 \end{bmatrix} I \tag{6.33}$$

Thus for each fixed time instant, y is a fixed vector in $\mathbf{R}^n$ and the four equations specifying v are

$$\frac{1}{9}\begin{bmatrix} 14 & 5 \\ 1 & 19 \end{bmatrix}\begin{bmatrix} v_1 \\ v_2 \end{bmatrix} + \frac{1}{9}\begin{bmatrix} 1 \\ 2 \end{bmatrix} = \begin{bmatrix} y_1 \\ y_2 \end{bmatrix} \tag{6.34}$$

for y in $f(R_1)$

$$\frac{1}{9}\begin{bmatrix} 15 & 4 \\ 3 & 17 \end{bmatrix}\begin{bmatrix} v_1 \\ v_2 \end{bmatrix} = \begin{bmatrix} y_1 \\ y_2 \end{bmatrix} \tag{6.35}$$

for y in $f(R_2)$

$$\frac{1}{9}\begin{bmatrix} 16 & 5 \\ 5 & 19 \end{bmatrix}\begin{bmatrix} v_1 \\ v_2 \end{bmatrix} + \frac{1}{9}\begin{bmatrix} 1 \\ 2 \end{bmatrix} = \begin{bmatrix} y_1 \\ y_2 \end{bmatrix} \tag{6.36}$$

for y in $f(R_3)$

$$\frac{1}{3}\begin{bmatrix} 5 & 2 \\ 1 & 7 \end{bmatrix}\begin{bmatrix} v_1 \\ v_2 \end{bmatrix} = \begin{bmatrix} y_1 \\ y_2 \end{bmatrix} \tag{6.37}$$

for y in $f(R_4)$, where the regions $f(R_i)$ are depicted in figure 6.40.

Determination of v follows by application of the Householder formula. Step one is to calculate the inverse of

$$\frac{1}{9}\begin{bmatrix} 14 & 5 \\ 1 & 19 \end{bmatrix} \tag{6.38}$$

which is

$$\frac{1}{29}\begin{bmatrix} 19 & -5 \\ -1 & 14 \end{bmatrix} \tag{6.39}$$

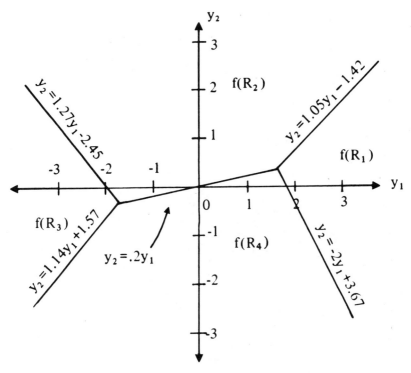

Figure 6.40. Image regions of the map $f(\cdot)$.

It was necessary to calculate this inverse by standard means. The next step is to calculate the inverse of

$$\frac{1}{9}\begin{bmatrix} 15 & 4 \\ 3 & 17 \end{bmatrix} \qquad (6.41)$$

via the Householder formula. For this observe that the common border of regions R_1 and R_2 is specified by the hyperplane

$$\begin{bmatrix} 1 & -1 \end{bmatrix}\begin{bmatrix} v_1 \\ v_2 \end{bmatrix} = 1 \qquad (6.42)$$

Therefore, there exists a vector "r" satisfying

$$\frac{1}{9}\begin{bmatrix} 15 & 4 \\ 3 & 17 \end{bmatrix} = \frac{1}{9}\begin{bmatrix} 14 & 5 \\ 1 & 19 \end{bmatrix} + r\begin{bmatrix} 1 & -1 \end{bmatrix} \qquad (6.43)$$

where $r = 1/9 \begin{bmatrix} 1 & 2 \end{bmatrix}^t$. Thus

$$\frac{1}{9}\begin{bmatrix} 15 & 4 \\ 3 & 17 \end{bmatrix} = \frac{1}{9}\begin{bmatrix} 14 & 5 \\ 1 & 19 \end{bmatrix} + \frac{1}{9}\begin{bmatrix} 1 & -1 \\ 2 & -2 \end{bmatrix} \tag{6.44}$$

Substituting these expressions into the formula yields

$$\frac{1}{9}\begin{bmatrix} 15 & 4 \\ 2 & 17 \end{bmatrix}^{-1} = \frac{1}{29}\left(\begin{bmatrix} 1 & 0 \\ 0 & 1 \end{bmatrix} - \frac{1}{261}\begin{bmatrix} 19 & -5 \\ -1 & 14 \end{bmatrix}\right. \tag{6.45}$$

$$\left. x \begin{bmatrix} 1 & -1 \\ 2 & -2 \end{bmatrix} \frac{261}{243}\right)\begin{bmatrix} 19 & -5 \\ -1 & 14 \end{bmatrix}$$

$$= \frac{1}{29}\frac{1}{243}\left(\begin{bmatrix} 243 & 0 \\ 0 & 243 \end{bmatrix} - \begin{bmatrix} 9 & -9 \\ 27 & -27 \end{bmatrix}\right)\begin{bmatrix} 19 & -5 \\ -1 & 14 \end{bmatrix} \tag{6.46}$$

$$= \frac{1}{27}\begin{bmatrix} 17 & -4 \\ 3 & 15 \end{bmatrix} \tag{6.47}$$

Computing the remaining inverses and then simplifying 6.34 through 6.37 yields the following specification of the implicit function, $k(x,u) = v$,

$$\begin{bmatrix} v_1 \\ v_2 \end{bmatrix} = \frac{1}{29}\begin{bmatrix} 19 & -5 \\ -1 & 14 \end{bmatrix}\begin{bmatrix} y_1 \\ y_2 \end{bmatrix} - \frac{1}{9}\begin{bmatrix} 1 \\ 2 \end{bmatrix} \tag{6.48}$$

for y in $f(R_1)$

$$\begin{bmatrix} v_1 \\ v_2 \end{bmatrix} = \frac{1}{27}\begin{bmatrix} 17 & -4 \\ 3 & 15 \end{bmatrix}\begin{bmatrix} y_1 \\ y_2 \end{bmatrix} \tag{6.49}$$

for y in $f(R_2)$

$$\begin{bmatrix} v_1 \\ v_2 \end{bmatrix} = \frac{1}{31}\begin{bmatrix} 19 & -5 \\ -5 & 16 \end{bmatrix}\begin{bmatrix} y_1 \\ y_2 \end{bmatrix} - \frac{1}{9}\begin{bmatrix} 1 \\ 2 \end{bmatrix} \tag{6.50}$$

for y in $f(R_3)$

$$\begin{bmatrix} v_1 \\ v_2 \end{bmatrix} = \frac{1}{11} \begin{bmatrix} 7 & -2 \\ -1 & 5 \end{bmatrix} \begin{bmatrix} y_1 \\ y_2 \end{bmatrix} \tag{6.51}$$

for y in $f(R_4)$.

Consequently, given a piecewise linear relationship for $x=f(x,a)$, the sought after equation becomes

$$\dot{x} = f(x,k(x,u)) \tag{6.52}$$

Although application of the Household formula for this example is exceedingly tedious, the numerical advantages for large matrices are great.

For the piecewise linear case, the implicit function becomes identified with a globally defined *inverse function*. Unfortunately for more general nonlinear systems such an identification is rare. However, the existence of the implicit function can be derived as a corollary to a *global inverse function theorem*. These details are illuminated in the next section. The reader interested in further discussion of piecewise linear problems will find such details at the end of section four of chapter IV.

7. GLOBALLY DEFINED IMPLICIT FUNCTIONS

In this section we undertake the task of determining the existence of the aforementioned *implicit function*, $k(x,u) = a$. First some general comments. If one were simulating a system, say similar to one of the previous section, one would use the initial state vector and the system inputs at the start-time to specify the composite component input vector, a, then one would sequentially update the numerical solution of the differential state equation and continue to construct a "trajectory" for the composite component input vector, a. The discussions of this section will provide *only a sufficient condition* for such a simulation procedure to exist. By necessity, a simulation uses the state trajectories in a solution. On the other hand will intentionally ignore state trajectories and attempt to uniquely determine "a" in terms of an arbitrary state vector and a given system input at each instant of time; thus the requirement that the implicit function be globally defined. An example of a nonlinear system not admitting a globally defined implicit function, but permitting a valid simulation, would be one containing a nonlinear inductor exhibiting hysteresis.

Moving on to the nitty-gritty, again consider the equation

$$h(a,x,u) \;=\; a - L_{11}g(x,a) - L_{12}u \;=\; 0 \qquad (7.1)$$

Stacking x and u together as $\mathrm{col}(x,u) = q$, we may rewrite $h(\cdot,\cdot,\cdot)$ as

$$h(a,q) \;=\; 0 \qquad (7.2)$$

where
$$h(\cdot,\cdot): \mathbf{R}^n x\, \mathbf{R}^m \to \mathbf{R}^n \qquad (7.3)$$

As such, the problem is to verify the existence of the implicit function, $k(\cdot):\mathbf{R}^m \to \mathbf{R}^n$, such that $k(q) = a$. The conditions for existence of this function will be a consequence of an *implicit function theorem* derived as a corollary to a *global inverse function theorem*.

For the inverse function theorem we will consider a map $f(\cdot):\mathbf{R}^n \to \mathbf{R}^n$ where $f(\cdot)$ is in the class, C^k, $K \geqslant 0$. By C_k we mean a set of maps between $\mathbf{R}^n$ and $\mathbf{R}^n$ having a least k continuous derivatives.[3] The theorem will enumerate conditions under which the inverse function, $f^{-1}(\cdot)$, will exist, be of class C^k, and satisfy

$$f^{-1} \circ f(x) \;=\; x \;=\; f \circ f^{-1}(x) \qquad (7.4)$$

for all x in $\mathbf{R}^n$. Basically we are describing the structure of those maps which are C^k-*diffeomorphisms* on $\mathbf{R}^n$ or, when k = 0 *homeomorphisms* on $\mathbf{R}^n$.

The implicit function theorem, derived as a corrollary to the above, will focus on a function, $h(\cdot,\cdot):\mathbf{R}^m x \mathbf{R}^n \to \mathbf{R}^n$ which is of class C^k. The desire here is to verify the existence of a function $k(\cdot):\mathbf{R}^m \to \mathbf{R}^n$ also of class C^k satisfying

$$h(k(q),q) \;=\; 0 \qquad (7.5)$$

for all q in $\mathbf{R}^m$.

The main theorem of the section is a variation of a result originally due to Palais.[10,20]

DEFINITION 7.6: A function $f(\cdot):\mathbf{R}^n \to \mathbf{R}^n$ is a *local C^k-diffeomorphism* ($k \geqslant 0$), at a point x if there exists an open neighborhood U of x and an open neighborhood V of f(x) such that (i) the restriction of $f(\cdot)$ to U is a one-to-one map of U onto V, and (ii) $f(\cdot)$ restricted to U and $f^{-1}(\cdot)$ restricted to V have k continuous derivatives.

Of course, if f is of class C^k ($k \geqslant 1$), the classical *local inverse function theorem*[3] assures that $f(\cdot)$ is a local C^k-diffeomorphism at x if and only if det($J_f(x)$) $\neq$ 0 where $J_f(x)$ is the *Jacobian matrix* of partial derivatives for f($\cdot$) evaluated at x.

Clearly for k = 0, no such test exists. By default a local C^0-diffeomorphism is a local homeomorphism. One final notion necessary for the discussion to follow is the notion of *proper function* by which we mean that for any *compact set* K(i.e., closed and bounded set in $\mathbf{R}^n$) f^{-1}(K) is also compact.

LEMMA 7.7: If f($\cdot$) is a continuous function mapping $\mathbf{R}^n$ into itself then the following statements are equivalent:

  (i) f($\cdot$) is proper,

  (ii) If B is a bounded set then f^{-1}(B) is a bounded set

  (iii) $|$ f(x) $|$ $\to\infty$ as $|$ x $|$ $\to\infty$.

THEOREM 7.8: Suppose f($\cdot$) is a continuous function such that f($\cdot$):$\mathbf{R}^n\to\mathbf{R}^n$. Then f($\cdot$) is a *global homeomorphism* if and only if f($\cdot$) is *proper* and a *local homoeomorphism* at each point.

The proofs of lemma 7.7 and theorem 7.8 are rather long and tedious and require the use of topological concepts not used elsewhere in the text. As such, they are relegated to appendix A.

To see the result intuitively consider f($\cdot$):$\mathbf{R}\to\mathbf{R}$ as a *piecewise linear function*. For f($\cdot$) to be a local homeomorphism, the slopes of the individual line segments must be non-zero. Now consider figure 7.9.

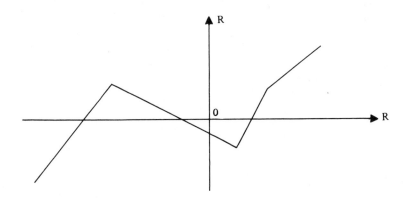

Figure 7.9. Illustration of a function not having a global inverse.

As can be deduced from the picture, two adjacent line segments must have slopes with the same sign either positive or negative. Otherwise, $f(\cdot)$ would not be a local homeomorphism at the point of intersection. In addition observe that since all slopes must have like signs, $|f(x)| \to \infty$ as $|x| \to \infty$. In summary if $f(\cdot): \mathbf{R} \to \mathbf{R}$ is a continuous piecewise linear function, then $f(\cdot)$ is a global homeomorphism if and only if all line segments have slopes with the same sign.

At last it is possible to formulate the desired *implicit function theorem* as a corollary to the above *inverse function theorem*.

THEOREM 7.10: Let $h(\cdot, \cdot)$ be a continuous function $h(\cdot, \cdot): \mathbf{R}^n \mathbf{x} \mathbf{R}^m \to \mathbf{R}^n$ such that for each q in $\mathbf{R}^m$, $h(\cdot, q)$ is proper and a local homeomorphism. Then there exists a continuous function $k(\cdot): \mathbf{R}^m \to \mathbf{R}^n$ such that

$$h(k(q), q) = 0 \tag{7.11}$$

for all q in $\mathbf{R}^n$.

PROOF: The required implicit function arises as the inverse of an appropriate auxilliary mapping, $f(\cdot, \cdot): \mathbf{R}^n \mathbf{x} \mathbf{R}^m \to \mathbf{R}^n \mathbf{x} \mathbf{R}^m$ defined as per 7.12.

$$f(x, q) = (h(x, q), q) \tag{7.12}$$

The function $f(\cdot, \cdot)$ is one to one and onto. To see this let (a, b) be an arbitrary vector in $\mathbf{R}^n \mathbf{x} \mathbf{R}^m$. For the equality of 7.12 to make sense it is necessary that

$$f(x, q) = (h(x, q), q) = (a, b) \tag{7.13}$$

and thus $q = b$. By theorem 7.8, $h(\ , b)$ is a global homeomorphism from $\mathbf{R}^n$ to $\mathbf{R}^n$. Thus there exists a unique x_0, such that $h(x_0, b) = a$. Consequently, equation 7.13 has the unique solution (x_0, b) for each (a, b). This verifies that $f(\cdot, \cdot)$ is one to one and onto.

Since $f(\cdot, \cdot)$ is continuous *invariance of the domain*[2] forces $f(\cdot, \cdot)$ to be a homeomorphism on $\mathbf{R}^n \mathbf{x} \mathbf{R}^m$. This permits $f(\cdot, \cdot)$ to admit an inverse satisfying

$$
\begin{aligned}
(a, b) &= f \circ f^{-1}(a, b) = (h \circ f^{-1}(a, b), b) \\
&= (h(f_1^{-1}(a, b), f_2^{-1}(a, b)), b)
\end{aligned}
\tag{7.14}
$$

By equation 7.12, $f_2^{-1}(a, b) = b$. Hence if we set $a = 0$, one concludes

$$(0, b) = (h(f_1^{-1}(0, b), b), b) \tag{7.15}$$

which verifies that

$$h(f_1^{-1}(0,b),b) \; = \; 0 \qquad\qquad (7.16)$$

Thus the desired implicit function $k(q) = f_1^{-1}(0,q)$. The proof is complete.

To extend these theorems to encompass functions of class C^k, one assumes $f(\cdot)$ is a proper and a local C^k-diffeomorphism. All C^k-diffeomorphisms are necessarily homeomorphisms; therefore the hypotheses of theorem 7.8 are satisfied. Basically then $f(\cdot)$ admits a continuous inverse which must coincide with the diffeomorphic local inverses at each point.[3] Hence $f^{-1}(\cdot)$ is a local C^k-diffeomorphism. Finally, noting that the classical local inverse function theorem[3] states that a function $f(\cdot)$ of class C^k is a C^k-diffeomorphism if and only $\det(J_f(x)) \neq 0$ brings us to the following corollary.

COROLLARY 7.17: Let $f(\cdot)$ be a class C^k function $k>0$, such that $f(\cdot):\mathbf{R}^n \to \mathbf{R}^n$. Then $f(\cdot)$ is a global C^k-diffeomorphism if and only if $\det(J_f(x)) \neq 0$ for all x in $\mathbf{R}^n$ and $f(\cdot)$ is proper.

Finally, in the modified implicit function case we have:

COROLLARY 7.18: Let $h(\cdot,\cdot):\mathbf{R}^n x\mathbf{R}^m \to \mathbf{R}^n$ be of class C^k, $k>0$, such that for each q, $h(\cdot,q):\mathbf{R}^n \to \mathbf{R}^n$ is proper and $\det(J_h(\cdot,q)(x)) \neq 0$. Then there exists a function $k(\cdot):\mathbf{R}^m \to \mathbf{R}^n$ of class C^k such that $h(k(q),q) = 0$ for all q in $\mathbf{R}^m$. As a final thought, consider the following example which illustrates the subtle distinctions between theorem 7.8 and corollary 7.17.

EXAMPLE 7.19: Consider the two curves in figure 7.20. In one case $z = x.^3$ Here $z = f(x) = x^3$ is clearly proper and a local homeomorphism. Thus it admits a globally continuous inverse $x = f^{-1}(z) = z^{1/3}$. However, since $\det(J_f(0)) = 0$, the conditions

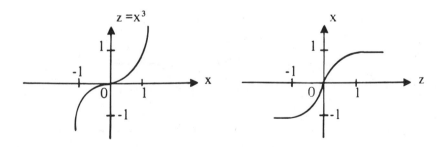

Figure 7.20. Illustration of $z = x^3$ and its inverse.

of corollary 7.17 fail to be satisfied. In particular $f^{-1}(\cdot)$ is continuous at zero but is not differentiable there. Of course, a similar phenomenon will occur at any *inflection point* of a function.

8. GENERALIZED COMPOSITE SYSTEM STATE MODELS

Can a *composite component state vector* also serve as the *composite system state vector*? No! Man in his quest for methodology designs by "sections." When fitting sections together, unforseen *loading effects* may crop up. A state of one black box may load down a state in an adjacent black box or the state vector of a particular component may become dependent on derivatives of the system input in addition to the usual dependence on the system input. Mathematically it would be impossible to specify arbitrary initial conditions for a composite system state vector which coincided with the composite component state vector. This follows since the value of one state could directly depend on the value of one or more of the other states. Even though a randomly interconnected dynamical system admits a state model as per theorem 5.7 with probability one, practically speaking numerous systems fail to satisfy the conditions of theorem 5.7.

EXAMPLE 8.1: Figure 8.2 shows a system illustrating the above pathology. Here a *π-section filter* drives a *capacitively coupled transistor amplifier*. Each component has a state model with the capacitor voltages taken as the component state vector. The resultant composite component state vector cannot, however, coincide with the state vector for the composite

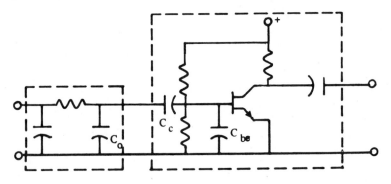

Figure 8.2: π-section filter driving a capacitively coupled amplifier.

system. Clearly the voltages on the output capacitor of the filter, C_0, the input coupling capacitor for the amplifier, C_c, and the base to emitter capacitance for the transistor, C_{be} are linearly dependent.

$$V_{C_0} = V_{C_c} + V_{C_{be}} \qquad (8.3)$$

If the composite component and composite system state vectors are to coincide to form the system model

$$\begin{aligned} \dot{x} &= Fx + Gu \\ y &= Hx + Ju \end{aligned} \qquad (8.4)$$

then it must be possible to arbitrarily specify a set of initial conditions. However equation 8.3 contradicts any such possibility. In particular, it is impossible to use the composite component state vector as a composite system state vector. However the system does posess a valid state model. In this example, simply deleting V_{C_0} from the state vector and replacing it algebraically by the sum $V_{C_c} + V_{C_{be}}$ produces a suitable state vector for the composite system. Obviously the dimension of the composite system state vector will, however, be less than the dimension of the composite component state vector, x.

EXAMPLE 8.5: Another sore spot occurs whenever $(I - L_{11}D) = 0$. Figure 8.6 illustrates this anomaly. Here an *all-pass network* is the open loop transfer function of a *unity gain positive feedback loop*.

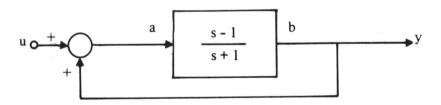

Figure 8.6. All-pass network in unity gain feedback loop.

The component connection model for the system is

$$\begin{aligned} \dot{x} &= [-1] x + [1] a \\ b &= [-2] x + [1] a \end{aligned} \qquad (8.7)$$

and

$$\left[\frac{a}{y}\right] = \left[\begin{array}{c|c} 1 & 1 \\ \hline 1 & 0 \end{array}\right]\left[\frac{b}{u}\right] \tag{8.8}$$

Calculating $(I-L_{11}D)$ yields

$$(I-L_{11}D) = 0 \tag{8.9}$$

which by theorem 5.7 disallows the existence of a composite system state model as per 8.4. However forgetting state models momentarily, the system has a perfectly well defined composite system transfer function

$$S(s) = \frac{s-1}{2} \tag{8.10}$$

Consequently equation I.4.67 implies that the composite system has a well defined *improper composite system state model.*

Examples 8.1 and 8.5 illustrate two dissimilar anomalies encountered in applying theorem 5.7. These examples suggest that a significantly strengthened theorem arises if the composite system state vector does not necessarily coincide with the composite component state vector and/ or the composite system is permitted to have an improper state model. Recall that an improper state model allows x to depend on x, a, and derivatives of the input a. It is possible to prove that any interconnected system with composite component state model

$$\begin{aligned} \dot{x} &= Ax + Ba \\ b &= Cx + Da \end{aligned} \tag{8.11}$$

and *generalized algebraic connection equations*

$$\left[\frac{a}{y}\right] = \left[\begin{array}{c|c} L_{11} & L_{12} \\ \hline L_{21} & L_{22} \end{array}\right]\left[\frac{b}{u}\right] + \left[\begin{array}{c|c} R_{11} & R_{12} \\ \hline R_{21} & R_{22} \end{array}\right]\left[\frac{a}{y}\right] \tag{8.12}$$

admits such an improper state model provided the system is *well defined.* Well defined means that for an initial state and any admissible composite system input vector, u, equations 8.11 and 8.12 uniquely determine the vectors, a, x, b, and y.

A word of caution: The connection equations of 8.12 are a

mathematical representation of the system or network. When using these generalized connection equations it is a simple matter to construct *degenerate* representations—i.e., representations which do not model all the connection or topological information of the network. For example, choose $L_{11} = 0$ and $R_{11} = I$ and you have built a degenerate connection model. Recall connection equations are akin to loop or node equations. Such equations represent the topology of a network if and only if all pertinent loops or nodes are accounted for. When using the usual algebraic connection equations such difficulties are not present. In this case the connection equations define an *explicit* relationship between a and b. In the generalized connection equation case, however, the relationship between a and b is implicitly defined.

The details surrounding the existence of a composite system state vector are lengthy, tedious, and somewhat obscure at this time. However, the usual plan of attack is to convert equations 8.11 and 8.12 into a form amenable to the theory of *inverse systems*.

Observe that equation 8.11 uniquely determines x and b regardless of the value of x(0). The problem is to show that under proper hypotheses, s, x(0) = 0, and equations 8.11 and 8.12 uniquely define a and y.

By suitable algebraic juggling of 8.11 and 8.12 we have the following set of equalities where v = col(a,y).

$$\begin{aligned} \dot{x} &= Ax + Pv \\ Qu &= Sx + Tv \end{aligned} \tag{8.13}$$

with

$$P = \begin{bmatrix} B & | & 0 \end{bmatrix} \tag{8.14}$$

$$Q = \begin{bmatrix} L_{12} \\ \hline L_{22} \end{bmatrix} \tag{8.15}$$

$$S = \begin{bmatrix} -L_{11}C \\ \hline -L_{21}C \end{bmatrix} \tag{8.16}$$

and

$$T = \begin{bmatrix} I - R_{11} -L_{11}D & | & -R_{12} \\ \hline -R_{21} - L_{21}D & | & I - R_{22} \end{bmatrix} \tag{8.17}$$

These equations convert the original problem into solving 8.13 for v = col(a,y) in terms of u and possibly its derivatives.

Basically we know x and Qu which we have manipulated into being an output of the artificially constructed system of 8.13. The goal is to find v = col(a,y). Clearly then this is an *inverse system* question. An alternate and very instructive point of view is this: Can we build a system whose input is Qu and whose output is v integrated L-times for some integer L. If so, the system is said to admit an *L-integral inverse*.[15] Recovering v amounts to taking the derivative of this output L-times. This is why the problem is to solve for v = col(a,y) in terms of u and possibly its derivatives.

Suppose that T is nonsingular. Then there is an explicit solution to 8.13 as

$$v = -T^{-1}Sx + T^{-1}QU \qquad (8.18)$$

Substituting this expression for v into the first equation of 8.13, produces the following composite system state equations.

$$\begin{aligned}
\dot{x} &= [A - PT^{-1}S]\, x + [PT^{-1}Q]\, u \\
v &= [-T^{-1}S]\, x + [T^{-1}Q]\, u
\end{aligned} \qquad (8.19)$$

Basically this case explicitly computes v = col(a,y) in terms of u. Hence the system is well defined. By deleting the rows corresponding to "a" in the equation for we produce the correct system state model. Unfortunately, the state model of 7.19 is for all intents and purposes the same as the one derived in theorem 5.7. In particular, suppose a standard algebraic connection model is used rather than a generalized algebraic connection model (i.e., $R_{ij} = 0$) then T is nonsingular if and only if $(I-L_{11}D)$ is nonsingular.

On the other hand we allow the possibility of an improper composite system state model. Thus if T is singular, differentiate the second equation in 8.13, Qu = SX + Tv, to obtain additional equations with the same number of unknowns. Differentiating this equation yields

$$Q\dot{u} = S\dot{x} + T\dot{v} = [SA]\, x + [SP]\, v + [\dot{T}]\, v \qquad (8.20)$$

Now we attempt to solve this equation simultaneously with

$$Qu = Sx + Tv \tag{8.21}$$

to find v as a function of u, u, and x. If this attempt is successful, one substitutes the solution into $x = Ax + Pv$. This results in the indicated improper composite system state model. If equations 8.20 and 8.21 fail to have a simultaneous solution, differentiate equation 8.20. This produces a set of three simultaneous equations which one must attempt to solve for v in terms of u, u, and u. Repeating the process through k-differentiations results in the set of simultaneous equations delineated in 8.22. It is unnecessary to consider $k \geqslant n$, the dimension of x, since if the equations have no solution for $k = n-1$, then they have no solution for $k \geqslant n$.[17]

$$
\begin{aligned}
Qu &= Sx + Tv \\
Qu^{(1)} &= SAx + SPv + Tv^{(1)} \\
Qu^{(2)} &= SA^2x + SAPv + SPv^{(1)} + Tv^{(2)} \\
Qu^{(3)} &= SA^3x + SA^2Pv + SAPv^{(1)} + SPv^{(2)} + Tv^{(3)}
\end{aligned}
\tag{8.22}
$$

$$\vdots \qquad\qquad \vdots$$

$$Qu^{(k)} = SA^kx + SA^{k-1}Pv + \ldots + SPv^{(k-1)} + Tv^{(k)}$$

In equation 8.22 the notation $u^{(i)}$ and $v^{(i)}$ denotes the i-th derivative of u and v respectively. In matrix notation, equation 8.22 becomes

$$Q_k\underline{u}^k = N_kX + M_k\underline{v}^k \tag{8.23}$$

where $Q_k = \text{diag}(Q,Q,Q,\ldots,Q)$, $u^k = \text{col}(u,u^{(1)},\ldots,u^{(k)})$, $v^k = \text{col}(v,v^{(1)},\ldots,v^{(k)})$, $N_k = \text{col}(S,SA,SA^2,\ldots,SA^k)$, and

$$
M_k =
\begin{bmatrix}
T & 0 & 0 & \cdots & 0 \\
SP & T & 0 & & 0 \\
SAP & SP & T & & 0 \\
SA^2P & SAP & SP & & 0 \\
\vdots & & & \ddots & \vdots \\
SA^{k-1}P & SA^{k-2}P & SA^{k-3}P & SP & T
\end{bmatrix}
\tag{8.24}
$$

Thus the construction of an improper composite system state model is equivalent to solving 8.23 for v in terms of u^k. This is a purely algebraic equation. Thus it can be solved by linear algebraic arguments. The following theorem is a statement of the details of existence.

THEOREM 8.25: Let an interconnected dynamical system have a component connection model as in 8.11 and 8.12. Then the following are equivalent.

    i) The system is well defined—i.e. for each admissible u, equations 8.11 and 8.12 uniquely define the vectors a,b,x, and y with $x(0) = 0$.

    ii) The composite system is characterized by a (possibly improper) state model.

    iii) There exists an integer $k \leqslant n-1$ such that rank $[Q^k \vert M_k] = $ rank $[M_{k-1}]$ + r where r is the dimension of the T matrix and $Q^k = \text{col}(Q,Q,....,Q)$.

    The proof of the theorem was originally given by Singh and Liu, so the details are referenced to the literature.[17] In order to render condition (iii) meaningful when k equals zero, define $M_0 = T$ and $M_{-1} = 0$.

EXAMPLE 8.26: Consider the electric network shown in figure 8.27.

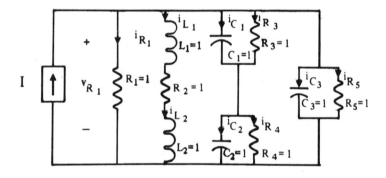

Figure 8.27. RLC Network whose composite component state vector and composite system state vector are different.

    By inspection it is obvious that the circuit admits a state description. In fact classical circuit theory[7] guarantees that such a state model exists. The method is to pick a *tree* with a maximal number of capacitors and a minimal number of inductors. The branch voltages corresponding to the capacitors in the tree and the branch currents corresponding to the inductors not in the tree become the state variables. In this circuit either inductor current, but not both, and any two capacitor voltages will serve as a suitable composite system state variables. However to illustrate the theorem we do the necessary algebraic manipulations. In the process we will construct an algorithmic framework to illustrate the procedure for more general systems.

To begin, pick v_R as the circuit output, I as the input, and embed the resistors into the connection equations. This leaves five components having the following composite component state model.

$$
\begin{bmatrix} \dot{i}_{L_1} \\ \dot{i}_{L_2} \\ \dot{v}_{C_1} \\ \dot{v}_{C_2} \\ \dot{v}_{C_3} \end{bmatrix} = \begin{bmatrix} 0 & 0 & 0 & 0 & 0 \\ 0 & 0 & 0 & 0 & 0 \\ 0 & 0 & 0 & 0 & 0 \\ 0 & 0 & 0 & 0 & 0 \\ 0 & 0 & 0 & 0 & 0 \end{bmatrix} \begin{bmatrix} i_{L_1} \\ i_{L_2} \\ v_{C_1} \\ v_{C_2} \\ v_{C_3} \end{bmatrix} + \begin{bmatrix} 1 & 0 & 0 & 0 & 0 \\ 0 & 1 & 0 & 0 & 0 \\ 0 & 0 & 1 & 0 & 0 \\ 0 & 0 & 0 & 1 & 0 \\ 0 & 0 & 0 & 0 & 1 \end{bmatrix} \begin{bmatrix} v_{L_1} \\ v_{L_2} \\ i_{C_1} \\ i_{C_2} \\ i_{C_3} \end{bmatrix} \quad (8.28)
$$

and

$$
\begin{bmatrix} i_{L_1} \\ i_{L_2} \\ v_{C_1} \\ v_{C_2} \\ v_{C_3} \end{bmatrix} = \begin{bmatrix} 1 & 0 & 0 & 0 & 0 \\ 0 & 1 & 0 & 0 & 0 \\ 0 & 0 & 1 & 0 & 0 \\ 0 & 0 & 0 & 1 & 0 \\ 0 & 0 & 0 & 0 & 1 \end{bmatrix} \begin{bmatrix} i_{L_1} \\ i_{L_1} \\ v_{C_1} \\ v_{C_2} \\ v_{C_3} \end{bmatrix} + \begin{bmatrix} 0 & 0 & 0 & 0 & 0 \\ 0 & 0 & 0 & 0 & 0 \\ 0 & 0 & 0 & 0 & 0 \\ 0 & 0 & 0 & 0 & 0 \\ 0 & 0 & 0 & 0 & 0 \end{bmatrix} \begin{bmatrix} v_{L_1} \\ v_{L_2} \\ i_{C_1} \\ i_{C_2} \\ i_{C_3} \end{bmatrix} \quad (8.29)
$$

As such, $A = D = 0$ and $B = C = I$. It is thus impossible to write the usual set of connection equations, necessitating the formulation of a suitable set of generalized connection equations. Because of the choice of parameter values, it is imperative to put as much information into the L_{11} matrix as possible. In particular, the L_{11} matrix must have maximal rank. Otherwise a degenerate representation of the circuit will result. A suitable set of generalized connection equations is given in 8.30.

$$
\begin{bmatrix} v_{L_1} \\ v_{L_2} \\ i_{C_1} \\ i_{C_2} \\ i_{C_3} \\ v_{R_1} \end{bmatrix} =
\left[\begin{array}{ccccc|c}
-1 & 0 & 0 & 0 & 1 & 0 \\
0 & -1 & 1 & 1 & 0 & 0 \\
-1 & 0 & -1 & 0 & -2 & 1 \\
0 & 0 & 1 & -1 & 0 & 0 \\
0 & -1 & -1 & 0 & -2 & 1 \\
\hline
0 & 0 & 0 & 0 & 1 & 0
\end{array} \right]
\begin{bmatrix} i_{L_1} \\ i_{L_2} \\ v_{C_1} \\ v_{C_2} \\ v_{C_3} \\ I \end{bmatrix}
\tag{8.30}
$$

$$
+
\left[\begin{array}{ccccc|c}
0 & -1 & 0 & 0 & 0 & 0 \\
-1 & 0 & 0 & 0 & 0 & 0 \\
0 & 0 & 0 & 0 & -1 & 0 \\
0 & 0 & 1 & 0 & 0 & 0 \\
0 & 0 & -1 & 0 & 0 & 0 \\
\hline
0 & 0 & 0 & 0 & 0 & 0
\end{array} \right]
\begin{bmatrix} v_{L_1} \\ v_{L_2} \\ i_{C_1} \\ i_{C_2} \\ i_{C_3} \\ v_{R_1} \end{bmatrix}
$$

The first step is to apply the theorem so as to decide if a composite system state model does exist. Thus we initially form P, Q, S, T, and thereby M_k. Consequently

$$
P = [B \vdots O] = [I \vdots 0] =
\left[\begin{array}{ccccc|c}
1 & 0 & 0 & 0 & 0 & 0 \\
0 & 1 & 0 & 0 & 0 & 0 \\
0 & 0 & 1 & 0 & 0 & 0 \\
0 & 0 & 0 & 1 & 0 & 0 \\
0 & 0 & 0 & 0 & 1 & 0
\end{array} \right]
\tag{8.31}
$$

$$
Q = \left[\begin{array}{c} L_{12} \\ \hline L_{22} \end{array} \right] =
\begin{bmatrix} 0 \\ 0 \\ 1 \\ 0 \\ \hline 1 \\ 0 \end{bmatrix}
\tag{8.32}
$$

$$S = \begin{bmatrix} -L_{11}C \\ \hline -L_{21}C \end{bmatrix} = \begin{bmatrix} -L_{11} \\ \hline -L_{21} \end{bmatrix} \qquad (8.33)$$

$$= \begin{bmatrix} 1 & 0 & 0 & 0 & -1 \\ 0 & 1 & -1 & -1 & 0 \\ 1 & 0 & 1 & 0 & 2 \\ 0 & 0 & -1 & 1 & 0 \\ 0 & 1 & 1 & 0 & 2 \\ \hline 0 & 0 & 0 & 0 & -1 \end{bmatrix}$$

and

$$T = \begin{bmatrix} I - R_{11} - L_{11}D & -R_{12} \\ \hline -R_{21} - L_{21}D & I\text{-}R_{22} \end{bmatrix} = \begin{bmatrix} I\text{-}R_{11} & -R_{12} \\ \hline -R_{21} & I\text{-}R_{22} \end{bmatrix} \qquad (8.34)$$

$$= \begin{bmatrix} 1 & 1 & 0 & 0 & 0 & 0 \\ 1 & 1 & 0 & 0 & 0 & 0 \\ 0 & 0 & 1 & 0 & 1 & 0 \\ 0 & 0 & -1 & 1 & 0 & 0 \\ 0 & 0 & 1 & 0 & 1 & 0 \\ \hline 0 & 0 & 0 & 0 & 0 & 1 \end{bmatrix}$$

To apply the theorem we must search for an integer k such that $M_k - M_{k-1} = r$ where r is the dimension of T. For this example $r = 6$. Consider $M_0 = T$. The rank of M_0 is four. Therefore, $\text{Rank}(M_0) - \text{Rank}(M_{-1}) \neq 6$. Consequently consider M_1 as given in equation 8.35.

$$M_1 = \begin{bmatrix} T & 0 \\ \hline SP & T \end{bmatrix} \qquad (8.35)$$

$$= \begin{bmatrix} 1 & 1 & 0 & 0 & 0 & 0 & 0 & 0 & 0 & 0 & 0 & 0 \\ 1 & 1 & 0 & 0 & 0 & 0 & 0 & 0 & 0 & 0 & 0 & 0 \\ 0 & 0 & 1 & 0 & 1 & 0 & 0 & 0 & 0 & 0 & 0 & 0 \\ 0 & 0 & -1 & 1 & 0 & 0 & 0 & 0 & 0 & 0 & 0 & 0 \\ 0 & 0 & 1 & 0 & 1 & 0 & 0 & 0 & 0 & 0 & 0 & 0 \\ 0 & 0 & 0 & 0 & 0 & 1 & 0 & 0 & 0 & 0 & 0 & 0 \\ \hline 1 & 0 & 0 & 0 & -1 & 0 & 1 & 1 & 0 & 0 & 0 & 0 \\ 0 & 1 & -1 & -1 & 0 & 0 & 1 & 1 & 0 & 0 & 0 & 0 \\ 1 & 0 & 1 & 0 & 2 & 0 & 0 & 0 & 1 & 0 & 1 & 0 \\ 0 & 0 & -1 & 1 & 0 & 0 & 0 & 0 & -1 & 1 & 0 & 0 \\ 0 & 1 & 1 & 0 & 2 & 0 & 0 & 0 & 1 & 0 & 1 & 0 \\ 0 & 0 & 0 & 0 & -1 & 0 & 0 & 0 & 0 & 0 & 0 & 1 \end{bmatrix}$$

However $\text{Rank}(M_1) = 6$ forcing $\text{Rank}(M_1) - \text{Rank}(M_0) = 6$. Thus if $\text{Rank}\,[Q^k|M_1]$ also equals six, the circuit has a composite state model as expected. Note that $Q^k = Q^2 = \text{col}(Q,Q)$. A little thought reveals that $\text{Rank}[Q^2|M_1] = 6$. Thus the circuit does have an improper composite system state model.

The next question is: how does one construct the state vector for the composite system and thereby the composite system state model? Mathematically, the goal of this task is to find a matrix G such that

$$x = Gz \tag{8.36}$$

which implies that

$$z = G^{-L}x \tag{8.37}$$

where G^{-L} is a left inverse of G. In other words we are determining the composite system state vector, z, in terms of the composite component state vector, x. The matrix G is one whose columns form a basis for the null space of the matrix, $S = \text{col}(S_0,...,S_{k-1})$, where k is the integer for which $\text{Rank}(M_k) - \text{Rank}(M_{k-1}) = r$. How then does one construct S or more appropriatley S_j. In this endeavor, first find a nonsingular matrix G_0 such that

$$G_0\,[Q|S|T] = \begin{bmatrix} Q_0 & \underline{S_0} & \underline{T_0} \\ \hline \tilde{Q}_0 & \tilde{S}_0 & 0 \end{bmatrix} \tag{8.38}$$

Basically G_0 is a matrix representing the set of elementary row opertions which row reduce the matrix, T, into the form $\text{col}(T_0,0)$ where T_0 has full row rank. Finding an appropriate G_0 allows one to compute S_0.

Now replace the second row of the right hand side of 8.38 as indicated in equation 8.39.

$$\begin{bmatrix} Q_0 & S_0 & T_0 \\ \hline 0 & \tilde{S}_0A & \tilde{S}_0P \end{bmatrix} \tag{8.39}$$

Now find another nonsingular matrix G_1 (representing a suitable set of elementary row operations) such that

$$G_1 \begin{bmatrix} Q_0 & S_0 & T_0 \\ \hline 0 & \tilde{S}_0A & \tilde{S}_0P \end{bmatrix} = \begin{bmatrix} Q_1 & S_1 & T_1 \\ \hline \tilde{Q}_1 & \tilde{S}_1 & 0 \end{bmatrix} \tag{8.40}$$

It is necessary to repeat the process k-times before being able to build S. The general formula is given in equation 8.41.

$$G_j \left[\begin{array}{c|c|c} Q_{j-1} & S_{j-1} & T_{j-1} \\ \hline 0 & \tilde{S}_{j-1}A & \tilde{S}_{j-1}P \end{array} \right] = \left[\begin{array}{c|c|c} Q_j & S_j & T_j \\ \hline \tilde{Q}_j & \tilde{S}_j & 0 \end{array} \right] \tag{8.41}$$

In our example k=1, therefore $S = S_0$. Thus we have

$$G_0 \quad [Q|S|T] \tag{8.42}$$

$$= \begin{bmatrix} 1 & 0 & 0 & 0 & 0 & 0 \\ 0 & 0 & 1 & 0 & 0 & 0 \\ 0 & 0 & 0 & 1 & 0 & 0 \\ 0 & 0 & 0 & 0 & 0 & 1 \\ -1 & 1 & 0 & 0 & 0 & 0 \\ 0 & 0 & -1 & 0 & 1 & 0 \end{bmatrix} \begin{bmatrix} 0 & 1 & 0 & 0 & 0 & -1 & 1 & 1 & 0 & 0 & 0 & 0 \\ 0 & 0 & 1 & -1 & -1 & 0 & 1 & 1 & 0 & 0 & 0 & 0 \\ 1 & 1 & 0 & 1 & 1 & 2 & 0 & 0 & 1 & 0 & 0 & 0 \\ 0 & 0 & 0 & -1 & 1 & 0 & 0 & 0 & -1 & 1 & 0 & 0 \\ 1 & 0 & 1 & -1 & 0 & 2 & 0 & 0 & 1 & 0 & 1 & 0 \\ 0 & 0 & 0 & 0 & 0 & -1 & 0 & 0 & 0 & 0 & 0 & 0 \end{bmatrix}$$

$$= \begin{bmatrix} 0 & 1 & 0 & 0 & 0 & -1 & 1 & 1 & 0 & 0 & 0 & 0 \\ 1 & 1 & 0 & 1 & 0 & 2 & 0 & 0 & 1 & 0 & 1 & 0 \\ 0 & 0 & 0 & -1 & 1 & 0 & 0 & 0 & -1 & 1 & 0 & 0 \\ 0 & 0 & 0 & 0 & 0 & -1 & 0 & 0 & 0 & 0 & 0 & 1 \\ \hline 0 & -1 & 1 & -1 & -1 & 1 & 0 & 0 & 0 & 0 & 0 & 0 \\ 0 & -1 & 1 & 0 & 0 & 0 & 0 & 0 & 0 & 0 & 0 & 0 \end{bmatrix}$$

Therefore we have that

$$\tilde{S}_0 = \begin{bmatrix} -1 & 1 & -1 & -1 & 1 \\ -1 & 1 & 0 & 0 & 0 \end{bmatrix} \tag{8.43}$$

A basis for the null space of S is given by the following three column vectors.

$$\begin{bmatrix} 1 \\ 1 \\ 0 \\ 0 \\ 0 \end{bmatrix} ; \begin{bmatrix} 0 \\ 0 \\ 1 \\ -1 \\ 0 \end{bmatrix} , \begin{bmatrix} 0 \\ 0 \\ 0 \\ 1 \\ 1 \end{bmatrix} \tag{8.44}$$

Thus $x = Gz$ becomes

$$\begin{bmatrix} i_{L_1} \\ i_{L_2} \\ v_{C_1} \\ v_{C_2} \\ v_{C_3} \end{bmatrix} = \begin{bmatrix} 1 & 0 & 0 \\ 1 & 0 & 0 \\ 0 & 1 & 0 \\ 0 & -1 & 1 \\ 0 & 0 & 1 \end{bmatrix} \begin{bmatrix} z_1 \\ z_2 \\ z_3 \end{bmatrix} \tag{8.45}$$

One of many left inverses of G is

$$G^{-L} = \begin{bmatrix} 1 & 0 & 0 & 0 & 0 \\ 0 & 0 & 1 & 0 & 0 \\ 0 & 0 & 0 & 0 & 1 \end{bmatrix} \tag{8.46}$$

Consequently the composite system state vector, z, becomes

$$z = \begin{bmatrix} z_1 \\ z_2 \\ z_3 \end{bmatrix} = \begin{bmatrix} i_{L_1} \\ v_{C_1} \\ v_{C_3} \end{bmatrix} \tag{8.47}$$

which agrees with the intuition developed earlier in the example. The only remaining loose end is the construction of the composite system state model. For this define

$$[Q_j \mid S_j \mid T_j] = G_j \begin{bmatrix} Q_{j-1} & \vdots & S_{j-1} & \vdots & T_{j-1} \\ \hline 0 & \vdots & \widetilde{S}_{j-1}A & \vdots & \widetilde{S}_{j-1}P \end{bmatrix} \tag{8.48}$$

The form of the composite system equations is:[17]

$$\dot{z} = [G^{-L}AG + G^{-L}PT_k^{-1}S_kG]z - [G^{-L}PT_{k-1}Q_k]u \tag{8.49}$$

and

$$\begin{bmatrix} a \\ \hline y \end{bmatrix} = -[T_k^{-1}S_kG]z - T_k^{-1}Q_ku \tag{8.50}$$

where again k is the integer chosen so that $\text{Rank}(M_k) - \text{Rank}(M_{k-1}) = r$. Note that for $j=k$ in equation 8.48, $G_j = I$ since $\text{col}(T_{j-1}, S_{j-1}P)$ is square and nonsingular. It is now just a matter of doing the necessary arithmetic to obtain the composite system state model. It is left as an exercise for the energetic student. The example is finished.

An improper state model will exist whenever the given interconnected system is well defined. This suffices for most purposes since we rarely have an interest in modeling an interconnected system which is not well defined. In the rare event that one wants to model such a system it can be done with the aid of the *Rosenbrock model* defined by equation 1.4.69. The Rosenbrock model description of the composite component equations is

$$W(s) \ x(s) \ = \ U(s) \ a(s)$$
$$b(s) \ = \ V(s)x(s) \ - \ Y(s)a(s) \tag{8.51}$$

which together with the generalized algebraic connection model forms our generalized component connection model. The goal is to formulate a Rosenbrock model for the composite system of the form

$$P(s) \ z(s) \ = \ Q(s)u(s)$$
$$y(s) \ = \ R(s)z(s) \ - \ T(s)u(s) \tag{8.52}$$

A little deep concentration reveals that a Rosenbrock model will always exist since the matrices P,Q,R, and T are all polynomial matrices. However the only way to determine interesting properties of the system is to make additional assumptions on the properties of the matrices.[14]

9. PROBLEMS

1. Verify theorem 2.8 using the matrices $X = [1 \quad 2]$ and $Y = [-1 \quad 2]^t$.
2. Compute the matrices $Z(I - L_{11}Z)^{-1}$ and $(I - ZL_{11})^{-1}Z$ for the *Butterworth filter* of example 2.13. Are they equal?
3. Calculate the *zero set* for the polynomial $P(x_1,x_2) = x_1^2 + x_1x_2 + 1$.
4. Let an interconnected dynamical system have composite component transfer function matrix

$$Z(s) = \begin{bmatrix} s+1 & 0 \\ 0 & 1/s \end{bmatrix}$$

and determine the class of all L_{11} matrices for which $(I - L_{11}Z)$ fails to admit a rational inverse.

5. Give examples of a function which has a *left inverse* but not a *right inverse* and a function which has a right inverse but not a left inverse.
6. Show that if a function has an *inverse* then it is unique.
7 Show that a function has a left inverse if and only if it is *one-to-one* and a right inverse if and only if it is *onto*.

8. Prove that the *connection function* is nonlinear (viewed as a mapping from a space of matrices, Z, to a space of matrices, S.

9. Compute

$$\text{vec} \begin{bmatrix} 1 & 0 & 7 \\ 2 & 1 & -1 \\ 0 & -4 & 1 \end{bmatrix}$$

10. Show that every *linear function* mapping $\mathbf{R}^n$ to $\mathbf{R}^m$ is *continuous*.

11. Compute the *Jacobian matrix* for the function f: $\mathbf{R}^3 \to \mathbf{R}^2$ defined by

$$\begin{bmatrix} f_1(x_1,x_2,x_3) \\ f_2(x_1,x_2,x_3) \end{bmatrix} = \begin{bmatrix} x_1 + e^{x_1+x_3} \\ \ln(x_2) + x_1^2 \end{bmatrix}$$

12. Compute the derivative of the matrix valued function of the scalar variable r

$$M(r) = \begin{bmatrix} e^{-r} & 1+r^2 \\ 3 & 7r \end{bmatrix}$$

using the notation of equation 3.7.

13. For the system of example 3.14 assume that the *internal composite system transfer function matrix* is of the form

$$Q(s) = \begin{bmatrix} 1 & 0 \\ -1 & 1/s \end{bmatrix}$$

and determine the composite component transfer function matrix.

14. Prove that the *"vec" operation* defined by equation 3.5 is a linear invertible transformation from the space of m by n matrices to $\mathbf{R}^{mn}$.

15. Verify the properties of the *matrix tensor product* given in 3.25.

16. Find a set of components which realize the Butterworth filter function

$$S(s) = \frac{1}{2s^2 + 2s + 1}$$

using the connection structure of example 3.31. Repeat the problem using the connection structure of example 3.14. Show that the latter is unique but that the former is not.

17. In example 3.45 assume that the integrator gain, r, varies with temperature as $r = T^{1/3}$ and compute dS/dT. (Hint: Use the *chain rule*).

18. Verify that the matrix of equation 4.3 is the inverse of the matrix of equation 4.2.

19. Show that the inverse of a *tri-diagonal* matrix is "almost always" non-sparse.

20. Use the *back substitution process* to compute the 9th row of the matrix M^{-1} of equation 4.13.

21. Prove that a nonsingular matrix has a unique *LU-factorization* in the form $M = LU$ where L is lower triangular and U is upper triangular with 1's on the diagonal.

22. Use the *Crout algorithm* to compute an LU-factorization for the matrix

$$M = \begin{bmatrix} 1 & 0 & 1 \\ 2 & 1 & 0 \\ 4 & 0 & 1 \end{bmatrix}$$

23. Formulate an algorithm based on the classical *Gaussian elimination* algorithm for generating the inverse of a sparse matrix in factored form. Hint: show that the *row operations* required to reduce an arbitrary matrix to an upper triangular matrix are all represented by lower triangular matrices. As such, at the intermediate stage of a classical Gaussian elimination algorithm where M has been reduced to an upper triangular matrix U we have the equality

$$T_1 T_2 T_3 \ldots T_k M = U$$

where the T_i are lower triangular matrices representing the elementary row operations used to transform M into U. Then M^{-1} may be represented in factored form as

$$M^{-1} = U^{-1}(T_1, T_2, T_3, \ldots T_k) = U^{-1}L^{-1}$$

24. Compute the *derivative of the matrix* $M(r)^{-1}$ at $r = 1$ via equation 4.31 where

$$M(r) = \begin{bmatrix} r & 0 & 0 \\ 1 & r^2 & 0 \\ r & 0 & r \end{bmatrix}$$

25. Prove that $^\ell M + {}^u M = M$ for any matrix M.

26. Derive a formula for M^{-1} in terms of the formulae for L^{-1} and U^{-1} of theorem 4.34 and use it on the matrix of problem 24. Do the results compare with those of that problem?

27. Repeat example 4.42 using the exact formula for the integration of the matrix valued functions. Does the result coincide with that of equation 4.51.

28 Let M be a *block lower triangular matrix* in which all of the diagonal blocks are square. Show that such a matrix is invertible if and only if each diagonal block is invertible and formulate a *block back substitution* process for the computation of M^{-1} in terms of the inverses if its diagonal blocks.

29. Does a *composite system state model* in the form of equation 5.3 exist for the system of example 3.14. If so, compute it via the equalities of theorem 5.7.

30. Show that the system described by the *adjacency graph* of figure 5.23 will have no *algebraic loops* if component 2 is an integrator.

31. Repeat example 5.38 with the diode replaced by a *hard limiter* characterized by the equalities

$$b_1 = \begin{cases} 1; \ a_1 \geqslant 1 \\ a_1; \ -1 \leqslant a_1 \leqslant 1 \\ -1; \ a_1 \leqslant -1 \end{cases}$$

32. Formulate a back substitution-like algorithm for computing the *implicit function* $k(x,u)$ of theorem 5.35 in an interconnected nonlinear system which has no algebraic loops.

33. Assume in equation 6.6 the component output is obtained by passing $x + a$ (which we assume to be scalars through the hard limiter of problem 31. Then compute the matrices P_i and Q_i, the vectors p_i and q_i, and their regions of definition.

34. Give a detailed proof of theorem 6.11.

35. Use *Householder's formula* to compute the inverse of the matrix

$$D = \begin{bmatrix} 1 & 0 & 1 \\ 0 & 1 & 1 \\ -1 & 1 & 2 \end{bmatrix}$$

given the inverse of the triangular matrix

$$E = \begin{bmatrix} 1 & 0 & 0 \\ 0 & 1 & 0 \\ -1 & 1 & 1 \end{bmatrix}$$

36. Give a proof of theorem 7.10 which does not use *invariance of domain* (which is a very powerful tool to be used in such a simple theorem).

37. Let f be a *proper* continuous function of class C^1 such that $\det(J_f(x)) \geq 0$ (or ≤ 0) for all x and $\det(J_f(x)) = 0$ on at most a set of isolated points. Then show that f has a C° globally defined inverse.

38. Show that an *RLC network* containing a loop of capacitors or a cut-set of inductors does not admit a state model in which all capacitor voltages (or charges) and inductor currents (or fluxes) are used in the *composite system state vector*.

39. Show that the system of figure II.2.17 is not well defined while the system of figure II.2.14 is well defined. Here, we say that an interconnected dynamical system is *well defined* if for $x(0) = 0$ and any admissible composite system input u there exists unique vectors a, b, x, and y satisfying the generalized component connection equations.

40. Prove theorem 8.25. Hint: transform equations 8.13 to the frequency domain and then expand the vectors u and v in a Laurant series to derive the result.

10. REFERENCES

1. Bellman, R., *An Introduction to Matrix Analysis*, New York, McGraw-Hill, 1960.

2. Bers, L., *Topology*, New York, Courant Inst. of Math. Sci., 1956.

3. Fleming, W., *Functions of Several Variables*, Reading, Addison-Wesley, 1965.

4. Householder, A. S., "A Survey of Some Closed Closed Methods for Inverting Matrices", SIAM Jour. on Appl. Math., Vol. 5, pp. 155-169, (1957).

5. Jacobson, *Lectures on Abstract Algebra*, Van Nostrand, Princeton, 1953.

6. Kevorkian, A. K., and J. Snoek, "Decomposition in Large-Scale Systems: Theory and Application of Structural Analysis in Partitioning, Disjointing and Constructing Hierarchical Systems", in *Decomposition of Large Scale Problems* (ed. D. M. Himmelblau), North Holland/American Elsevier, Amsterdam, 1973.

7. Kuh, E. S., and R. A. Rorher, "The State Variable Approach to Network Analysis", Proc. Proc. of the IEEE, Vol. 53, pp. 672-686, (1965).

8. Loeve, M., *Probability Theory*, Van Nostrand, Princeton, 1954.

9. Massey, W. S., *Algebraic Topology; An Introduction*, Harcourt, Brace, and World, New York, 1967.

10. Palais, R. S., "Natural Operations on Differential Forms", Trans. of the AMS, Vol. 92, pp. 125-141, (1959).

11. Pease, W., *Matrix Algebra*, Academic Press, New York, 1967.

12. Ransom, M. N. and R. Saeks, "The Connection Function — Theory and Application", Inter. Jour. on Circuit Theory and Its Application, Vol. 3, pp. 5-21, (1975).

13. Rao, C. R., and S. K. Mitra, *Generalized Inverses of Matrices and its Applications*, New York, Wiley, 1971.
14. Rosenbrock, H. H., and A. C. Pugh, "Contributions to a Hierarchical Theory of Systems", Int. Jour. on Cont., Vol. 19, pp. 845-867, (1974).
15. Sain, M. K., and J. L. Massey, "Invertibility on Linear Time-Invariant Dynamical Systems", IEEE Trans. on Auto. Cont., Vol. AC-14, pp. 141-149, (1969).
16. Silverman, L. M., "Inversion of Multivariable Linear Systems", IEEE Trans. on Auto. Cont., Vol. AC-14, pp. 270-276, (1969).
17. Singh, S. P., and R. W. Liu, "Existence of State Equation Representation of Linear Large-Scale Dynamical Systems", IEEE Trans. on Circuit Theory, Vol. CT-20, pp. 239-246, (1973).
18. Spivak, M., *Calculus on Manifolds*, Benjamin, Amsterdam, 1967.
19. Tewerson, R. P., *Sparse Matrices*, Academic Press, New York, 1972.
20. Wu, F. F., and C. A. Desoer, "Global Inverse Function Theorem", IEEE Trans. on Circuit Theory, Vol. CT-19, pp. 199-201, (1972).

IV. SIMULATION

1. INTRODUCTION

System *simulation* refers to the process of computing the *composite system output vector*, y, and possibly the *composite component vectors*, a and b, given the *component connection model* of the system, the *composite system input vector*, u, and *initialization* information. Historically, system simulation involves a two step process. First one computes a *composite system model* such as

$$y = Su \tag{1.1}$$

Then one uses this model to compute y from u and the appropriate initial conditions. However, for a large system, a typical interconnected dynamical system, such a procedure is grossly inefficient. In particular, such a modus operandi requires the solution of two sets of equations: one set computes S and the other computes y. Since S is generally non-sparse, it is impossible to exploit the efficient sparse matrix techniques applicable to the component connection model. Finally, in nonlinear systems the relatively simple *component nonlinearities* of the component connection model generally combine to form extremely complex *composite system nonlinearities*.

171

To circumvent the above obstacles it is important to design numerical techniques and simulation schemes directly applicable to the component connection model. Clearly this alleviates any need to even formulate a composite system model. Two such techniques, a tableau algorithm and a relaxation algorithm, are the foundation of the present chapter. Both algorithms are applicable to *linear and nonlinear systems.*

In keeping with the numerical flavor of the chapter, we focus on quantitive aspects of computing system responses as opposed to studying qualitative behavior of the particular algorithms. However, our goal, here, is not to specifically document numerical procedures, but more simply to affirm the numerical realities constraining the formulation of simulation algorithms. Consequently, the chapter includes material on *numerical integration*, the numerical solution of *simultaneous nonlinear equations*, and the *fast Fourier transform*. Again these developments have a motivational aim and are appropriately cursory in nature.

The following section outlines the *sparse tableau approach* to simulation. In general terms, the algorithm "stacks" the various component equations together with the connection model to form a large, highly sparse set of simultaneous equations. Given an input vector, u, and a set of initial conditions, one then solves for a, b, and y via *sparse matrix inversion.*

The concept of a *relaxation algorithm* fills the pages of section 3. The relaxation algorithm builds around a *predictor-corrector integration scheme*. It can cope with both linear and nonlinear systems. The algorithm allows one to employ a different *variable order and/or step-size integration routine* for each component of the system.

Section 4 presents a brief critique of the various simulation algorithms. Specifically, an *error analysis* of numerical integration and differentiation schemes is discussed. Then we develop a correspondence between the various integration formulas and the numerical schemes devoted to solving nonlinear simultaneous equations.

The fifth section describes a tableau algorithm devoted to *frequency domain simulation*. An alternative algorithm based on the "*tearing concept*" is also outlined.

The final section of this chapter discusses the *discrete Fourier transform* implemented via the *fast Fourier transform algorithm.*

2. THE SPARSE TABLEAU ALGORITHM

The sparse *tableau algorithm*[5,6,14] adapts flawlessly to the component connection model whenever the individual component models are *state* equations coupled through the usual connection equations. For the linear case

$$\dot{x} = Ax + Ba$$
$$b = Cx + Da \tag{2.1}$$

and

$$a = L_{11}b + L_{12}u$$
$$y = L_{21}b + L_{22}u \tag{2.2}$$

where x is the *composite component state vector*, a and b are the *composite component input and output vectors*, and u and y are the *composite system input and output vectors*. The immediate chore is to solve equations 2.1 and 2.2 simultaneously given u and x(0) in a manner exploiting the sparseness and the matrices.

Since a computer executes the simulation of a system discretely, step one requires replacing the derivative, $\dot{x}$, by a discrete approximation. There are myriad discrete approximations for $\dot{x}$.[8] Basically, all such techniques approximate x(t) by a polynomial. The derivative of this polynomial approximates $\dot{x}(t)$. Specific approximations vary depending upon two hypotheses: the number of values over which the approximating polynomial interpolates x(t) and the possibly nonuniform spacing between these values. For instance approximate x(t) as a straight line through two points, $x(t_1)$ and $x(t_2)$ qith spacing $h = t_2-t_1$. The unique first order approximating polynomial is the straight line having slope $(x(t_2) - x(t_1))/h$. Clearly the first order approximation for $\dot{x}(t)$ is

$$\dot{x}(t_2) \approx \frac{x(t_2) - x(t_1)}{h} \tag{2.3}$$

Suppose x(t) takes on the value x_k at t_k. Suppose we interpolate x(t) over the points $x_k, x_{k-1}, x_{k-2},...,x_{k-r}$. Then the polynomial which approximates x, has the form

$$\dot{x}_k \approx \sum_{i=0}^{r} d_i x_{k-i} \tag{2.4}$$

A further discussion of *numerical differentiation* takes place in section 3.

EXAMPLE 2.5: Consider a second order approximation of $x(t)$ with an unequal spacing between interpolation points. Let $t_0=0$, $t_1=1$, and $t_2=3$. The standard *Lagrange interpolation*[8] formula yields the following degree-two *polynomial approximation*

$$x(t) \approx \frac{(t-1)(t-2)}{3} x_0 + \frac{t(t-3)}{-2} x_1 + \frac{t(t-1)}{6} x_2$$

$$= (\frac{x_2}{6} - \frac{x_1}{2} + \frac{x_0}{3}) t^2 + (-\frac{x_2}{6} + \frac{3x_1}{2} - \frac{4x_0}{3}) t + x_0$$

(2.6)

Differentiating this polynomial produces the approximation to $\dot{x}(t)$.

$$\dot{x}(t) \approx 2 \frac{x_2}{6} - \frac{x_1}{2} + \frac{x_0}{3} t + -\frac{x_2}{6} + \frac{3x_1}{2} - \frac{4x_0}{3}$$

(2.7)

The discrete approximation of $\dot{x}(t_2)$ results by evaluating this formula at $t = t_2$:

$$\dot{x}_2 \approx \frac{5}{6} x_2 - \frac{3}{2} x_1 + \frac{2}{3} x_0$$

(2.8)

Returning to a more general discussion, let equation 2.4 approximate $\dot{x}(t_k)$. The *discretized model* of equations 2.1 and 2.2 is:

$$\sum_{i=0}^{r} d_i x_{k-i} = A x_k + B a_k$$
$$b_k = C x_k + D a_k$$

(2.9)

and

$$a_k = L_{11} b_k + L_{12} u_k$$
$$y_k = L_{21} b_k + L_{22} u_k$$

(2.10)

where $x_k = x(t_k)$ and similarly for a_k, b_k, y_k, and u_k. Again the distance between t_k and t_{k+1} need not be the same for various k's. This is a prerequisite for using *variable step size simulation techniques*.

Since the second equation of 2.10 is an expression for explicitly calculating y_k from b_k and u_k, it may be neglected in the actual simulation algorithm. Consequently formulation of the tableau algorithm requires us to simultaneously solve 2.9 and the first equation of 2.10 for x_k, a_k, and b_k. Of course the solution depends on the given matrices, the input vector, u_k, and the prior state values, x_{k-i}, $i \geq 1$. A tractable form results after writing the three equations as a simple, highly sparse, matrix equation.

$$
\begin{bmatrix} \sum\limits_{i=1}^{r} d_i x_{k-i} \\ \hline 0 \\ \hline -L_{12} u_k \end{bmatrix} = \begin{bmatrix} A - d_0 I & B & 0 \\ \hline C & D & I \\ \hline 0 & -I & L_{11} \end{bmatrix} \begin{bmatrix} x_k \\ \hline a_k \\ \hline b_k \end{bmatrix} = \begin{bmatrix} T \end{bmatrix} \begin{bmatrix} x_k \\ \hline a_k \\ \hline b_k \end{bmatrix} \qquad (2.11)
$$

Interpreting these equations it is clear that the simulation of the system rests in the inversion of the *sparse tableau matrix*, T. Of course, the initial condition on x initiates the process.

Dimensionally, T is an (n+m+j) by (n+m+j) matrix where n is the dimension of x. The matrix T is highly *sparse* since all of the non-zero matrices in T are block diagonal excepting L_{11} which is typically sparse anyway. Fortunately the inversion of T usually proves more facile than inverting a smaller j by j matrix which would result when simulating the same system using a composite system state model as per III.5.6. Indeed, in this case it is also necessary to invert an nxn matrix in the actual construction of the composite system state model. On the other hand, since T is independent of k, it is only necessary to invert T once and then use the same matrix inverse repeatedly at each time instant, t_k. However, if a user desired to change the order of the approximating polynomial or to change the step size of the discrete differentiation formula, a new T would result requiring a second inversion. But a change in step size only changes the d_0 coefficient in the definition of T. Changing d_0 only causes a rank n perturbation in the tableau matrix T. Hence a new inverse can be obtained by inverting a j by j matrix as per lemma III.6.15, the *Householder formula*. Finally, observe it is possible to select different order, numerical, differentiation formulas for each component of the system simply by using different order approximations for different entries of x, say a first order approximation for by-pass capacitors and a higher order approximation for a tuned circuit.

EXAMPLE 2.12: Consider the single loop feedback system illustrated by figure 2.13. The component connection equations for this system are given in equations 2.14 and 2.15.

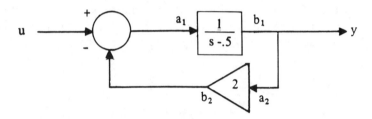

Figure 2.13. Single loop feedback system of example 2.12.

$$\begin{aligned} \dot{x} \Big] &= \Big[.5\Big]x\Big] + \Big[1\ \ 0\Big]\begin{matrix}a_1\\a_2\end{matrix}\Big] \\[2mm] \begin{matrix}b_1\\b_2\end{matrix}\Big] &= \begin{bmatrix}1\\0\end{bmatrix}x\Big] + \begin{bmatrix}0&0\\0&2\end{bmatrix}\begin{matrix}a_1\\a_2\end{matrix}\Big] \end{aligned} \tag{2.14}$$

and

$$\begin{matrix}a_1\\a_2\\y\end{matrix}\Bigg] = \left[\begin{array}{cc|c}0 & -1 & 1\\ 1 & 0 & 0\\ \hline 1 & 0 & 0\end{array}\right]\begin{matrix}b_1\\b_2\\u\end{matrix}\Bigg] \tag{2.15}$$

Assume the system input $u(t) = t$ and all initial conditions are zero. Let us use the first order approximation of 2.3 for $x(t)$ assuming a uniform step size $h=1$. The tableau matrix T is:

$$T = \left[\begin{array}{cc|cc|cc}-.5 & 1 & 1 & 0 & 0 & 0\\ \hline 1 & 0 & 0 & -1 & 0\\ 0 & 0 & 2 & 0 & -1\\ \hline 0 & -1 & 0 & 0 & -1\\ 0 & 0 & -1 & 1 & 0\end{array}\right] \tag{2.16}$$

Upon invoking the *Crout algorithm* of section 4 of chapter III this matrix is found to have the *LU factorization*

$$T = \begin{bmatrix}-.5 & 0 & 0 & 0 & 0\\ 1 & 2 & 0 & 0 & 0\\ 0 & 0 & 2 & 0 & 0\\ 0 & -1 & 0 & -.5 & 0\\ 0 & 0 & -1 & 1 & -2.5\end{bmatrix}\begin{bmatrix}1 & -2 & 0 & 0 & 0\\ 0 & 1 & 0 & -.5 & 0\\ 0 & 0 & 1 & 0 & -.5\\ 0 & 0 & 0 & 1 & 2\\ 0 & 0 & 0 & 0 & 1\end{bmatrix} \tag{2.17}$$

from which back substitution yields the factored inverse in the form

$$
T^{-1} = \begin{bmatrix} 1 & 2 & 0 & 1 & -.2 \\ 0 & 1 & 0 & .5 & -1 \\ 0 & 0 & 1 & 0 & .5 \\ 0 & 0 & 0 & 1 & -2 \\ 0 & 0 & 0 & 0 & 1 \end{bmatrix} \begin{bmatrix} -2 & 0 & 0 & 0 & 0 \\ 1 & .5 & 0 & 0 & 0 \\ 0 & 0 & .5 & 0 & 0 \\ -2 & -1 & 0 & -2 & 0 \\ -.8 & -.4 & -.2 & -.8 & -.4 \end{bmatrix}
\tag{2.18}
$$

Calculating the solution for the network at t=1 and t=2 gives

$$
\begin{bmatrix} x \\ a_1 \\ a_2 \\ b_1 \\ b_2 \end{bmatrix}_1 = T^{-1} \begin{bmatrix} 0 \\ 0 \\ 0 \\ -1 \\ 0 \end{bmatrix} = \begin{bmatrix} .4 \\ .2 \\ .4 \\ .4 \\ .8 \end{bmatrix}
\tag{2.19}
$$

and

$$
\begin{bmatrix} x \\ a_1 \\ a_2 \\ b_1 \\ b_2 \end{bmatrix}_2 = T^{-1} \begin{bmatrix} -.2 \\ 0 \\ 0 \\ -2 \\ 0 \end{bmatrix} = \begin{bmatrix} .88 \\ .24 \\ .88 \\ .88 \\ 1.76 \end{bmatrix}
$$

Substituting these values into the last equation of 2.15 yields the composite **system output values $y_1 = .4$ and $y_2 = .88$. Comparing these values with the** exact solutions $y_1 = .28$ and $y_2 = .74$ indicates an error of .12 in y_1 and .13 in y_2 due to the use of the overly large step size, h = 1.

This example serves only to illustrate the use of the sparse tableau algorithm. As the system becomes large, the benefits of manipulating T heavily outweigh the complexity of increased dimensionality involved in the method. For the nonlinear case the advantages are further compounded when one deals exclusively with the *component nonlinearities* as opposed to the usually intractable *composite system nonlinearities*.

For the nonlinear case the component connection model has the symbolic form

$$
\begin{aligned}
\dot{x} &= f(x,a) \\
b &= g(x,a)
\end{aligned}
\tag{2.20}
$$

and

$$a = L_{11}b + L_{12}u$$
$$y = L_{21}b + L_{22}u$$

(2.21)

In line with the linear case approximate $\dot{x}$ by a numerical differentiation formula as per equation 2.4. Such a *discretized model* is

$$\sum_{i=0}^{r} d_i x_{k-i} = f(x_k, a_k)$$
$$b_k = g(x_k, a_k)$$

(2.22)

and

$$a_k = L_{11}b_k + L_{12}u_k$$
$$y_k = L_{21}b_k + L_{12}u_k$$

(2.23)

Once again, the last equation of 2.23 is neglected. The task is to simultaneously solve the first three equations of 2.33 and 2.23. This necessitates construction of the *nonlinear sparse tableau equation*

$$T(x_k, a_k, b_k) \overset{\Delta}{=} \begin{bmatrix} f(x_k, a_k) - d_0 x_k \\ \hline g(x_k, a_k) - b_k \\ \hline -a_k + L_{11}b_k \end{bmatrix} = \begin{bmatrix} \sum_{i=1}^{r} d_i x_{k-i} \\ \hline 0 \\ \hline -L_{12}u_k \end{bmatrix} \overset{\Delta}{=} \underline{v}$$

(2.24)

This equation reduces the simulation procedure to solving a set of $(n+m+j)$ algebraic nonlinear equations in $(n+m+j)$ unknowns. Indeed, this is a formidable task, as is any technique for the simulation of a nonlinear system. Fortunately, the *Jacobian matrix* for the tableau function, T, is both sparse and explicitly computable in terms of the component nonlinearities. This is advantageous because most numerical techniques for solving a set of *simultaneous nonlinear equations* iteratively invert the Jacobian matrix to obtain an approximate solution of the equation as per equation 2.25. Clearly sparseness of the Jacobian matrix is fundamental to computational simplicity in nonlinear simulation.

For instance, a *Newton-Raphson algorithm*[8] simulation of 2.24 begins with an "initial guess" say (x_k^o, a_k^o, b_k^o) and updates this guess via the equation

$$\begin{bmatrix} x_k^{i+1} \\ \hline a_k^{i+1} \\ \hline b_k^{i+1} \end{bmatrix} = \begin{bmatrix} x_k^i \\ \hline a_k^i \\ \hline b_k^i \end{bmatrix} - J_T^{-1}(x_k^i, a_k^i, b_k^i) [T(x_k^i, a_k^i, b_k^i) - v]$$

(2.25)

Here $J_T(x_k^i, a_k^i, b_+^i)$ is the *Jacobian matrix* of T at the point (x_k^i, a_k^i, b_k^i), i.e., the matrix of partial derivatives defined by equation 2.26.

$$
J_T = \begin{bmatrix}
\dfrac{df}{dx} - d_0 I & \dfrac{df}{da} & 0 \\[2mm]
\dfrac{dg}{dx} & \dfrac{dg}{da} & -I \\[2mm]
0 & -I & L_{11}
\end{bmatrix}
\tag{2.26}
$$

It is evaluated at the particular point in question.

The reason the Jacobian is sparse is because L_{11} is sparse and because the sparse decoupled structure of f and g makes their partial derivatives sparse. Moreover, the partial derivatives of f and g are typically easily computed from the less complex component nonlinearities, not the composite system nonlinearities. Finally, the sparsity pattern of J_T is independent of the particular point of evaluation. Hence it is possible to design efficient algorithms for *repeatedly inverting* J_T. Again this follows from the fixed sparsity structure of J_T, independent of the point of evaluation. For these reasons, the sparse tableau approach for nonlinear simulation proves to be highly efficient for large scale systems.

EXAMPLE 2.27: Consider the earlier feedback system of figure 2.13 with the linear feedback gain replaced by a saturating nonlinear gain

$$
b_2 = 2\tanh(a_2) \tag{2.28}
$$

The component connection equations for the modified system are

$$
\begin{aligned}
\dot{x} &= [.5]\, x + [1 \; 0] \begin{bmatrix} a_1 \\ a_2 \end{bmatrix} \\[2mm]
\begin{bmatrix} b_1 \\ b_2 \end{bmatrix} &= \begin{bmatrix} 1 \\ 0 \end{bmatrix} x + \begin{bmatrix} 0 \\ 2\tanh(a_2) \end{bmatrix}
\end{aligned}
\tag{2.29}
$$

and

$$
\begin{bmatrix} a_1 \\ a_2 \\ y \end{bmatrix} = \begin{bmatrix} 0 & -1 & 1 \\ 1 & 0 & 0 \\ 1 & 0 & 0 \end{bmatrix} \begin{bmatrix} b_1 \\ b_2 \\ u \end{bmatrix}
\tag{2.30}
$$

The sparse tableau function for this nonlinear system is:

$$T(x_k,a_k,b_k) \quad = \quad \begin{bmatrix} .5x_k+a_1-x_k \\ x_k-b_k \\ 2\tanh(a_2)-b_2 \\ -a_1-b_2 \\ -a_2+b_1 \end{bmatrix} \qquad (2.31)$$

The Jacobian matrix for T is

$$J_T(x_k,a_k,b_k) \quad = \quad \begin{bmatrix} -.5 & 1 & 0 & 0 & 0 \\ 1 & 0 & 0 & 1 & 0 \\ 0 & 0 & 2\operatorname{sech}^2(a_2) & 0 & -1 \\ 0 & -1 & 0 & 0 & -1 \\ 0 & 0 & -1 & 1 & 0 \end{bmatrix} \qquad (2.32)$$

To simulate the system use this Jacobian matrix in any standard numerical algorithm in conjunction with a sparse matrix inversion algorithm which exploits the fixed sparsity pattern of J_T.

Unlike the linear case where T^{-1} is independent of k, a nonlinear simulation requires one to reevaluate J_T at each step. Sometimes several reevaluations are necessary if an iterative procedure for solving the equations is employed. Thus one encounters no additional computational cost when changing numerical differentiation formulas and/or step sizes at each k. The tableau algorithm is ideally suited for self adaptive change of order and/or step size during simulation. A criterion for self adaptive change in order and/or step size depends on an *error analysis* of the numerical processes. Discussion of such an analysis occurs in section 4.

EXAMPLE 2.33: This example is a sparse tableau simulation of a nonlinear RC network. It is strongly suggested that the student explore all facets of the example.

Consider the circuit containing a nonlinear resistor sketched in figure 2.34. The goal is to calculate all circuit variables at times t=0, 1, and 2. The simulation will use a first order differentiation formula to approximate the derivative of the capacitor voltage. The nonlinear simulation utilizes a *Newton-Raphson* approach. The example is broken down into five steps. Step 1 sets up the component connection equation. Step 2 computes the Tableau matrix, the Jacobian matrix, etc. Steps 3, 4, and 5 compute the network variables at the desired times.

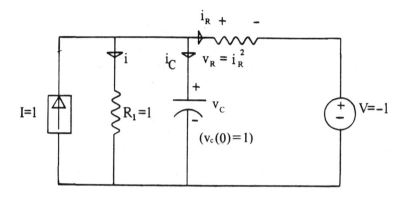

Figure 2.34. Nonlinear RC network with $v_c(0)=1$.

STEP 1: Let the capacitor be component one. Let v_c be the state variable x in addition to the component output and let i_c be the component input, then

$$\dot{v}_C = [0]\ v_C + [1]\ i_C = f(v_C, i_C)$$
$$v_C = [1]\ v_C + [0]\ i_C = g_1(v_C, i_C)$$

(2.35)

Let the second component be the nonlinear resistor with component input v_R and component output i_R. For this component there is no state equation. However equation 2.36 gives the input-output relationship.

$$i_R = \sqrt{v_R} = g_2(v_R)$$

(2.36)

Clearly the system inputs are V and I. The simulation does not require us to specify the system outputs. Thus we neglect this part of the connection equations. The component interconnections are given in 2.37.

$$\begin{bmatrix} i_C \\ v_R \end{bmatrix} = \begin{bmatrix} -1 & -1 \\ 1 & 0 \end{bmatrix} \begin{bmatrix} v_C \\ i_R \end{bmatrix} + \begin{bmatrix} 0 & 1 \\ -1 & 0 \end{bmatrix} \begin{bmatrix} V \\ I \end{bmatrix}$$

(2.37)

Observe that R_1 is submerged into the connection equations. For this example *only* we modify notation as follows: q(k) will denote the value of the variable q(·) at the k-th time instant.

Using a first order approximation with step size h=1 for the derivative of the capacitor voltage and writing the equation in discrete form gives

$$v_C(k) \; - \; v_C(k-1) \; \approx \; v_C(k) \; = \; [0] \, v_C(k) + [1] \, i_C(k) \qquad (2.38)$$
$$= \; f(v_C(k), i_C(k))$$

Recall that we assume that

$$\dot{v}_C(k) \; = \; \sum_{i=0}^{2} \; d_i v_C(k-i) \; = \; d_0 v_C(k) + d_1 v_C(k-1) \quad (2.39)$$

This implies that $d_0 = 1$ and $d_1 = -1$. These values are needed to construct the tableau function. Writing these equations together in the proper form gives

$$v_C(k) \; - \; v_C(k-1) \; \approx \; [0] \, v_C(k) + [1] \, i_C(k)$$
$$= \; f(v_C(k), i_C(k)) \qquad\qquad (2.40)$$

$$\begin{bmatrix} v_C \\ i_R \end{bmatrix} = \begin{bmatrix} v_C \\ \sqrt{v_R} \end{bmatrix} = g(v_C, v_R) \qquad (2.41)$$

$$\begin{bmatrix} i_C(k) \\ v_R(k) \end{bmatrix} = \begin{bmatrix} -1 & -1 \\ 1 & 0 \end{bmatrix} \begin{bmatrix} v_C(k) \\ i_R(k) \end{bmatrix} + \begin{bmatrix} 0 & 1 \\ -1 & 0 \end{bmatrix} \begin{bmatrix} V \\ I \end{bmatrix} \qquad (2.42)$$

STEP 2: Construction of the various simulation matrices. Substituting the equations of step 1 into equation 2.24 gives

$$T \; \overset{\Delta}{=} \; \begin{bmatrix} i_C(k) - v_C(k) \\ v_C(k) - v_C(0) \\ \sqrt{v_R(k)} - i_R(k) \\ -v_C(k) - i_R(k) - i_C(k) \\ v_C(k) - v_R(k) \end{bmatrix} \; = \; \begin{bmatrix} -v_C(k-1) \\ 0 \\ 0 \\ -1 \\ V \end{bmatrix} \; \overset{\Delta}{=} \; v \quad (2.43)$$

In many simulations it is necessary to carefully understand where the different values come from. In general the pertinent quantities should be labeled a, b, x, etc. Now define $T^S = T-v$ as given in equation 2.44.

$$T^S = \begin{bmatrix} i_C(k)^i - v_C(k)^i + v_C(k-1)^i \\ v_C(k)^i - v_C(0)^i \\ \sqrt{v_R(k)}^i - i_R(k)^i \\ -v_C(k)^i - i_R(k)^i - i_C(k)^i + I \\ v_C(k)^i - v_R(k)^i - V \end{bmatrix} \qquad (2.44)$$

where the superscript "i" denotes the i-th iteration of the Newton-Raphson algorithm.

Now it is necessary to calculate the Jacobian matrix. For pedagogical reasons we restate the definition in equation 2.45.

$$J_1 = \begin{bmatrix} \dfrac{df}{dx} - d_0 I & \dfrac{df}{da} & 0 \\ \dfrac{dg}{dx} & \dfrac{dg}{da} & -I \\ 0 & -I & L_{11} \end{bmatrix} \qquad (2.45)$$

Calculating the 1-1 entry for this example gives

$$\frac{df}{dx} - d_0 I = -1 \qquad (2.46)$$

Taking the partial derivative of f with respect to the two-vector, a, gives

$$\frac{df}{da} = \begin{bmatrix} \dfrac{df}{di_C} & \dfrac{df}{dv_R} \end{bmatrix} = [1 \ 0] \qquad (2.47)$$

Continuing in this manner produces the following relationships

$$\frac{dg}{dx} = \begin{bmatrix} \dfrac{dg_1}{dv_C} \\ \dfrac{dg_2}{dv_C} \end{bmatrix} = \begin{bmatrix} 0 \\ 0 \end{bmatrix} \qquad (2.48)$$

and

$$\frac{dg}{da} = \begin{bmatrix} \dfrac{dg_1}{di_C} & \dfrac{dg_1}{dv_R} \\[2mm] \dfrac{dg_2}{di_C} & \dfrac{dg_2}{dv_R} \end{bmatrix} = \begin{bmatrix} 0 & 0 \\[2mm] 0 & \dfrac{1}{2\sqrt{v_R}} \end{bmatrix} \tag{2.49}$$

We therefore conclude that

$$J_T = \begin{bmatrix} -1 & 1 & 0 & 0 & 0 \\ 1 & 0 & 0 & -1 & 0 \\ 0 & 0 & \dfrac{1}{2\sqrt{v_R}} & 0 & -1 \\ 0 & -1 & 0 & -1 & -1 \\ 0 & 0 & -1 & 1 & 0 \end{bmatrix} \tag{2.50}$$

Performing an LU FACTORIZATION on J_T allows us to obtain the following matrix product.

$$J_T = LU = \begin{bmatrix} -1 & 0 & 0 & 0 & 0 \\ 1 & 1 & 0 & 0 & 0 \\ 0 & 0 & \dfrac{1}{2\sqrt{v_R}} & 0 & 0 \\ 0 & -1 & 0 & -2 & 0 \\ 0 & 0 & -1 & 1 & -.5-2\sqrt{v_R} \end{bmatrix} \begin{bmatrix} 1 & -1 & 0 & 0 & 0 \\ 0 & 1 & 0 & -1 & 0 \\ 0 & 0 & 1 & 0 & -2\sqrt{v_R} \\ 0 & 0 & 0 & 1 & .5 \\ 0 & 0 & 0 & 0 & 1 \end{bmatrix} \tag{2.51}$$

Appropriately then $J_T^{-1}=U^{-1}L^{-1}$ where

$$L^{-1} = \begin{bmatrix} -1 & 0 & 0 & 0 & 0 \\ 1 & 1 & 0 & 0 & 0 \\ 0 & 0 & 2\sqrt{v_R} & 0 & 0 \\ -.5 & -.5 & 0 & -.5 & 0 \\ \dfrac{-1}{1+4\sqrt{v_R}} & \dfrac{-1}{1+4\sqrt{v_R}} & \dfrac{2\sqrt{v_R}}{.5+2\sqrt{v_R}} & \dfrac{-1}{1+4\sqrt{v_R}} & \dfrac{-1}{.5+2\sqrt{v_R}} \end{bmatrix} \tag{2.52}$$

and

$$U^{-1} = \begin{bmatrix} 1 & 1 & 0 & 1 & -.5 \\ 0 & 1 & 0 & 1 & -.5 \\ 0 & 0 & 1 & 0 & 2\sqrt{v_R} \\ 0 & 0 & 0 & 1 & -.5 \\ 0 & 0 & 0 & 0 & 1 \end{bmatrix} \tag{2.53}$$

With these tedious preliminaries completed the actual simulation begins.

STEP 3: The initialization or calculation of system parameters at t=0. For t=0, $v_C(0)=1$. The connection equations or equivalently the KVL and KCL yield

$$
\begin{bmatrix}
v_C(0) \\
i_C(0) \\
v_R(0) \\
v_C(0) \\
i_R(0)
\end{bmatrix}
=
\begin{bmatrix}
1.0 \\
-\sqrt{2} \\
2.0 \\
1.0 \\
\sqrt{2}
\end{bmatrix}
\tag{2.54}
$$

So ends the third step. It is always good practice to finish the trivial before attempting the exasperating.

STEP 4: The problem here is to iteratively estimate the system variables at t=1, using the Newton-Raphson method. As a first approximation suppose

$$
\begin{bmatrix}
v_C(1) \\
i_C(1) \\
v_R(1) \\
v_C(1) \\
i_R(1)
\end{bmatrix}^{0}
=
\begin{bmatrix}
1.0 \\
-\sqrt{2} \\
2.0 \\
1.0 \\
\sqrt{2}
\end{bmatrix}
\tag{2.55}
$$

Computing the initial estimate of $T^{\$}(1)$ we have

$$
T^{\$}(1)^{0}
=
\begin{bmatrix}
-\sqrt{2} \\
0 \\
0 \\
0 \\
0
\end{bmatrix}
\tag{2.56}
$$

To compute the first updated vector, we must solve the equation given in 2.57.

$$
\begin{bmatrix}
v_C(1) \\
i_C(1) \\
v_R(1) \\
v_C(1) \\
i_R(1)
\end{bmatrix}^{1}
=
\begin{bmatrix}
1.0 \\
-\sqrt{2} \\
2.0 \\
1.0 \\
\sqrt{2}
\end{bmatrix}^{0}
- \; U^{-1}L^{-1}T^{\$}
\tag{2.57}
$$

where $J_T^{-1}=U^{-1}L^{-1}$ and $T^{\$}$ are evaluated at the initial guess. After a few hours of heavy arithmetic, the solution becomes

$$
\begin{bmatrix} v_C(1) \\ i_C(1) \\ v_R(1) \\ v_C(1) \\ i_R(1) \end{bmatrix}^{1} = \begin{bmatrix} 1.0 \\ -\sqrt{2} \\ 2. \\ 1.0 \\ \sqrt{2} \end{bmatrix}^{0} - \begin{bmatrix} .601 \\ -.813 \\ .6 \\ .601 \\ .212 \end{bmatrix} = \begin{bmatrix} 0.3990 \\ -.6012 \\ 1.4000 \\ 0.3990 \\ 1.2020 \end{bmatrix} \tag{2.58}
$$

To obtain a second updated guess, it is necessary to solve

$$
\begin{bmatrix} v_C(1) \\ i_C(1) \\ v_R(1) \\ v_C(1) \\ i_R(1) \end{bmatrix}^{2} = \begin{bmatrix} 0.3990 \\ -.6012 \\ 1.4000 \\ 0.3990 \\ 1.2020 \end{bmatrix} - J_T^{-1}T^{\$} \tag{2.59}
$$

where J_T^{-1} and $T^{\$}$ are updated by the solutions of equation 2.58. The updated T becomes

$$
T^{\$}(1)^2 = \begin{bmatrix} -.0002 \\ .0000 \\ .0188 \\ .0002 \\ -.0010 \end{bmatrix} \tag{2.60}
$$

At this point the calculation of the remaining system parameters is a straightforward task.

$$
\begin{bmatrix} v_C(1) \\ i_C(1) \\ i_R(1) \\ v_C(1) \\ i_R(1) \end{bmatrix}^{2} = \begin{bmatrix} 0.3990 \\ -.6012 \\ 1.4000 \\ 0.3990 \\ 1.2020 \end{bmatrix} - \begin{bmatrix} .00794 \\ .00814 \\ .00692 \\ .00794 \\ .01587 \end{bmatrix} = \begin{bmatrix} 0.4069 \\ .5931 \\ 1.4069 \\ 0.4069 \\ 1.1860 \end{bmatrix} \tag{2.61}
$$

Thus we proceed to step 5.

This concludes the example and the discussion of the sparse tableau algorithm. The convergence of the algorithm in this example is very fast. Observe that the nonlinearity was less than complicated.

3. A RELAXATION ALGORITHM

The *sparse tableau algorithm* is built around a single high dimensional set of equations. Sparse matrix techniques permit computational feasibility. The *relaxation algorithm* simulates the individual system components one at a time. The arrays of zeros needed as in the tableau algorithm for a single high dimensional equation are never written down. The necessity for sparse matrix techniques is therefore bypassed. Basically the algorithm simulates a very large system with a minimal amount of *computer memory* by bringing only a single component equation and the appropriate part of the connection equations into core at a time. Assuming state models for each component, each equation is integrated individually with the results iterated through the connection equations. This generates a simultaneous solution of the entire set of component connection equations for the system.[11][13]

The theory for the linear and nonlinear cases is more or less equivalent so the remainder focuses on the nonlinear case. The starting point for constructing a relaxation algorithm is a set of *strictly proper state equations* describing each component of the system.

$$\begin{aligned} \dot{x}_i &= f_i(x_i, a_i) \\ b_i &= g_i(x_i) \end{aligned} \qquad i = 1, 2, \ldots, r \qquad (3.1)$$

together with the usual algebraic connection equations

$$\begin{aligned} a &= L_{11}b + L_{12}u \\ y &= L_{21}b + L_{22}u \end{aligned} \qquad (3.2)$$

The thrust of the algorithm is to explicitly deal with the individual component state equations and neglect the *composite component state model*. Also, it is necessary to decompose the L_{11} matrix into distinct submatrices corresponding to the different b_i's as per equation 3.3.

$$a = \sum_{i=1}^{r} L_{11}^{i} b_i + L_{12}u \qquad (3.3)$$

ere L_{11}^{i} represents the columns of L_{11} corresponding to the subvector, b_i, b.

STEP 5: Calculation of the system parameters at t=2. For t=2, w‹
our initial guess, the vector

$$
\begin{bmatrix}
v_C(2) \\
i_C(2) \\
v_R(2) \\
v_C(2) \\
i_R(2)
\end{bmatrix}^0
=
\begin{bmatrix}
0.4069 \\
-.5931 \\
1.4069 \\
0.4069 \\
1.1860
\end{bmatrix}
$$

The first updated guess is the solution to equation 2.63.

$$
\begin{bmatrix}
v_C(2) \\
i_C(2) \\
v_R(2) \\
v_C(2) \\
i_R(2)
\end{bmatrix}^1
=
\begin{bmatrix}
0.4069 \\
-.5931 \\
1.4069 \\
0.4069 \\
1.1860
\end{bmatrix}
- J_T^{-1} T^{\$}
$$

where J_T^{-1} and $T\$$ are evaluated at the initial guess values. T‹
vector is

$$
T^{\$}(2)^0 =
\begin{bmatrix}
i_C(2)^0 - v_C(2)^0 + v_C(1) \\
v_C(2)^0 - v_C(2)^0 \\
\sqrt{v_R(2)^0} - i_R(2)^0 \\
-v_C(2)^0 - i_R(2)^0 - i_C(2)^0 + I \\
v_C(2)^0 - v_R(2)^0 - V
\end{bmatrix}
=
\begin{bmatrix}
-.5931 \\
0 \\
0 \\
0 \\
0
\end{bmatrix}
$$

The solution to 2.64 is

$$
\begin{bmatrix}
v_C(2) \\
i_C(2) \\
v_R(2) \\
v_C(2) \\
i_R(2)
\end{bmatrix}^1
=
\begin{bmatrix}
0.4069 \\
-.5931 \\
1.4069 \\
0.4069 \\
1.1860
\end{bmatrix}
-
\begin{bmatrix}
.2449 \\
-.3483 \\
.2451 \\
.2449 \\
.1033
\end{bmatrix}
=
\begin{bmatrix}
0.1620 \\
-.2448 \\
1.1618 \\
0.1620 \\
1.082⁷
\end{bmatrix}
$$

After performing a second iteration we obtain the f‹
variable values.

$$
\begin{bmatrix}
v_C(2) \\
i_C(2) \\
v_R(2) \\
v_C(2) \\
i_R(2)
\end{bmatrix}^2
=
\begin{bmatrix}
0.1637 \\
-.2432 \\
1.1652 \\
0.1637 \\
1.0795
\end{bmatrix}
$$

As before we ignore the second equation of 3.2 in the simulation process. The goal is to simultaneously solve equations 3.1 and 3.3. Given the *composite system input* and the appropriate *initial conditions* we first evaluate the *component output vectors* at t=0 as

$$b_i(0) = g_i(x_i(0)) \tag{3.4}$$

and then we evaluate the composite component input vector at t=0 as

$$a(0) = \sum_{i=1}^{r} L_{11}^i b_i(0) + L_{12}u(0) \tag{3.5}$$

These equations construct the initial values for all of the system variables under consideration. These ideas more or less complete phase one of the simulation.

The next step is to solve the individual component differential equations by first converting them to equivalent integral equations as per 3.6.

$$x_i(t) = x_i(t_0) + \int_{t_0}^{t} f_i(x_i(q), a_i(q))dq \tag{3.6}$$

The solution proceeds by first approximating $f_i(x_i(q), a_i(q))$ by polynomial. In this endeavor let $t_0, t_1, ..., t_{k-1}, t_k = t$ denote a discrete set of points over which $f_i(x_i(q), a_i(q))$ is to be interpolated with values

$$f_i^j = f_i(x_i(t_j), a_i(t_j)) \tag{3.7}$$

The interpolation scheme may depend on any subset of the points, t_j: j=0,1,...,k, to obtain a polynomial approximating $f_i(x_i(q), a_i(q))$ over the interval $t_0 \leq q \leq t$. One then integrates this polynomial and substitutes the result into equation 3.6 to numerically solve for $x_i(t)$.

At this point we will take a lengthy digression to discuss the *numerical integration* of equation 3.6. As a beginning, we qualitatively catalog the various integration routines into *explicit* and *implicit* schemes. Suppose the nonlinear function f_i takes on the values f_i^j ($j \leq k$) over k different points. An *explicit* routine does not use the point f_i^k in the approximating formula for the integral of f_i over the interval $[t_0, t_k]$. Therefore an explicit integration scheme for solving 3.6 has the form integration scheme for solving 3.6 has the form

$$x_i(t_k) = x_i(t_0) + \sum_{i=0}^{k-1} c_j f_i^j \qquad (3.8)$$

where the coefficient c_j depends on the particular algorithm and the chosen step sizes. The solution then proceeds iteratively beginning at some initial point t_0. It is a "bootstrapping" type of operation valid over an arbitrarily long interval. Obviously one first evaluates f_i^0 in terms of $x_i(t_1)$, etc. An advantage of the procedure is that it leaves one free to choose the step size and the degree of the interpolating polynomial at each iteration.

Implicit integration schemes include the point f_i^k in the numerical integration formula. This forces one to simultaneously solve

$$x_i(t_k) = x_i(t_0) - \sum_{j=0}^{k} c_j f_i^j \qquad (3.9)$$

in conjunction with

$$f_i^k = f_i(x_i(t_k), a_i(t_k)) \qquad (3.10)$$

for $x_i(t_k)$ and f_i^k before using the "bootstrapping" effect to obtain $x_i(t_{k+1})$. In other words, an implicit scheme may force one to solve a set of simultaneous nonlinear equations at each step of the algorithm. This unfortunately has a gluttonous effect on computer time, but usually improves numerical accuracy over explicit schemes in which error may compound at each iteration.

A compromise between the two approaches is a *predictor-corrector algorithm* offering high numerical accuracy without having to solve simultaneous nonlinear equations. Basically the algorithm first uses an explicit integration routine to compute $x_i(t_k)$ and f_i^k. These values then initialize an "implicit-like" routine which iteratively converges to values for $x_i(t_k)$ and f_i^k obtained by using an implicit formula to begin with. The essential trick is to compute $a_i(t_k)$ for all components via the connection equations. This produces a more accurate new estimate of f_i^k at each corrector step. The number, f_i^k, then appears in the implicit-like corrector scheme which computes a better approximation for $x_i(t_k)$. The process continues until suitable convergence occurs. This guarantees that the final result of the sequence of corrections satisfies the connection equations in addition to the component differential equations.

There are many explicit and implicit integration formulas.[8] Two common ones are the *Euler formula* and the *trapezoidal formula*. The Euler formula is a zeroth order explicit approximation of $f_i(x_i(q), a_i(q))$ on the interval $t_0 \leqslant q \leqslant t_1 = t$ by a constant function, f_i^0. Thus the Euler formula in the form of equation 3.8 is:

the interval $t_0 \leqslant q \leqslant t_1 = t$ by a constant function, f_i^0. Thus the Euler formula in the form of equation 3.8 is:

$$x_i(t_1) = x_i(t_0) + (t_1-t_0) f_i^0 \tag{3.11}$$

On the other hand, the trapezoidal formula is a first order implicit approximation of $f_i(x_i(q), a_i(q))$ on the interval $t_0 \leqslant q \leqslant t_1$ by a straight line passing through f_i^0 and f_i^1. This then yields a state vector for the i-th component as

$$x_i(t_1) = x_i(t_0) + .5(t_1-t_0)(f_i^0+f_i^1) \tag{3.12}$$

EXAMPLE 3.13: Consider the differential equation

$$\dot{x} = x_i^2 \quad \text{with} \quad x(0)=1.0 \tag{3.14}$$

Assume $t_0 = 0$ and $t_1 = .1$. The implicit trapezoidal integration formula of 3.12 produces

$$x(.1) = x(0) + .05(1 + x(.1)^2) \tag{3.15}$$

Solving via the quadratic formula yields

$$x(.1) = 1.111806 \tag{3.16}$$

On the other hand utilization of the explicit Euler formula gives

$$x(.1) = x(0) + .1 x(0)^2 = 1.1 \tag{3.17}$$

To see what we mean by an implicit-like formula, substitute this value into the right hand side of equation 3.15. Doing the indicated arithmetic results in a "corrected" approximation of $x(.1) = 1.11050$. Repeating the process produces successive corrections of $x(.1) = 1.11166$, $x(.1) = 1.11179$, $x(.1) = 1.11180$, and $x(.1) = 1.111805$ which converge to the solution given in 3.16. This procedure avoids having to solve the implicit routine, but offers similar accuracy.

EXAMPLE 3.18: In practice, a "small" step size usually compensates for errors growing from low order integration schemes. On the other hand, a scheme like the classical *Simpson's rule* is a higher order implicit integration formula obtained by interpolating $f_i(x_i(a), a_i(q))$ at three equally spaced points with a second order polynomial. Assume the points are $t_0 = 0$, $t_1 = 1$, and $t_2 = 2$. The *Lagrange interpolation formula* produces the quadratic approximation defined in 3.19.

$$P_s(t) \;=\; .5\,(t^2-t)\,f_i^2 \;-\; (t^2-2t)\,f_i^1 \;+\; .5\,(t^2-3t+2)f_i^0 \qquad (3.19)$$

The definite integral over [0,2] of $P_s(t)$ is:

$$\int_0^2 P_s(q)dq \;=\; .333(f_i^0 + 4\,f_i^1 + f_i^2) \qquad (3.20)$$

which is the classical Simpson's formula. Using the same interpolating polynomial but integrating from zero to three produces the following explicit integration formula

$$\int_0^3 P_s(q)(dq \;=\; .75(f_i^0 + 3\,f_i^2) \qquad (3.21)$$

This approximates the integral of $f_i(x_i(q), a_i(q))$ over the interval [0,3]. Suppose we now wish to solve the following differential equation

$$\dot{x}_i(t) \;=\; f_i(x_i(t), a_i(t)) \qquad (3.22)$$

Using the formula given by 3.21 produces the approximate solution

$$x(3) \;=\; x(0) \;+\; .75\,(f_i^0 + 3\,f_i^2) \qquad (3.23)$$

At this point let us return to the original task by properly fitting these notions into the framework of the relaxation algorithm. Before the discussion on numerical integration, we were solving equation 3.6. Recall that we assume u is known for all t and that the initial state $x_i(0)$ is specified for each component. With this information we computed the initial values for $b_i(0)$ and $a(0)$.

In the relaxation algorithm we will use the predictor corrector approach to solve 3.6. The next step then is to use an explicit integration scheme to obtain an initial estimate for the component state vectors at time t_1. Equation 3.24 provides the means for this end.

t_1. Equation 3.24 provides the means for this end.

$$x_i(t_1) = x_i(t_0) + c_0 f_i^0 \qquad (3.24)$$

This is the aforementioned Euler formula. Since only initial condition information is available, its use is mandatory at this stage.

After computing the first estimate of $x_i(t_1)$, initial approximate values for $b_i(t_1)$ and $a(t_1)$ follow by solving equations 3.25 and 3.26.

$$b_i(t_1) = g_i(x_i(t_1)) \qquad (3.25)$$

and

$$a(t_1) = \sum_{i=1}^{r} L_{11}^i b_i(t_1) + L_{12} u(t_1) \qquad (3.26)$$

The coarseness of the explicit integration formula and the disregard of the connection structure in the initial estimates make these approximations inaccurate. Improvement occurs by a corrector step implemented through an implicit-like integration formula. The rough estimates of $x_i(t_1)$ and $a_i(t_1)$ (the approximate sub-vector of $a(t_1)$) are used in conjunction with the connection information to approximate f_i^1.

$$f_i^1 = f_i(x_i(t_1), a_i(t_1)) \qquad (3.27)$$

Plugging this information into an implicit integration formula leads to an improved estimate of $x_i(t_1)$ implemented through 3.28.

$$x_i(t_1) = x_i(t_0) + c_0 f_i^0 + c_1 f_i^1 \qquad (3.28)$$

This allows us to produce another set of estimates for $b_i(t_1) a(t_1)$ and f_i^1. The process is repeated until the estimates converge to a solution at $t = t_1$ of the set of simultaneous equations characterizing the interconnected dynamical system.

The initial conditions and the estimates at $t = t_1$ now initialize the same type of procedure to estimate the system variables at $t = t_2$. This provides values for $b_i(t_2), a_i(t_2)$, and f_i^2, etc. Of course, for computing system variables at t_k, $k \leqslant 2$ one has already computed a large number of the prior values of the system variables. This permits use of *higher order integration formulas* and/or *unequal step sizes*.

EXAMPLE 3.29: Consider an *LC tank circuit* driven by a DC source as illustrated by figure 3.30.

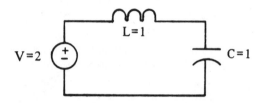

Figure 3.30. LC tank circuit.

The component connection equations for this circuit take the form

$$\dot{v}_C = [0]\,v_C + [1]\,i_C$$
$$v_C = [1]\,v_C \tag{3.31}$$

$$\dot{i}_L = [0]\,i_L + [1]\,v_L$$
$$i_L = [1]\,i_L \tag{3.32}$$

and

$$\begin{bmatrix} i_C \\ v_L \\ \hline i_C \end{bmatrix} = \begin{bmatrix} 0 & 1 & 0 \\ -1 & 0 & 1 \\ \hline 0 & 1 & 0 \end{bmatrix} \begin{bmatrix} v_C \\ i_L \\ \hline E \end{bmatrix} \tag{3.33}$$

Suppose the battery voltage E equals 2 volts. Assume all initial conditions are zero, i.e., $v_C(0) = i_L(0) = 0$. Equations 3.4 and 3.5 lead to the initial values for all variables.

$$\begin{aligned} v_C(0) &= 0 \\ i_L(0) &= 0 \\ i_C(0) &= 0 \\ v_L(0) &= 2. \end{aligned} \tag{3.34}$$

Using the Euler formula as a predictor we estimate $v_C(.1)$ and $i_L(.1)$ via

$$\begin{aligned} v_C(.1) &= 0 + .1 \times 0.0 = 0.0 \\ i_L(.1) &= 0 + .1 \times 2.0 = .2 \\ i_C(.1) &= .2 \\ v_L(.1) &= 2.0 \end{aligned} \tag{3.35}$$

Here $v_C(.1)$ and $i_L(.1)$ follow by applying the Euler approximation to the solution of 3.31 and 3.32. Also $i_c(.1)$ and $v_L(.1)$ are computed via 3.33 and the estimates of $v_C(.1)$ and $i_L(.1)$.

To improve these estimates we utilize a trapezoidal corrector formula

$$X(.1) \approx X(0) + .1 (f_i^1 + f_i^0)/2 \qquad (3.36)$$

Equation 3.36 together with equation 3.33 results in

$$\begin{aligned}
v_C(.1) &= 0 + .05(.2 + 0) = .01 \\
i_L(.1) &= 0 + .05(2 + 2) = .2 \\
i_c(.1) &= .2 \\
v_L(.1) &= 1.99
\end{aligned} \qquad (3.37)$$

Using these new data points and successive data points in repeated iterations of the corrector scheme produces system parameter values obtainable by simulating the system via the implicit trapezoidal formula to begin with. A table of the corrector values appears in 3.38.

	v_C	i_L	i_c	v_L
initial values	0.0	0.0	0.0	2.0
predicted values	0.0	0.2	0.2	2.0
1st corr. values	.01	0.2	0.2	1.99
2nd corr. values	.01	.1995	.1995	1.99
3rd corr. values	.00998	.1995	.1995	1.99002
4th corr. values	.00998	.199501	.199501	1.99002
5th corr. values	.00997505	.199501	.199501	1.99002495

Table 3.38. Results of the simulation of the circuit of figure 3.30.

At first glance the relaxation algorithm seems more intricate than the tableau approach. However, it is possible to implement the relaxation **algorithm in a minimal amount of random access memory provided proper care is taken in managing the data required for the simulation.**[7] **Basically** everything is stored on disk (or some other type of bulk computer memory). The equations for a single component, f_i and g_i, and the corresponding columns of the connection matrix, L_{i1}^i are individually **brought into memory. Using only this data, one computes a new value of** b_i **as** $b_i=g_i(x_i)$ **and a new value of "a" via the formula**

$$a^n = a^o + L_{11}^i (x_i^n - x_i^o) \qquad (3.39)$$

where the superscript "n" denotes the new value of the variable and the superscript "o" denotes the old value.

Another interesting and somewhat unique attribute of the relaxation algorithm is the freedom to choose various *step-sizes* for different components of the system. Suppose a component state vector x_i is changing very slowly. It is then possible to update x_i say every other iteration or even less. One simply bypasses the i-th component in the numerical integration process, never even bringing the equations into memory. Basically this is equivalent to having $x_i^n = x_i^o$. Thus the procedure introduces no error into the system other than the approximation error in x_i^n. Moreover, no new bookkeeping costs arise when skipping simulation of the i-th component at a given time instant.[7] Any of the standard schemes for adjusting step sizes is usable in the relaxation algorithm.[4] In addition, the number of corrector steps is a measure for reducing or lengthening step size.[7] If the number of corrector steps is large, then it is advisable to reduce the step size and conversely. In practice it is advantageous to adjust step sizes so that the variables converge to the desired degree of accuracy with one or two correction steps. If the scheme requires more than two correction steps, comparable accuracy would result with less computer time used by implementing a *higher order numerical integration* routine.[4]

At this point, the main disadvantage of the algorithm is the necessity of having strictly proper state models for components. In practice many systems have a large number of purely algebraic components. The presence of components with strictly *proper state models* is limited in applicability, even though it is sometimes possible to embed linear algebraic components into the connection structure, for instance load and source resistors. However, the general applicability of the algorithm depends on its capability of coping with components characterized by *proper state models* as per equation 3.40.

as per equation 3.40.

$$\begin{aligned} \dot{x}_i &= f_i(x_i, a_i) \\ b_i &= g_i(x_i, a_i) \end{aligned} \quad ; \ i=1,2,\dots,r \qquad (3.40)$$

When the components have proper state models, the algorithm remains the same except the part which computes a and b. For each new value of x, it is necessary to solve the pair of simultaneous equations in 3.41

$$\begin{aligned} b &= g(x,a) \\ a &= L_{11}b + L_{12}u \end{aligned} \qquad (3.41)$$

for the vectors a and b. If the system has no *algebraic loops* it is always possible to *reorder* the connection equations so that they still have an explicit solution.[10] If the system has algebraic loops, however, it is always necessary to solve a set of simultaneous equations whose dimensionality depends on the number of algebraic loops in the system. Since 3.41 must be solved after each correction, it is important to design numerical integration formulas and choose step sizes which minimize the number of required corrector steps. If the system has no algebraic loops, such difficulties never materialize.

4. ANALYSIS OF NUMERICAL METHODS

The last two sections briefly overviewed various *numerical differentiation and integration* schemes utilized for the simulation of interconnected dynamical systems. All such techniques basically evolve along the same path. A polynomial interpolates data points associated with a functional equation. Either differentiating or integrating the interpolation polynomial will provide the appropriate approximation. The evolution of the schemes along the same path occurs because there is a *unique* interpolating polynomial for each set of (n+1) data points.[8]

Similarly *error analysis* for the various schemes are also equivalent. The uniqueness of the interpolating polynomial, therefore, justifies basing the foregoing analysis on the more widely understood *Lagrange interpolating formula*. This formula has the freedom to allow non-uniform distance between successive data points. The analysis, then, is sufficiently general for most purposes.

Consider the interpolation of the function $f(\cdot)$ at the points $x_0, x_1, x_2, \ldots, x_n$. Define $f_i = f(x_i)$. The interpolating polynomial as per the Lagrange formula is

$$p_n(x) = \sum_{i=0}^{n} \frac{(x-x_0) \cdots (x-x_{i-1})(x-x_{i+1}) \cdots (x-x_n)}{(x_i-x_0) \cdots (x_i-x_{i-1})(x_i-x_{i+1}) \cdots (x_i-x_n)} f_i \quad (4.1)$$

Mathematically, the function $f(\cdot)$ may be expressed as the sum of $p_n(\cdot)$ and an error term as per 4.2

$$f(x) = p_n(x) + r(x) \quad (4.2)$$

The thrust of the first part of this part of this section is to provide a detailed analysis of the *remainder or error term*, $r(x)$. As a first step consider the following theorem.

THEOREM 4.3: Suppose $f(\cdot)$ has at least $(n+1)$ continuous derivatives and that $p_n(\cdot)$ is the interpolating polynomial of $f(\cdot)$ as above. Then for each x there exists a real number, q, contained in the closed interval $[x_0, x_n]$ such that

$$r(x) = (x-x_0)(x-x_1) \cdots (x-x_n) \frac{f^{(n+1)}(q)}{(n+1)!} \quad (4.4)$$

where $f^{(n+1)}(q)$ is the $(n+1)$-st derivative of $f(\cdot)$ evaluated at q.

PROOF: By construction, $f(x_i) = p_n(x_i) = f_i$, which forces $r(x_i) = 0$. Hence without loss of generality it is possible to express $r(x)$ as

$$r(x) = (x-x_0) \cdots (x-x_n) k(x) \quad (4.5)$$

The problem now reduces to evaluating $k(x)$. For each x define a new function, $W(\cdot)$, as

$$\begin{aligned} W(t) &= f(t) - p_n(t) - (t-x_0) \cdots (t-x_n) k(x) \\ &= f(t) - p_n(t) - (t^{n+1} + a_n t^n + \cdots + a_0) k(x) \end{aligned} \quad (4.6)$$

which clearly has $(n+1)$ zeros at $t = x_i$, $i = 0, 1, \ldots, n$. In addition $W(x)$ is zero since by assumption $f(x) = p_n(x) + r(x)$. Thus $W(t)$ has $(n+2)$ zeros in the interval $[x_0, x_n]$. Computing the $(n+1)$-st derivative of $W(\cdot)$ results in the equality

$$W^{(n+1)}(t) = f^{(n+1)}(t) - (n+1)! \, k(x) \qquad (4.7)$$

This follows because the $(n+1)$-st derivative of $p_n(t)$ is zero whereas the $(n+1)$-st derivative of $r(t)$ is $(n+1)!$.

Since $W(\cdot)$ has $(n+2)$ zeros over $[x_0, x_n]$, we have that $W^{(n+1)}(\cdot)$ has at least one zero, say at q, over the interval. *Rolle's Theorem*[2] guarantees this fact. In particular

$$W^{(n+1)}(q) = f^{(n+1)}(q) - (n+1)! \, k(x) = 0 \qquad (4.8)$$

which implies that

$$k(x) = \frac{f^{(n+1)}(q)}{(n+1)!} \qquad (4.9)$$

as was to be shown.

Again, q depends on the particular x. However, by bounding $f^{(n+1)}(\cdot)$ over $[x_0, x_n]$ it is possible to bound $r(x)$, thus obtaining a general measure of the accuracy of the interpolation.

Given the *remainder formula* as per equation 4.4, it is a simple matter to compute error or remainder formulas for the various numerical differentiation and integration formulas

$$\dot{f}(\cdot) = \dot{p}_n(\cdot) + \dot{r}(\cdot) \qquad (4.10)$$

and

$$\int_{x_0}^{x} f(y)dy = \int_{x_0}^{x} p_n(y)dy + \int_{x_0}^{x} r(y)dy \qquad (4.11)$$

COROLLARY 4.12: Suppose $f(\cdot)$ has at least n continuous derivatives. Suppose the derivative of $f(\cdot)$ at x_k is approximated by the derivative of $p_n(\cdot)$ at x_k. Then there exists a real number q, $x_0 \leqslant q \leqslant x_n$, such that

$$\dot{r}(x_k) = (x_k - x_0) \cdots (x_k - x_{k-1})(x_k - x_{k+1}) \cdots (x_k - x_n) \frac{f^{(n+1)}(q)}{(n+1)!} \qquad (4.13)$$

PROOF: The proof is a direct consequence of theorem 4.3 by differentiating the expression for $r(x)$ and evaluating it at x_k. Observe that equation 4.13 has a further simplification if one uses *equal step sizes*. Under such restrictions

$$\dot{r}(x_k) = (-1)^{(n-k)} \frac{(k!) (n-k)!}{(n+1)!} h^n f^{(n+1)}(q) \tag{4.14}$$

This formula indicates that in general the *step size* and the *order of the interpolating polynomial* determine the error of approximation. A similar although much more difficult procedure will result in a like-formula for the error analysis of integration schemes. If the step sizes are of equal length, then the remainder term for an integration scheme has the form

$$\int_{x_0}^{x_k} r(y)dy = C(p_n)h^{n+2} f^{(n+1)}(q) \tag{4.15}$$

where $C(p_n)$ is a constant dependent on the interpolating polynomial p_n. Also q is an unknown real number in the interval $[x_0, x_n]$. To illustrate these ideas consider the following example.

EXAMPLE 4.16: Let us interpolate $f(x) = \sin(\pi x)$ over the set of points [0.0, 0.5, 1.5, 2.0] by a cubic polynomial. The Lagrange interpolation formula of equation 4.1 gives

$$p_3(x) = \frac{8}{3} (x^3 - 3x^2 + 2x) \tag{4.17}$$

It follows that for some unknown, q, the remainder or error approximation is

$$r(x) = x(x-.5) (x-1.5) (x-2.0) \frac{\pi^4 \sin(\pi q)}{4!} \tag{4.18}$$

Clearly $r(x)$ can be bounded as follows

$$| r(x) | \leq | x(x-.5) (x-1.5) (x-.0) | \frac{\pi^4}{4!} \tag{4.19}$$

Now observe that at $x = .25$, $\sin(\pi/4) = .707$, $p_3(.25) = .88$, and the actual error, $r(.25) = -.17$. This is well within the upper bound given by equation 4.19 as .55. By viewing the plots sketched at the end of this example, one can see an accurate picture of the quality of the approximations.

In the case of approximating $f(x) = \pi\cos(\pi x)$, we have

$$\dot{p}_3(x) \;=\; \frac{8}{3}\,[3x^2 - 6x + 2] \qquad\qquad (4.20)$$

At $x = .5$, we have $f(.5) = \pi\,\cos(\pi/2) = 0.0$ whereas $p_3(.5) = .67$. Thus the actual error is $r(.5) = .67$. Now bounding equation 4.13 yields

$$|\,\dot{r}\,(1.0)\,| \;\leq\; |\;\;.5(.5{-}1.5)\,(.5{-}2.0)\;|\;\frac{\pi^4}{24} \;=\; 3.04 \qquad (4.21)$$

Thus, the error is much smaller than its upper bound.

Finally, let us consider the error due to integration schemes. For this case consider that

$$\int_0^x f(q)dq \;=\; \int_0^x \sin(q)dq \;=\; \frac{1}{\pi}\,(1 - \cos(\pi x)) \quad (4.22)$$

$$\int_0^x p_3(q)dq \;=\; \frac{2}{3}\,x^2(x{-}2)^2 \qquad\qquad (4.23)$$

At $x = .5$ the integral of f is $.318$ and the integral of p_3 is $.38$ which produces an actual error of $-.06$. By taking a bound for $r(x)$ and integrating the resultant expression we obtain a bound for the integral of $r(x)$ as

$$|\,\textstyle\int r(x)dx\,| \;\leq\; \int_0^{.5} |\,r(x)\,|\;dx \;=\; \frac{\pi^4}{24}\,\int_0^{.5} |\,x(x{-}.5)\,(x{-}1.5)\,(x{-}2.0)\,|\;dx$$
$$\qquad\qquad\qquad (4.24)$$
$$\qquad\quad =\; \frac{\pi^4}{24}\,(.0458) \;=\; .186$$

Again the actual error is well within the bound. Notice that the approximation error associated with integration schemes is in general smaller than that for differentiation schemes. Intuitively this is because integration is a smoothing process as opposed to the "unsmoothing" process of differentiation. The following three figures illustrate the approximations of this example.

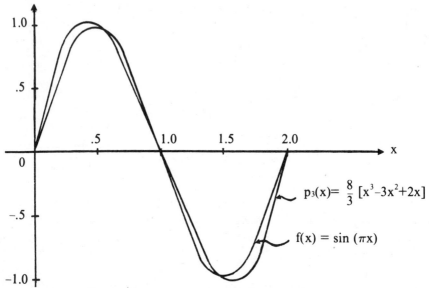

Figure 4.25. Plot of f(x) and its approximation p₃(x).

As can be seen the approximation is remarkably accurate. In the following figure we see that the derivative of the approximation is a fair look-alike also.

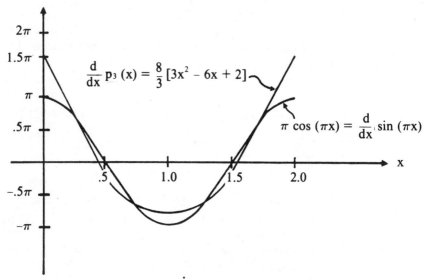

Figure 4.26. Plot of $\dot{f}$(x) and its approximation $\dot{p}_3$(x).

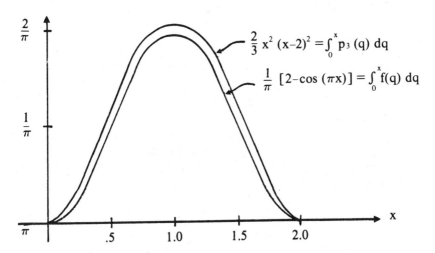

Figure 4.27. Plot of the integral of f(x) and its approximation.

The error formulas needed in measuring the accuracy of a simulation are either of the derivative nature or integration nature. A considerable amount of computation is required to evaluate these entities. On the other hand these formulas prove invaluable to algorithms which *adaptively choose step sizes* in either the tableau or relaxation formats.

To discuss this use of the error formulas, define E to be the maximum allowable error over the entire time interval, say T, in which a simulation is to be made. For the i-th step in the simulation process, choose a numerical technique for which

$$\frac{E_i}{h_i} \leq \frac{E}{T} \tag{4.28}$$

where h_i is the step-size of the i-th step and E^i is the estimated error for the step. One of the above remainder formulas will provide an estimate of E_i. Basically these estimates should arise from prior computations

In addition to numerical differentiation and integration techniques we must also analyze the various techniques for solving nonlinear equations required by the simulation algorithms. The classical approach for solution is by way of the *Newton-Raphson* algorithm.

Suppose the problem is to solve the continuous non-linear equation 4.29 for x, where x is a vector in some Euclidean space. One picks an initial

$$f(x) = 0 \qquad (4.29)$$

guess and uses the Newton-Raphson algorithm to update the guess via the formula

$$x^{i+1} = x^i - J_f^{-1}(x^i)f(x^i) \qquad (4.30)$$

Assuming x^0 is sufficiently close to the solution, say x, the algorithm is assured to converge.[2, 8] Here $J_f(x^i)$ is the *Jacobian matrix* of partial derivatives of $f(\cdot)$ evaluated at x^i where $f(\cdot)$ is a vector valued function of a vector as given in 4.32.

$$f(\cdot) = \begin{bmatrix} f_1(\cdot) \\ f_1(\cdot) \\ \vdots \\ f_n(\cdot) \end{bmatrix} ; \; x = \begin{bmatrix} x_1 \\ x_1 \\ \vdots \\ x_n \end{bmatrix} \qquad (4.31)$$

Justification for the Newton-Raphson approach and most other numerical processes designed to solve continuous nonlinear equations as 4.29 may be obtained by converting the solution to an equivalent problem. The equivalent problem amounts to computing the *stable singular points of a nonlinear differential equation*. After such a conversion, any of the standard integration routines provide a means for calculating the desired *stable singular points*. In particular, these stable limiting points are the limiting value of any trajector of the differential equation.

THEOREM 4.32: Assume $f(\cdot)$ has continuous partial derivatives. Then the stable singular points of the nonlinear differential equation

$$\dot{x} = -J_f^{-1}(x)f(x) \qquad (4.33)$$

are the solutions to the nonlinear equation $f(x) = 0$.

PROOF: The thrust of the proof is to show that

$$f(x(t)) = f(x(0))\exp(-t) \qquad (4.34)$$

Therefore as t approaches ∞, we obtain the required solution, $f(x(\infty)) = 0$.

As a first step rewrite equation 4.33 as:

$$J_f(x)\dot{x} = -f(x) \tag{4.35}$$

Now observe that by definition

$$\frac{df}{dx} = J_f(x) \tag{4.36}$$

where d/dx is the total derivative with respect to the vector x. Thus applying the *chain rule* we have

$$J_f(x) \dot{x} = \frac{df}{dx}\frac{dx}{dt} = \dot{f}(x) \tag{4.37}$$

which implies, by 4.35, that

$$\dot{f}(x) = -f(x) \tag{4.38}$$

This is a linear first order equation in $f(\cdot)$ whose solution is

$$f(x(t)) = f(x(0))\exp(-t) \tag{4.39}$$

As t approaches infinity, $f(x(t))$ approaches zero, which implies that $x(t)$ approaches the solution of the equation, $f(x) = 0$. By construction, $x(t)$ satisfies equation 4.33, so its limiting value is nothing more than a stable singular point of the equation. The proof is now complete. Note that without the continuity assumption, the above constructions are ridiculous.

Equation 4.33 provides a catalyst for a whole slew of numerical schemes capable of solving $f(x) = 0$. Basically one picks an initial condition (an initial guess) and integrates the equation to infinity by any of the myriad integration schemes found in any supermarket text on numerical methods. Each different routine supplies a different mode of solution.

This point of view is essentially the *continuations approach* discussed previously. It is interesting to realize that the continuations approach defines a one to one correspondence between classical integration routines and classical equation solving algorithms.[2,8] For instance if the *Euler* method of integration with step size, $h = 1$, is employed to solve 4.33, then the resulting iterative formula is

$$x^{i+1} = x^i - J_i^{-1}(x^i) \; f(x^i) \tag{4.40}$$

This coincides precisely with the *Newton-Raphson Algorithm*. More generally if one employs an arbitrary h with the Euler method, the iterative formula for solution becomes

$$x^{i+1} = x^i - h \; J_f^{-1}(x^i) \; f(x^i) \tag{4.41}$$

EXAMPLE 4.42: Consider the following nonlinear equation

$$f(x,y) = \begin{bmatrix} f_1(x,y) \\ f_2(x,y) \end{bmatrix} = \begin{bmatrix} \exp(x) - 2 \\ \exp(xy) - .1 \end{bmatrix} = \begin{bmatrix} 0 \\ 0 \end{bmatrix} \tag{4.43}$$

The Jacobian matrix for f(x,y) is

$$J_f(x,y) = \begin{bmatrix} \exp(x) & 0 \\ y \exp(xy) & x \exp(xy) \end{bmatrix} \tag{4.44}$$

The inverse of the Jacobian matrix is

$$J_f^{-1}(x,y) = \begin{bmatrix} \exp(-x) & 0 \\ -\dfrac{y}{x} \exp(-x) & \dfrac{1}{x} \exp(-xy) \end{bmatrix} \tag{4.45}$$

The solution of the example is the solution to the following differential equation.

$$\begin{bmatrix} x \\ y \end{bmatrix} = - \begin{bmatrix} \exp(-x) & 0 \\ \dfrac{-y}{x}\exp(-x) & \dfrac{\exp(-xy)}{x} \end{bmatrix} \begin{bmatrix} \exp(x) - 2.0 \\ \exp(xy) - .1 \end{bmatrix} \tag{4.46}$$

Using Euler integration with step size h = 1, we obtain the iterative formula

$$\begin{bmatrix} x^{i+1} \\ y^{i+1} \end{bmatrix} = \begin{bmatrix} x^i \\ y^i \end{bmatrix} - \begin{bmatrix} 1 - 2\exp(-x) \\ \dfrac{1}{x}(2y\exp(-x) - y + 1 - .1 \exp(-xy)) \end{bmatrix} \tag{4.47}$$

Given an initial guess of $[x,y]^t = [1,1]^t$, we have the table of iterations as given in **4.49** which shows the solution to be

$$[x,y]^t = [.6931, -3.3219]^t \qquad (4.48)$$

i	1	2	3	4	5	6	7
x^i	1	.73576	.69404	.69315	.69315	.69315	.69315
y^i	1	.30103	-.93213	-2.099	-2.9236	-3.2717	-3.3211

Table **4.49**. Table of iterations for example 4.43.

The stable singular points of **4.33** depend on the differential equation and not the type of integration scheme employed to solve the equation. Thus one may choose any integration formula to work with. For example after x^0 and x^1 are computed it is possible to use a second order formula wherein the two previous values, x^i and x^{i-1}, help compute x^2.

EXAMPLE **4.50**: Suppose it is desired to solve the same differential equation as in example **4.42**. In this case let us use a *second order trapezoidal predictor* as given in equation **4.51**.

$$x^{i+1} = x^i + \frac{h}{2} 3\left[g(x^i) - g(x^{i-1})\right] \qquad (4.51)$$

where h is the step size and $g(x^i) = -J_f^{-1}(x^i)f(x^i)$. Choosing a step size of h = .5, the iterative formula becomes

$$\begin{bmatrix} x^{i+1} \\ y^{i+1} \end{bmatrix} = \begin{bmatrix} x^i \\ y^i \end{bmatrix} + .25 \begin{bmatrix} 2 - 6\exp(-x^i) + 2\exp(-x^{i-1}) \\ \frac{3}{x^i} 2y^i\exp(-x^i) - y^i + 1 - .1\exp(-x^i y^i) \end{bmatrix} \qquad (4.52)$$

$$+ .25 \begin{bmatrix} 0 \\ \frac{1}{x^{i-1}} 2y^{i-1}\exp(-x^{i-1}) - y^{i-1} + 1 - .1\exp(-x^{i-1} y^{i-1}) \end{bmatrix}$$

The result of several steps of the iteration are tabulated in 4.53.

i	x^i	y^i
0	1	1
1	.73576	.30103
2	.77053	-.44910
3	.72511	-1.0091
4	.72013	-1.5718
5	.70803	-2.0498
6	.70361	-2.4350
7	.69950	-2.7236
8	.69735	-2.9265

Table 4.53. Table of iterations for trapezoidal predictor scheme.

If we had chosen $h = 1$, the process would converge at an even slower rate. It is instructive to note that increasing the order of the numerical routine does not necessarily improve performance. It is important to match the numerical routine to the problem. Many times better overall efficiency occurs through use of a lower order numerical routine with smaller step size, than through use of a higher order routine and more complicated programming.

The final comments of the section focus on the nonlinear equation

$$f(x) = 0 \qquad (4.54)$$

when $f(\cdot)$ is *continuous piecewise linear*. If $f(\cdot)$ is continuous piecewise linear, the solution of the equation reduces to the simpler solution of a finite number of linear algebraic equations. *Householder's formula* adds a further bit of simplicity to the solution procedure. Note that we assume nothing about partial derivatives (which don't exist at the boundary of the regions which define f). Thus theorem 4.32 does not apply in this instance.

Suppose $f(\cdot)$ is a continuous piecewise linear function mapping $\mathbf{R}^n$ into $\mathbf{R}^n$. Suppose $[R_1,...,R_m]$ is a partition of $\mathbf{R}^n$ into m-disjoint *regions* which cover $\mathbf{R}^n$ such that each "piece" of the map $f(\cdot)$ has domain over one of the R_i's. Note that the regions R_i are determined by an appropriate set of *hyperplanes*. In particular suppose $f(\cdot)$ is specified by

$$f(x) = f_i(x) = D_i x + d_i \qquad i=1,...,m \qquad (4.55)$$

for x in R_i. Furthermore assume that each $f_i(\cdot)$ is one to one. Therefore D_i is one to one and **nonlinear**. Thus continuity and the affine nature of $f_i(\cdot)$ make $f_i^{-1}(\cdot)$ well defined.

To solve the equation $f(x) = 0$, for the set of points $\{x\}$ which satisfy it, one computes the set of points $S = \{x_i\}$ as follows.

$$x_i = -D_i^{-1} d_i \tag{4.56}$$

for each i. Remember from **chapter III** that after computing D_1^{-1} Householder's formula eases the computation of the remaining inverses. Now form the set $\underline{S} = \{x_i \epsilon S \mid x_i \epsilon R_i\}$. The set $\underline{S}$ is the set of solutions to the equation $f(x) = 0$.

To see this observe that each x_i in S satisfies

$$D_i x_i - d_i = 0 \tag{4.57}$$

However this equation coincides with the specification of the continuous piecewise linear function, $f(\cdot)$, if and only if x_i is in R_i. By eliminating the points x_i which are not in the domain of $f_i(\cdot)$, we eliminate the meaningless solutions. Consequently $\underline{S}$ is the desired set of solutions.

EXAMPLE 4.58: Consider the following continuous piecewise linear equation

$$f(x) = 0 \tag{4.59}$$

specified by the following set of seven affine equations.

$$\begin{bmatrix} 1 & 0 \\ 1 & 1 \end{bmatrix} x \tag{4.60}$$

for x in R_1,

$$\begin{bmatrix} 2 & 1 \\ -1 & -1 \end{bmatrix} x + \begin{bmatrix} -1 \\ 2 \end{bmatrix} \tag{4.61}$$

for x in R_2,

$$\begin{bmatrix} 3 & 0 \\ -1 & -1 \end{bmatrix} x + \begin{bmatrix} 0 \\ 2 \end{bmatrix} \tag{4.62}$$

for x in R₃,

$$\begin{bmatrix} 2 & -1 \\ 1 & 1 \end{bmatrix} x + \begin{bmatrix} 1 \\ 0 \end{bmatrix} \tag{4.63}$$

for x in R₄,

$$\begin{bmatrix} 2 & 0 \\ 1 & -1 \end{bmatrix} x + \begin{bmatrix} 2 \\ -2 \end{bmatrix} \tag{4.64}$$

for x in R₅,

$$\begin{bmatrix} 1 & 1 \\ 1 & -1 \end{bmatrix} x + \begin{bmatrix} 1 \\ -2 \end{bmatrix} \tag{4.65}$$

for x in R₆,

$$\begin{bmatrix} 2 & 2 \\ -1 & -3 \end{bmatrix} x \tag{4.66}$$

for x in R₇, where $x = (x_1, x_2)^t$, and the regions R_i are depicted in figure 4.67.

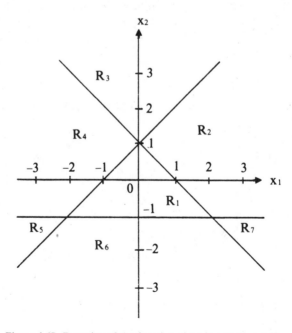

Figure 4.67. Domains of the function given in equation 4.59.

To compute the set S, we solve the following set of equations where the inverses are computed with the aid of Householder's formula.

$$x_1 = \begin{bmatrix} 1 & 0 \\ 1 & 1 \end{bmatrix}^{-1} \begin{bmatrix} 0 \\ 0 \end{bmatrix} \tag{4.68}$$

$$x_2 = \begin{bmatrix} 2 & 1 \\ -1 & -1 \end{bmatrix}^{-1} \begin{bmatrix} 1 \\ -2 \end{bmatrix} \tag{4.69}$$

$$x_3 = \begin{bmatrix} 3 & 0 \\ -1 & -1 \end{bmatrix}^{-1} \begin{bmatrix} 0 \\ -2 \end{bmatrix} \tag{4.70}$$

$$x_4 = \begin{bmatrix} 2 & -1 \\ 1 & 1 \end{bmatrix}^{-1} \begin{bmatrix} -1 \\ 0 \end{bmatrix} \tag{4.71}$$

$$x_5 = \begin{bmatrix} 2 & 0 \\ 1 & -1 \end{bmatrix}^{-1} \begin{bmatrix} -2 \\ 2 \end{bmatrix} \tag{4.72}$$

$$x_6 = \begin{bmatrix} 1 & 1 \\ 1 & -1 \end{bmatrix}^{-1} \begin{bmatrix} -1 \\ 2 \end{bmatrix} \tag{4.73}$$

$$x_7 = \begin{bmatrix} 2 & 2 \\ -1 & -3 \end{bmatrix}^{-1} \begin{bmatrix} 0 \\ 0 \end{bmatrix} \tag{4.74}$$

After completing the necessary arithmetic, we find that

$$x_1 = \begin{bmatrix} 0 \\ 0 \end{bmatrix} ; x_2 = \begin{bmatrix} -1 \\ 3 \end{bmatrix} ; x_3 = \begin{bmatrix} 0 \\ 2 \end{bmatrix} ; x_4 = \begin{bmatrix} -1/3 \\ 1/3 \end{bmatrix} ;$$

$$x_5 = \begin{bmatrix} -1 \\ -3 \end{bmatrix} x_6 = .5 \begin{bmatrix} 1 \\ -3 \end{bmatrix} ; x_7 = \begin{bmatrix} 0 \\ 0 \end{bmatrix} \tag{4.75}$$

By checking the appropriate domains we find that the solution set is

$$S = x_1 = \begin{bmatrix} 0 \\ 0 \end{bmatrix} ; x_3 = \begin{bmatrix} 0 \\ 2 \end{bmatrix} ; x_6 = .5 \begin{bmatrix} 1 \\ -3 \end{bmatrix} \tag{4.76}$$

Consequently these are the only three solutions to equation 4.59. This completes the example.

Finally observe that the above discussion is easily extendable to a wider class of continuous piecewise linear functions. In particular, suppose $f(\cdot)$: $\mathbf{R}^m \to \mathbf{R}^n$ where $n \geq m$ is a continuous piecewise linear map. Then the same algorithm for solving the equation $f(x) = 0$ is applicable provided each of the pieces $f_i(\cdot)$ admits a left inverse. Although the left inverse of a particular $f_i(\cdot)$ is nonunique, the solution to the equation is independent of the choice of left inverse.

There is another technique for solving the equation

$$f(x) \;=\; y \tag{4.77}$$

where $f(\cdot)$: $\mathbf{R}^n \to \mathbf{R}^m$ is continuous piecewise linear.[13] Basically take an initial guess, say x_0 and calculate $y_0 = f(x_0)$. Draw a straight line from y_0 to y in $\mathbf{R}^n$. To find the solution x such that $f(x) = y$, one inverts this line. Specifically one calculates a continuous piecewise linear trajectory in the domain space. The initial point of the trajectory is the initial guess, x_0, and the end point is the solution, x. These notions are illustrated in figure 4.78.

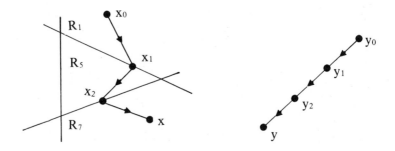

Figure 4.78. Illustration of a technique to solve piecewise linear equations.

At this point we refer the reader to the literature[13] where ample discussion of these ideas exists. Note that there are numerous restrictions on the kind of permissible map, $f(\cdot)$, and that the algorithm for constructing the solution is rather involved.

5. FREQUENCY DOMAIN ALGORITHMS

Simulation of a *linear time invariant system* permits the direct use of frequency domain techniques in lieu of the various general purpose algorithms described previously. One such technique is a variation of the *sparse tableau algorithm*. To construct this scheme, one begins with the usual component connection equations

$$b = Z(s)a \qquad (5.1)$$

and

$$\begin{aligned} a &= L_{11}b + L_{12}u \\ y &= L_{21}b + L_{22}u \end{aligned} \qquad (5.2)$$

where all system variables depend on the frequency parameters. As in our previous discussion we neglect the system output equation. Equation 5.1 and the first connection equation combine to form the following tableau format

$$\left[T(s)\right]\left[\begin{array}{c} a \\ \hline b \end{array}\right] = \left[\begin{array}{c|c} Z(s) & -I \\ \hline -I & L_{11} \end{array}\right]\left[\begin{array}{c} a \\ \hline b \end{array}\right] = \left[\begin{array}{c} 0 \\ \hline -L_{12}u \end{array}\right] \qquad (5.3)$$

To carry out the simulation it is necessary to solve for a and b. The block diagonal structure of $Z(s)$ and the typical sparseness of L_{11} make $T(s)$ sparse. Consequently any of the standard *sparse matrix inversion* algorithms may be used to compute $T^{-1}(s)$ and thereby solve equation 5.3. Since inversion must take place over a discrete set of different frequencies, it is important to exploit the sparsity structure of $T(s)$ which fortunately is independent of s. An efficient simulation algorithm for the inversion of $T(s)$ would **incorporate** a subroutine to automatically generate a program for the inversion of matrices with the same sparsity pattern.[1] A *continuations algorithm* offers an alternative procedure for computing the inverse of $T(s)$ at multiple frequencies.[13] Recall that the technique was outlined in theorem III.4.34. Here, one computes an *LU-factorization* of $T(s_p)$ and the inverses of these factors at this frequency. The next step is to integrate the differential equations of theorem III.4.34 to obtain the LU-factors of $T(s)$ and their inverses at other frequencies.

EXAMPLE 5.4: Consider the *LC tank circuit* of figure 3.30. Assume that the two volt battery switches on at $t = 0$ and that all *initial conditions* are zero. The component connection equations in the frequency domain format are

$$\left[\begin{array}{c} v_c \\ i_L \end{array}\right] = \left[\begin{array}{cc} \dfrac{1}{s} & 0 \\ 0 & \dfrac{1}{s} \end{array}\right]\left[\begin{array}{c} i_c \\ v_L \end{array}\right] \qquad (5.5)$$

and

$$
\begin{bmatrix} i_c \\ \hline v_L \\ \hline i_s \end{bmatrix} = \begin{bmatrix} 0 & 1 & 0 \\ -1 & 0 & 1 \\ 0 & 1 & 0 \end{bmatrix} \begin{bmatrix} v_c \\ i_L \\ \dfrac{2}{s} \end{bmatrix}
\tag{5.6}
$$

The *frequency domain tableau equation* is

$$
\begin{bmatrix} T(s) & a \\ & b \end{bmatrix} = \begin{bmatrix} \dfrac{1}{s} & 0 & -1 & 0 \\ 0 & \dfrac{1}{s} & 0 & -1 \\ -1 & 0 & 0 & 1 \\ 0 & -1 & -1 & 0 \end{bmatrix} \begin{bmatrix} i_c \\ v_L \\ v_c \\ i_L \end{bmatrix} = \begin{bmatrix} 0 \\ 0 \\ 0 \\ \dfrac{2}{s} \end{bmatrix}
\tag{5.7}
$$

Inverting $T(s)$ produces the solution vector for the system. The solution is given in equation 5.8.

$$
\begin{bmatrix} i_c \\ v_L \\ v_c \\ i_L \end{bmatrix} = \frac{1}{s^2+1} \begin{bmatrix} s & -s^2 & -s^2 & -s & 0 \\ s^2 & s & s & -s^2 & 0 \\ -s^2 & -s & -s & -1 & 0 \\ s & -s^2 & 1 & -s & \dfrac{2}{s} \end{bmatrix} = \begin{bmatrix} \dfrac{-2}{s^2+1} \\ \dfrac{-2s}{s^2+1} \\ \dfrac{-2}{s^3+1} \\ \dfrac{-2}{s^2+1} \end{bmatrix}
\tag{5.8}
$$

    The above example permits hand manipulations. Extremely large matrices do not. However, the highly sparse nature of the particular matrices permits fabrication of highly efficient numerical simulation routines for *frequency domain analysis*.

    A second technique sometimes applicable to frequency domain simulation is the notion of *tearing*. In many cases it makes the computation of the *composite system transfer function matrix* feasible. Assuming efficient computation of $(I-Z(s)L_{11})^{-1}$, then the following equations will specify the various system variables.

$$
a = (I - L_{11}Z(s))^{-1}L_{12}\, u(s)
\tag{5.9}
$$

$$
b = Z(s)(I - L_{11}Z(s))^{-1}L_{12}\, u(s)
\tag{5.10}
$$

and

$$y = [L_{21}Z(s)(I - L_{11}Z(s))^{-1}L_{12} + L_{22}]\, u(s) \qquad (5.11)$$

Although smaller in dimension than $T(s)$, the matrix $(I-L_{11}Z(s))$ is often less sparse than $T(s)$. However, if the system is *hierarchical* $(I-L_{11}Z(s))$ will be *block triangular*. Its inversion proceeds by inverting its diagonal blocks together with a *back substitution* process similar to that discussed in section III.4 Unfortunately, few real world systems are hierarchical. Many practical systems are, however, *almost hierarchical*. By almost hierarchical, we mean that if one removes a "small" number of branches the remaining system will be hierarchical. The concept of *tearing* is applicable to such systems. To apply the concept, divide the components of the system into two classes. Class one comprises the *hierarchical components* and class two the *tearing components*. This decomposes the composite component transfer function matrix into a sum

$$Z(s) = Z_h(s) + Z_t(s) \qquad (5.12)$$

Here, $Z_h(s)$ is a block diagonal matrix. This matrix is what remains of $Z(s)$ after the blocks corresponding to tearing components have been zeroed out. The matrix $Z_t(s)$ is what remains of $Z(s)$ after the blocks corresponding to the hierarchical components are zeroed out. Clearly then

$$(I - L_{11}Z(s)) = (I - L_{11}Z_h(s)) + L_{11}Z_t(s) \qquad (5.13)$$

The inversion of this matrix then becomes a two step process. First, exploit the hierarchical structure to invert $(I-L_{11}Z_h(s))$ by inverting its diagonal blocks. Then apply the *Householder formula* of lemma III.6.15 to incorporate the, usually *low rank, perturbation*, $L_{11}Z_t(s)$, into the inverse. Obviously, the power of the technique depends on the structure of the specific system. In many cases the "tearing-out" of a relatively small number of components will decompose the system into a hierarchical structure with small diagonal blocks. This reduces the inversion of a large matrix to the inversion of many small matrices. Unfortunately, tearing out too many components renders the rank of $L_{11}Z_t(s)$ too large for efficient use of the Householder formula. However, when doing a frequency domain analysis, in which it is necessary to invert $(I-L_{11}Z(s))$ for many different values of s, the computational overhead for determining an "optimal" tearing of the network, is often well worth the cost.

EXAMPLE 5.14: Consider the *feedback system* shown in figure 5.15.

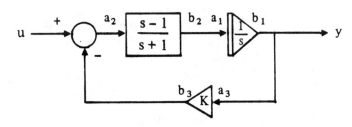

Figure 5.15. Feedback system of example 5.14.

In the frequency domain format, the component connection equations for this system are

$$
\begin{bmatrix} b_1 \\ b_2 \\ b_3 \end{bmatrix} = \begin{bmatrix} \dfrac{1}{s} & 0 & 0 \\ 0 & \dfrac{s-1}{s+1} & 0 \\ 0 & 0 & K \end{bmatrix} \begin{bmatrix} a_1 \\ a_2 \\ a_3 \end{bmatrix}
\tag{5.16}
$$

and

$$
\begin{bmatrix} a_1 \\ a_2 \\ a_3 \\ \hline y \end{bmatrix} = \left[\begin{array}{ccc|c} 0 & 1 & 0 & 0 \\ 0 & 0 & -1 & 1 \\ 1 & 0 & 0 & 0 \\ \hline 1 & 0 & 0 & 0 \end{array} \right] \begin{bmatrix} b_1 \\ b_2 \\ b_3 \\ \hline u \end{bmatrix}
\tag{5.17}
$$

A simulation requires us to invert

$$
[I - L_{11}Z(s)] = \begin{bmatrix} 1 & -\dfrac{s-1}{s+1} & 0 \\ 0 & 1 & K \\ \dfrac{-1}{s} & 0 & 1 \end{bmatrix}
\tag{5.18}
$$

which has no hierarchical structure regardless of the component rearrangement. If, however, we delete the feedback component, and appropriately *reorder* the components as 2, 1, 3, then we have

$$(I - \hat{L}_{11}\hat{Z}(s)) = (I - \hat{L}_{11}\hat{Z}_h(s)) - \hat{L}_{11}\hat{Z}_t(s) \qquad (5.19)$$

$$= \begin{bmatrix} 1 & 0 & 0 \\ -\dfrac{s-1}{s+1} & 1 & 0 \\ 0 & -\dfrac{1}{s} & 1 \end{bmatrix} + \begin{bmatrix} 0 & 0 & k \\ 0 & 0 & 0 \\ 0 & 0 & 0 \end{bmatrix}$$

where the "hats" denote the reordered Z and L_{11} matrices. Note that the first matrix is triangular making it amenable to inversion by *back substitution*. The second matrix is of rank one. With the Householder formula incorporating the effect of the rank one perturbation in the inverse, the problem of inverting a non-hierarchical three by three matrix has been converted to the much easier problem of inverting four one by one matrices.

In regard to feedback systems, similar results are achievable by removing all feedback elements. The remaining feedforward structure will always be *series-parallel*. The inversion of the matrix $(I-L_{11}Z(s))$ for a feedback system, then reduces to the "trivial" inversion of a triangular matrix with unit diagonal and the inversion of a rank "r" matrix where r is the number of feedback elements in the system.

6. DISCRETE AND FAST FOURIER TRANSFORMS

The past growth and applicability of circuit and system theory directly stems from the evolution of the concepts of frequency. From the origins of circuit response or analysis along the imaginary axis, to analysis in the entire complex plane, to the present beginnings of multidimensional frequency responses (image processing) and several complex variables. With the advent of computer technology, researchers explored the numerical realms of efficient frequency domain simulation techniques for *linear time invariant interconnected dynamical systems*. The utility of these techniques depended on designing efficient numerical algorithms to transform, the *composite component and system variables*, a, b, y, and u from the time domain to the frequency domain, and back. The research began with the less efficient *discrete Fourier transform* and progressed to the highly efficient *fast Fourier transform* or FFT. The FFT is *not* a new-fangled transform but merely a very efficient means of computing the discrete Fourier transform.

Recall that the classical Fourier transform has the definition

$$F(\omega) = \int_{-\infty}^{\infty} f(t)\exp. (-j\omega t)dt \tag{6.1}$$

Most often $f(t)$ is either negligible or uninteresting outside a finite interval. Moreover the prevalence of digital techniques, discrete time systems, and real time digital control made the idea of a discrete approximation to the continuous Fourier integral highly attractive. By sampling a time function $f(t)$ at N points and using N steps of an *Euler integration* scheme with step size, h, one obtains the following discrete approximation to the Fourier integral

$$F(\omega) \approx h \sum_{n=0}^{N-1} f(nh)\exp(-j\omega h) \tag{6.2}$$

where we have normalized the sampling interval to have starting point zero. These ideas foster the rigorization of the *discrete Fourier transform* as follows.

If one samples the time function, $f(t)$, every h seconds, then the frequency domain representation of $f(\cdot)$ will be a discrete set of points F_k. Assuming an accurate approximation to the continuous integral, F_k should approximate $F(\omega_k)$. The actual Fourier transform evaluated at the appropriate discrete frequencies. Assuming one takes N samples of $f(t)$, then it is necessary to pick the number v to satisfy

$$N = \frac{2\pi}{hv} \tag{6.3}$$

The number v is the distance between the frequency samples as h is the distance between time samples. Now define

$$W_N = \exp(-j\frac{2\pi}{N}) = \exp(-jvh) \tag{6.4}$$

Also define $f_n = f(t_n) = f(nh)$ for $n = 0,1,...,N-1$. The *discrete Fourier transform* of the sequence of points, $[f_n | n = 0,1,...,N-1]$ is again a sequence of points, $[F_k | k = 0,1,...,N-1]$ computed via the formula

$$F_k = h \sum_{n=0}^{N-1} f_n \exp(-j\frac{2\pi}{N}nk) \tag{6.5}$$

$$= h \sum_{n=0}^{N-1} f_n [W_N]^{nk}$$

Using the relationship in equation 6.6 below

$$\sum_{p=0}^{N-1} [W_N]^{pk} [W_N]^{-pn} = \begin{cases} N & \text{if } k=n \\ 0 & \text{if } k \neq n \end{cases} \tag{6.6}$$

it is possible to establish the *inverse discrete Fourier* relationship[9] as

$$f_n = \frac{1}{hN} \sum_{k=0}^{N-1} F_k \exp(j \frac{2\pi}{N} nk) \tag{6.7}$$

$$= \frac{1}{hN} \sum_{k=0}^{N-1} F_k [W_N]^{-nk}$$

for $n = 0,1...,N-1$. The relationships of 6.5 and 6.7 have all the usual symmetry properties associated with their classical continuous time counterparts. Before stating and proving a theorem outlining the kinship between the continuous and discrete transforms, it is helpful to root the ideas with a simple example.

EXAMPLE 6.8: Consider the *one sided exponential function* defined by

$$f(t) = \begin{cases} 0 & t<0 \\ .5 & t=0 \\ \exp(-t) & t>0 \end{cases} \tag{6.9}$$

This function has the well known Fourier transform

$$F(j\omega) = \frac{1}{1 + j\omega} \tag{6.10}$$

Assuming one samples $f(t)$ every .5 seconds and takes a total of eight samples, the defining formulas reduce to

$$F_k = .5 \sum_{n=0}^{7} f_n \exp(-j \frac{\pi}{4} nk) \tag{6.11}$$

where $f_0 = .5$ and $f_n = \exp(-.5n)$ for $n = 1,2,...,7$. The discrete Fourier transform then is the discrete sequence of eight complex numbers listed in table 6.12.

F_0	.9974
F_1	$.2996 - j.4127$
F_2	$.1008 - j.2176$
F_3	$.0651 - j.0946$
F_4	$.0556$
F_5	$.0651 + j.0946$
F_6	$.1088 + j.2176$
F_7	$.2996 + j.4127$

Table 6.12. Values of discrete Fourier transform for example 6.8.

The discrete Fourier transform, in contrast to its continuous counterpart, is *always* periodic. This becomes apparent when one observes that $F_7 = F_1$, $F_6 = F_2$, $F_5 = F_3$. As such we define $F_{-1} = F_7, F_{-2} = F_6$, and $F_{-3} = F_5$. These relationships mirror the *symmetry* of the magnitude of the continuous Fourier transform and the anti-symmetry of the phase of the continuous Fourier transform. A comparison of the discrete approximation with the actual continuous Fourier transform is shown in the two following plots of magnitude and phase where the x's indicate the appropriate property of the F_k's and the continuous curve the actual magnitude and phase of $F(\omega) = 1/(1-j\omega)$.

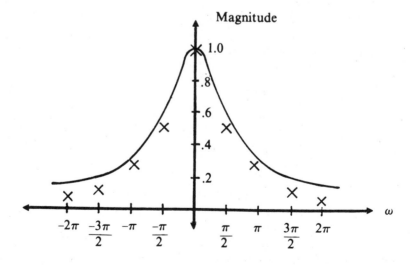

Figure 6.13. Magnitude plots for discrete and continuous cases.

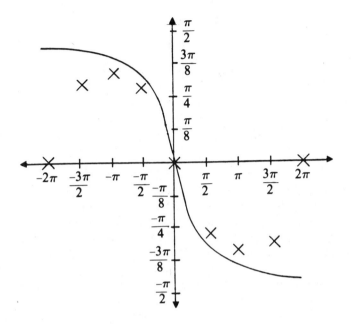

Figure 6.14. Phase, plots for discrete and continous cases.

With the intuition of this example we may now discuss the extent to which the discrete Fourier transform approximates the classical Fourier transform.

THEOREM 6.15: Let f(t) be a function of time and $F(\omega)$ its Fourier transform as per equation 6.1. Define sequences of real and complex numbers respectively by the following two formulas:

$$f_n = \sum_{m=-\infty}^{\infty} f(nh + mhN) \qquad (6.16)$$

for n = 0,1,...,N-1, and

$$F_k = \sum_{m=-\infty}^{\infty} f(kv + mvN) \qquad (6.17)$$

for k = 0,1,...,N-1. The symbols v, N, and h have their earlier meaning.

Given these relationships, then

$$F_k = h \sum_{n=0}^{N-1} f_n [W_N]^{nk}$$ (6.18)

for $k = 0,1,...,N-1$, and

$$f_n = \frac{1}{hN} \sum_{k=0}^{N-1} F_k [W_N]^{-nk}$$ (6.19)

for $n = 0,1,...,N-1$.

Intuitively this theorem says that if we infinitely sample $f(t)$ with step size h and sum the resultant sample values to form f_n and if we similarly infinitely sample $F(\omega)$ with step size v to form F_k, then the two resulting sums of samples are precisely related by the discrete Fourier transform. Fortunately, in practice, $f(t)$ is often negligible outside of a finite time interval. **By properly choosing h and v, all terms except the m=0 term in 6.6** and 6.17 become negligible. Basically then the theorem establishes an exact relationship between the samples of $f(t)$ and $F(\omega)$ which indirectly describes the kinship between the approximations, formulas 6.16 and 6.17 when m=0. Bounds on $f(t)$ and $F(\omega)$ outside the "interesting" intervals permits an *error analysis of the formulas.*[9]

PROOF OF THEOREM 6.15: The derivations for equations 6.18 and 6.19 are dual. Consequently the proof will construct only 6.18. Taking the inverse transform of $F(\omega)$ and playing games with the limits of integration, we obtain

$$f(nh) = \frac{1}{2\pi} \int_{-\infty}^{\infty} F(\omega)\exp(j\omega nh)d\omega$$

$$= \frac{1}{2\pi} \sum_{n=-\infty}^{\infty} \int_{\frac{2\pi n}{h}}^{\frac{2\pi(n+1)}{h}} F(\omega + \frac{2\pi n}{h})\exp(jnh\omega)d\omega$$ (6.20)

$$= \frac{1}{2\pi} \sum_{n=-\infty}^{\infty} \int_{0}^{\frac{2\pi}{h}} F(\omega + \frac{2\pi n}{h})\exp(jnh(\omega + \frac{2\pi n}{h}))d\omega$$

$$= \frac{1}{2\pi} \int_{0}^{\frac{2\pi}{h}} \sum_{n=-\infty}^{\infty} F(\omega + \frac{2\pi n}{h})\exp(jnh\omega)d\omega$$

With these magical manipulations over, a moment's thought reveals that except for a scale factor, h, the last line of 6.20 is the formula for computing the n-th Fourier series coefficient of the periodic function, $p(\omega)$, defined as

$$p(\omega) = \sum_{m=-\infty}^{\infty} F(\omega + \frac{2\pi m}{h}) \tag{6.21}$$

for $0 \leqslant \omega \leqslant 2\pi/h$. The period of $p(\omega)$ is clearly $2\pi/h$ and it has the Fourier series expansion

$$p(\omega) = \sum_{m=-\infty}^{\infty} F(\omega + \frac{2\pi m}{h}) = h \sum_{n=-\infty}^{\infty} f(nh)\exp(-jnh\omega) \tag{6.22}$$

Observe that $2\pi m/h = mNv$ and choose $\omega = kv$ where k is an integer ranging as $k = 0, 1, ..., N-1$; the middle term of 6.22 is just F_k. Consequently we have the following equality.

$$F_k = \sum_{m=-\infty}^{\infty} F(kv + mNv) = h \sum_{n=-\infty}^{\infty} f(nh)\exp(-j\frac{kn2\pi}{N}) \tag{6.23}$$

Finally converting the right side of 6.23 to a double summation by letting $n = p + mN$ with p ranging from zero to N-1 and m ranging from $-\infty$ to ∞ we obtain

$$p(\omega) = \sum_{m=-\infty}^{\infty} F(\omega + \frac{2\pi m}{h}) \tag{6.24}$$

which is the desired formula.

Numerically speaking, the computational complexity of the discrete Fourier transform grows with the square of N. In point there are N^2 multiplications ignoring any additions. For large systems the drain on computer time becomes much too large. However if $N = 2^L$ for some L, i.e., N is a power of 2, then the FFT algorithm greatly increases the efficiency of a frequency domain simulation.

The key feature of the scheme is that it first reduces the computation of a discrete Fourier transform with N sample points to the computation of two discrete Fourier transforms, each with $N/2$. Each of these then is transformed into two discrete transforms of length $N/4$, etc. Repeating the process (L-1) times eventually reduces the computation of a discrete Fourier transform of length $N = 2^L$ to the computation of $N/2$ transforms of length 2. Observe that since $W_2 = \exp(-\pi) = -1$, computation of a length two transform involves no multiplications. The major part of the arithmetic of the FFT centers around transforming the *half length transforms* into full length transforms.

THEOREM 6.25: Let $[f_n| n = 0,1,...,N-1]$ be a sequence of length N with discrete Fourier transform $[F_k| k = 0,1,...,N-1]$. Define two new sequences of length N/2 corresponding to the even and odd terms of $[f_n]$ respectively as

$$f_n^e = f_{2n} \quad \text{for } n=0,1,..., \frac{N}{2} - 1, \tag{6.26}$$

and

$$f_n^0 = f_{2n+1} \quad \text{for } n=0,1,..., \frac{N}{2} - 1. \tag{6.27}$$

Let F_n^e and F_k^0 be their discrete Fourier transforms for $k=0,1,...,(N/2)-1$. Then the conclusion of the theorem is that

$$F_k = F_k^e + [W_N]^k F_k^0 \tag{6.28}$$

and

$$F_{k+\frac{N}{2}} = F_k^e - [W_N]^k F_k^0 \tag{6.29}$$

for $k=0,1,...,(N/2)-1$.

PROOF: First observe that $[W_N]^2 = W_{N/2}$, hence the following manipulations are legitimate

$$F_k = \sum_{n=0}^{N-1} f_n [W_N]^{nk}$$

$$= \sum_{n=0}^{\frac{N}{2} - 1} f_{2n} [W_N]^{2nk} + f_{2n+1} [W_N]^{2nk-k} \tag{6.30}$$

$$= \sum_{n=0}^{\frac{N}{2} - 1} f_n^e [W_{\frac{N}{2}}]^{nk} + W_N^k \sum_{n=0}^{\frac{N}{2} - 1} f_n^0 [W_{\frac{N}{2}}]^{nk}$$

Now for $k=0,1,...,(N/2)-1$, the two summations in the last line of 6.30 are just the discrete Fourier transforms of f_n^e and f_n^0 respectively. Hence the conclusion follows that

$$F_k = F_k^e + [W_N]^k F_k^0 \qquad (6.31)$$

for $k=0,1,...,(N/2)-1$, verifying equation 6.28. To verify 6.29 replace k by $k+(N/2)$ in equation 6.30. The proof is complete.

Again no multiplications are required to compute the length-two discrete Fourier transforms; two multiplications are required to compute the length-four transforms from the length-two transforms; four multiplications are required to compute the length-eight transforms, etc. The number of multiplications for implementing the FFT is a grand total of $(N/2)L=(N/2)\log_2 N$. Needless to say, for large N this constitutes a major computational savings over the N^2 multiplications required in standard methods. This then is the doorway to practical implementation of frequency domain simulation techniques.

EXAMPLE 6.32: In this example we will use the FFT algorithm to calculate the DFT of

$$f(t) = \begin{cases} 0 & t < 0 \\ .5 & t = 0 \\ \exp(-t) & t > 0 \end{cases} \qquad (6.33)$$

for $8 = 2^3$ samples taken at .5 second intervals as in example 6.8. To apply the FFT algorithm we will break the eight samples down into three levels as illustrated in table 6.34

Level 1 $f = [f_0 \quad f_1 \quad f \quad f_3 \quad f_4 \quad f_5 \quad f_6 \quad f_7]$

Level 2 $f_e = [f_0 \quad f_2 \quad f_4 \quad f_6] \qquad f_0 = [f_1 \quad f_3 \quad f_5 \quad f_7]$

Level 3 $f^{ee} = [f_0 \quad f_4] \quad f^{eo} = [f_2 \quad f_6] \quad f^{oe} = [f_1 \quad f_5] \quad f^{00} = 0 [f_3 \quad f_7]$

Table 6.34. Sample levels.

The first step is to compute four length-two DFT's associated with level 3 of Table 6.34. These values are given in table 6.35, where the left and right sides of the table correspond to the left and right sides of table 6.34.

$$F_0^{ee} = h[f_0 + [W_2]^0 f_4]$$
$$= .318$$

$$F_0^{e0} = h[f_2 + [W_2]^0 f_6]$$
$$= .209$$

$$F_0^{0e} = h[f_1 + [W_2]^0 f_5]$$
$$= .345$$

$$F_0^{00} = h[f_3 + [W_2]^0 f_7]$$
$$= .127$$

$$F_1^{ee} = h[f_0 + [W_2]^1 f_4]$$
$$= .182$$

$$F_1^{e0} = h[f_2 + [W_2]^1 f_6]$$

$$F_1^{0e} = h[f_1 + [W_2]^1 f_5]$$
$$= .263$$

$$F_1^{00} = h[f_3 + [W_N]^1 f_7]$$
$$= .097$$

Table 6.35. Table of four length-two transforms.

Now we combine these four length-two transforms into two length-four transforms as per the formulas in equations 6.28 and 6.29. These are the transforms of the right and left sides of level 2 of table 6.34 respectively. Table 6.36 lists these values.

$$F_0^e = F_0^{ee} + [W_4]^0 F_0^{e0}$$
$$= .527$$

$$F_1^e = F_1^{ee} + [W_4]^1 F_1^{e0}$$
$$= .182 - j.159$$

$$F_2^e = F_3^{ee} [W_4]^0 F_0^{e0}$$
$$= .109$$

$$F_3^e = F_1^{ee} - [W_4]^1 F_1^{e0}$$
$$= .182 + j.159$$

$$F_0^0 = F_0^{0e} + [W_4]^0 F_0^{00}$$
$$= .472$$

$$F_1^0 = F_1^{0e} + [W_4]^1 F_1^{00}$$
$$= .263 - j.097$$

$$F_2^0 = F_0^{0e} - [W_4]^0 F_0^{00}$$
$$= .218$$

$$F_3^0 = F_1^{0e} - [W_4] F_1^{00}$$
$$= .263 + j.097$$

Table 6.36. The values of the two length-four transforms corresponding to level 2 of table 6.34.

Finally we obtain the desired length-eight DFT by suitably combining the values of table 6.36. The answer is given in table 6.37.

$$F_0 = F_0^c + [W_8]^0 F_0^0 = .999$$

$$F_1 = F_1^c + [W_8]^1 F_1^0 = .300 - j.413$$

$$F_2 = F_2^c + [W_8]^2 F_2^0 = .109 - j.218$$

$$F_3 = F_3^c + [W_j]^3 F_3^0 = .065 - j.095$$

$$F_4 = F_0^3 - [W_8]^0 F_0^0 = .056$$

$$F_5 = F_1^c - [W_8]^1 F_1^0 = .065 + j.095$$

$$F_6 = F_2^c - [W_8]^2 F_1^0 = .109 + j.218$$

$$F_7 = F_3^c - [W_8]^3 F_3^0 = .200 + j.413$$

Table 6.37. Table listing the length-eight DFT of example 6.32.

This coincides, excepting minor *roundoff error*, with the answers of example 6.8. Thus concludes this section and the chapter.

7. Problems

1. Derive a *numerical integration formula* for a function which is to be interpolated a five equally spaced points; 0,h,2h,3h, and 4h; with the derivative being evaluated at 4h.
2. Formulate the *sparse tableau* for the system of example II.2.18 using the numerical differentiation formula derived in example 2.5.
3. Formulate the *sparse tableau equations* for the system of example II.2.24 using the numerical differentiation formula of equation 2.3.
4. Formulate the *sparse tableau matrix* for the circuit of example II.3.53 and invert this matrix via LU factorization.
5. Compute the *Jacobian matrix* for the *sparse tableau equations* derived in problem 3. for the system of example II.2.24 using the numerical differentiation formula of equation 2.3.
6. Compute the solution of the nonlinear system whose tableau equations are formulated in example 2.27 at $t = 1$ assuming that $u(t) = t$ and $x(0) = 0$.
7. Integrate the equation

$$x = x^3; \quad x(0) = 1$$

for one time step of length h using *trapezoidal integration*. Using the result of this integration for $x(1)$ integrate the equation a second step to $2h = 2$ using *Simpson's rule*.

8. Derive a second order implicit integration formula for a function interpolated at $t_0 = 0$, $t_1 = 3/2$, and $t_2 = 2$.

9. Derive a third order explicit integration formula for a function interpolated at $t_0 = 0$, $t_1 = 1$, $t = 2$, $t_3 = 3$, and $t_4 = 4$.

10. Use an *Euler predictor* and a *trapezoidal corrector* to integrate the differential equation.

$$\dot{x} = \sin(\pi x) \; ; \; x(0) = 1$$

one step of length $h = .1$.

11. Use an *Euler predictor* and a *trapezoidal corrector* to compute the solutions of the system of example 2.27 at $t_1 = 1$. Assume that all initial conditions are zero and that $u(t) = t$.

12. The function e^{-x} is to be interpolated at points $x_0 = 0$, $x_1 = .5$ and $x_2 = 1$ by a second order polynomial. Determine an upper bound on the error in the interpolation formula at any point in the interval from zero to one.

13. Assume that the function e^{-x} is to be differentiated at the point $x_1 = .5$. Using the interpolation formula of problem 12 estimate the error resulting from the numerical differentiation formula and compare this estimate with the actual error.

14. Compute the *remainder term* $r(x)$ resulting from the use of *Simpson's rule* to approximate an integral.

15. Solve the set of two nonlinear equations in two unknowns

$$0 = f(X_1, X_2) = \begin{cases} 2\tanh(X_1) - 1 \\ \\ X_1 X_2 - 1 \end{cases}$$

using the *Newton-Raphson* algorithm.

16. Formulate the *frequency domain sparse tableau* for the *Butterworth filter* of example III.2.13 and compute the transforms of the network variables with a step input applied to the filter by inverting the tableau matrix.

17. Show that if all feedback elements are torn out of a feedback system then the remaining feedforward system will be *series-parallel*.

18. If in the system of example 5.14 component one is torn out of the system rather than component three will the remaining system be *hierarchical*? If so, formulate the *tearing decomposition* of the matrix $(I - Z(s)L_{11})$.

19. Invert the matrix of equation 5.18 by exploiting the *tearing decomposition* of equation 5.19.
20. Verify equation 6.6 and 6.7.
21. Show that the F_k computed by equation 6.5 are periodic in N if the k are allowed to range beyond the interval 0 to N-1.
22. Compute the *discrete Fourier transform* for the function $e^{-t}U(t)$ (U(t) the unit step function) with N = 8 and h = .5.
23. Repeat problem 22 using the *fast Fourier transform* algorithm.

8. References

1. Chao, K. S., and R. J. deFigueiredo, "Optimally Controlled Iterative Schemes for Obtaining the Solution of a Nonlinear Equation", *Int. Jour. on Cont.*, Vol. 18, pp. 377-384, (1973).
2. Fleming, W., *Functions of Several Variables*, Addison-Wesley, Reading, 1965.
3. Fujisawa, T., and E. S. Kuh, "Piecewise-Linear Theory of Nonlinear Networks", SIAM *Jour. on Appl. Math.*, Vol. 22, pp. 307-328, (1972).
4. Gear, C. W., *Numerical Initial Value Problems in Ordinary Differential Equations*, Prentice-Hall, Englewood Cliffs, 1971.
5. Hachtel, G., Brayton, R. K., and F. Gustavson, "The Sparse Tableau Approach to Network Analysis and Design," *IEEE Trans. on Circuit Theory*, Vol. CT-18, pp. 101-13, (1971).
6. Hachtel, G., Gustavson, F., Brayton, R. K., and T. Grapes, "A Sparse Matrix Approach to Network Analysis," Proc. of the Cornell Conf. on Computerized Electronics, 1969.
7. Knox, B., and R. Saeks, "A Componentwise-Adaptive Step-size Algorithm for the Simulation of LSDS", Proc. of the 8th Asilomar Conf. on Circuits, Systems, and Computers, pp. 580-583, 1974.
8. Kunz, K., *Numerical Analysis*, McGraw-Hill, New York, 1957.
9. Peled, A., and B. Liu, *Digital Signal Processing*, Wiley, New York, 1976.
10. Prasad, N., "Graph Theoretic and Combinatorial Algorithms in Digital System Simulation," Proc. of the 7th Asilomar Conf. on Circuits, Systems, and Computers, pp. 29-32, 1973.
11. Prasad, N., Smith, J., Reiss, J., and S., Robert, "MARSYAS—A new Language for the Digital Simulation of Systems," Research Rpt., Computer Applications Inc., 1969.
12. Saeks, R., and K. S. Chao, "Continuations Approach to Large Change Sensitivity Analysis" IEE (London) *Jour. on Electronic Circuits and Systems*, Vol. 1, pp. 11-16, (1976).
13. Trauboth, H., and N. Prasad, "MARSYAS—A Software Package for Digital Simulation of Physical Systems," Proc of the Spg. Joint Computer Conf., pp. 223-235, 1970.
14. Weeks, W. T., Jiminez, A. J., Mahoney, G. W., Metho, D., Quassemzedeh, H., and T. R. Scott, "Algorithms for ASTAP—A Network Analysis Program," *IEEE Trans. on Circuit Theory*, Vol. CT-20, pp. 623-634, (1973).

V. SENSITIVITY

1. INTRODUCTION

Classical *sensitivity analysis* describes the effect small component parameter variations have on *composite system* performance. In the present chapter we extend this concept to include the study of the relationship between component parameters and composite system performance, whether or not the parameter variations are small. As such, in addition to the classical small perturbation sensitivity analysis, we consider the problems of *large change sensitivity analysis, computer-aided design,* and *system diagnosis.* As with the classical theory, large change sensitivity analysis still investigates the effects of random parameter change on composite system performance although, in this case, large parameter variations are permitted. On the other hand, computer-aided design is concerned with the determination of (large or small) parameter variations which can be intentionally implemented to tune a system to some optimal performance level. Finally, system diagnosis may be viewed as an "*inverse sensitivity problem*" wherein we desire to determine the parameter changes which caused a measured change in composite system performance.

The following section focuses on transfer function matrix sensitivity. The vehicle for carrying out the investigation is the formula for the derivative of the connection function derived in **chapter III**. This formula permits one to study only *small change sensitivity*.

Section 3 explores two means for calculating large change sensitivity in the frequency domain. The first approach uses a modified small change sensitivity formula to set up a *continuations algorithm*. Basically, we set up a differential equation dependent on the parameter in question. The solution-trajectories then determine the transfer function change. The second approach rests upon the *Householder formula*. This approach is advantageous when considering problems in which several component parameters simultaneously undergo large variation.

Nonlinear systems lack frequency domain theory, to say the least. However, by defining one or more *integral performance measures* on the system, it is possible to characterize pertinent system behavior. As such, section 4 discusses a technique for computing the sensitivity of an arbitrary integral performance measure with respect to component parameter variations of the nonlinear system. As before, the technique has direct use for investigating small change sensitivity or in a *continuations algorithm* for large change sensitivity analyses.

The viability of both the frequency domain sensitivity analysis and the nonlinear performance measure sensitivity analysis is due to the fact that the majority of the computation required to compute the sensitivities duplicates computations required to evaluate the transfer function or performance index under study. In some sense the sensitivity computations come for "free." This, in turn, implies that an iterative process for optimizing internal parameters of an interconnected dynamical system relative to a given performance/sensitivity measure will be computationally feasible. This follows because a numerical implementation of such a procedure requires no additional computation over that necessary for system analysis. These ideas are the crux of the computer-aided design problem sketched in a more general framework in section 5. Specifically one iteratively searches for a parameter vector so as to optimize a given *performance index*. The analytical theory underlying the resultant optimization problem is also developed in section 5 while numerical techniques for implementing the required *optimization algorithm* are discussed in section 6.

Our study of system diagnosis is the topic of the final two sections of the chapter. Section 7 is concerned with the formulation of the *fault diagnosis equations* for linear and nonlinear systems whereas section 8 describes two algorithms for the solution of the fault diagnosis equations.

2. TRANSFER FUNCTION SENSITIVITY

Resistance, capacitance, inductance, etc. as labeled on real world components are merely nominal values guaranteed, say, within plus or minus 5%. The theoretical model then is only nominal. Consequently, there is a striking need to know the effect parameter variations have on nominal system response. Moreover, with time, components age, causing further degradation in system behavior.

There are several objectives of this section. One is to theoretically account for transfer function matrix variations as a function of parameter variations; in other words classical small change sensitivity analysis. However, we propose to show the explicit way parameter changes are woven into the sensitivity of the transfer function. In addition we claim a numerical feasibility of the approach. Presumably, one knows dZ/dr, the change of component transfer function models with respect to some parameter, r. Such information can be garnered from manufacturers' data, measurements, etc. Our goal then is to express dS/dr, the composite system transfer function variation, in terms of dZ/dr. Therefore for small changes in r, Δr, one can approximate the perturbed transfer function matrix as $S+(dS/dr)\,\Delta r$.

Recall that a frequency domain component connection model is given by the set of equations as per 2.1 and 2.2.

$$b = Za \tag{2.1}$$

and

$$a = L_{11}b + L_{12}u$$

$$\tag{2.2}$$

$$y = L_{21}b + L_{22}u$$

The composite system transfer function matrix S is given by

$$y = Su = f(Z)u = (L_{22}+L_{21}(I-ZL_{11})^{-1}ZL_{12})u \tag{2.3}$$

where the explicit dependence of Z and S on the complex frequency variable, s, is suppressed. Recall f is the *connection function* defined in section III.3. The objective is to compute dS/dr relative to some parameter, r, affecting the component transfer functions as modeled by Z. The variable, r, may represent an explicit component parameter such as a capacitance or gain, or it may account for an external parameter, such as temperature, affecting one or more of the components. An integrated circuit process parameter is another external parameter. The only assumption made in the section is that the component parameter variation with respect to r is known via the matrix dZ/dr.

The desired *small change sensitivity* formula is coincident with the derivative of the connection function derived in theorem III.3.43. It is rewritten in equation 2.4.

$$\frac{dS}{dr} = [L_{21} (I-ZL_{11})^{-1}] \frac{dZ}{dr} [I+L_{11}(I-ZL_{11})^{-1}Z]L_{12} \qquad (2.4)$$

Note that 2.4 is a universal frequency domain sensitivity formula. It is useful for the analysis of transfer function sensitivity with respect to any parameter affecting *only* the components of the given system. The only inverse required in computing dS/dr is $(I-ZL_{11})^{-1}$. This is also necessary for computing S itself. Thus, if one needs to compute S, no meaningful additional computation arises in calculating dS/dr. This fact sets up the possibility for an algorithm to optimally choose r by evaluating S for a given value of r. Essentially one uses dS/dr to produce an improved value of r. An efficient numerical implementation of the idea is possible through any of the standard numerical *optimization algorithms* such as *steepest descent*.

EXAMPLE 2.5: Consider the simple RC-voltage divider circuit of figure 2.6.

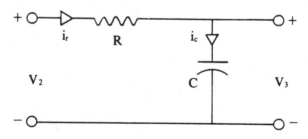

Figure 2.6. RC-voltage divider circuit.

The component connection model for the circuit is

$$\begin{bmatrix} i_r \\ \\ V_c \end{bmatrix} = \begin{bmatrix} \dfrac{1}{R} & 0 \\ \\ 0 & \dfrac{1}{Cs} \end{bmatrix} \begin{bmatrix} v_r \\ \\ ic \end{bmatrix} \qquad (2.7)$$

and

$$\begin{bmatrix} v_r \\ i_c \\ \hline v_o \end{bmatrix} = \left[\begin{array}{cc|c} 0 & -1 & 1 \\ 1 & 0 & 0 \\ \hline 0 & 1 & 0 \end{array}\right] \begin{bmatrix} i_r \\ v_c \\ \hline v_i \end{bmatrix} \tag{2.8}$$

For the nominal values of R=C=1.0 we have

$$(I-ZL_{11}) = \begin{bmatrix} 1 & 1 \\ \dfrac{-1}{s} & 1 \end{bmatrix} \tag{2.9}$$

and

$$(I-ZL_{11})^{-1} = \begin{bmatrix} \dfrac{s}{s+1} & \dfrac{-s}{s+1} \\ \dfrac{1}{s+1} & \dfrac{s}{s+1} \end{bmatrix} \tag{2.10}$$

The composite system transfer function is given by the formula

$$S(s) = \begin{bmatrix} 0 & 1 \end{bmatrix} \begin{bmatrix} \dfrac{s}{s+1} & \dfrac{-s}{s+1} \\ \dfrac{1}{s+1} & \dfrac{s}{s+1} \end{bmatrix} \begin{bmatrix} 1 & 0 \\ 0 & \dfrac{1}{s} \end{bmatrix} \begin{bmatrix} 1 \\ 0 \end{bmatrix} = \dfrac{1}{s+1} \tag{2.11}$$

The sensitivity of the composite system transfer function matrix is given by equation 2.12.

$$\dfrac{dS}{dr} = \begin{bmatrix} \dfrac{1}{s+1} & \dfrac{s}{s+1} \end{bmatrix} \dfrac{dZ}{dr} \begin{bmatrix} \dfrac{s}{s+1} \\ \dfrac{s}{s+1} \end{bmatrix} \tag{2.12}$$

Thus, to compute the sensitivity of the circuit transfer function with respect to the resistance value, we substitute

$$\frac{dZ}{dR} = \begin{bmatrix} -1 & 0 \\ 0 & 0 \end{bmatrix} \tag{2.13}$$

into 2.12 obtaining

$$\frac{dS}{dR} = \frac{-s}{(s+1)^2} \tag{2.14}$$

Similarly, to compute the circuit's sensitivity to its capacitance value, we use

$$\frac{dZ}{dC}\bigg|_{C=1} = \begin{bmatrix} 0 & 0 \\ 0 & \frac{-1}{C^2 s} \end{bmatrix}\bigg|_{C=1} = \begin{bmatrix} 0 & 0 \\ 0 & \frac{-1}{s} \end{bmatrix} \tag{2.15}$$

This implies that

$$\frac{dS}{dC} = \frac{-s}{(s+1)^2} \tag{2.16}$$

Finally, assuming that capacitance and resistance vary simultaneously with temperature as $C = \sqrt{T}$ and $R = T$ with the nominal value of temperature normalized to one we have

$$\frac{dZ}{dT} = \begin{bmatrix} -1 & 0 \\ 0 & \frac{-.5}{s} \end{bmatrix} \tag{2.17}$$

Substituting 2.17 into 2.12 yields

$$\frac{dS}{dT} = \frac{-3s}{s(s+1)^2} \tag{2.18}$$

as the sensitivity of the composite system transfer function with respect to temperature.

Equation 2.4 suffices for variations of S with respect to a single parameter, r. Note that the sensitivity matrix, dS/dr, is basically a matrix whose ij-th entry is the derivative of the ij-th entry of S with respect to r. However, suppose the object is to describe the sensitivities of the various entries in S relative to several simultaneous parameter variations. Equation 2.4 becomes intractable. Curing this difficulty requires a reintroduction of the *"vec" operation* defined earlier by III.3.15. Applying this transformation to S and invoking equation III.3.25 allows equation 2.4 to be written equivalently as

$$(2.19)$$

$$\frac{dvec(S)}{dr} = \left[I + L_{11}(I-ZL_{11})^{-1}Z \ L_{12} \right]^t \otimes \left[L_{21}(I-ZL_{11})^{-1} \right] \frac{dvec(Z)}{dr}$$

Note that $dvec(S)/dr$ is a column vector whose i-th entry is the derivative of the i-th entry in $vec(S)$. To study the sensitivity of the composite system transfer function with respect to several parameters simultaneously simply stack the column vectors $dvec(S)/dr_i$ for i=1,...,n, as defined in 2.19, side by side, to form the *Jacobian matrix* of $vec(S)$ with respect to the vector of parameters $r = col(r_i)$. In particular, let $dvec(Z)/dr$ denote the Jacobian matrix of $vec(Z)$ with respect to the vector of parameters, r. A little algebra results in the expected:

$$(2.20)$$

$$\frac{d(vec(S))}{dr} = \left[I + L_{11}(I-ZL_{11})^{-1}Z \ L_{12} \right]^t \otimes \left[L_{21}(I-ZL_{11})^{-1} \right] \frac{d(vec(Z))}{dr}$$

EXAMPLE 2.21: For this example let us reformulate the sensitivity analysis described in example 2.7. Writing the sensitivity matrix dS/dr of equation 2.12 in the guise of equation 2.20 yields

$$\frac{d(vec(S))}{dr} = \left[\frac{s}{(s+1)^2} \quad \frac{s^2}{(s+1)^2} \quad \frac{s}{(s+1)^2} \quad \frac{s^2}{(s+1)^2} \right] \frac{d(vec(Z).)}{dr} \quad (2.22)$$

Since S is a scalar transfer function, $vec(S)$ coincides with S. Thus, 2.22 is merely an alternate way to express 2.12. To compute the sensitivity matrix, dS/dr, with respect to the parameter vector, $col(R,C)$, we first calculate the component Jacobian matrix as per 2.23.

$$\frac{d(vec(Z))}{d(R,C)} = \begin{bmatrix} 1 & 0 \\ 0 & 0 \\ 0 & 0 \\ 0 & \frac{-1}{s} \end{bmatrix} \quad (2.23)$$

Substituting this equation into **2.22** and simplifying produces the composite system small change sensitivity Jacobian matrix.

$$\frac{d(vec(S))}{d(R,C)} = \begin{bmatrix} \dfrac{-s}{(s+1)^2} & \dfrac{-s}{(s+1)^2} \end{bmatrix} \tag{2.24}$$

Often one desires that the sensitivity matrix reflect a *per-unit normalization*. If S is a scalar transfer function matrix and r is a scalar parameter, the per-unit normalized sensitivity matrix takes the form

$$\frac{d(\ln(S))}{d(\ln(r))} = \frac{r}{S}\frac{dS}{dr} = \mathbf{S}_r^S \tag{2.25}$$

In the general case it is possible to formulate a corresponding per-unit normalized sensitivity matrix by multiplying the Jacobian matrix of equation 2.22 on the right and left by appropriate diagonal **matrices**. The entries of these diagonal matrices are the nominal values of S and r. The details are, however, more confusing than helpful.

3. LARGE CHANGE TRANSFER FUNCTION SENSITIVITY

Suppose a resistor in a circuit burns out, becomes open circuited. Suppose a capacitor shorts. What is the effect of such radical parameter changes on the circuit transfer function matrix? This question requires a structured answer since radical parameter changes frequently occur in real world systems. The *power line fault* and its effect on *system stability* is a common example. The analysis of such phenomena falls into the realm of *large change sensitivity analysis*. In this section we describe two distinct approaches to this topic.

The first is a *continuations approach*. Here the underlying hypothesis is that only one parameter undergoes a large variation, although the parameter may affect numerous components of the system. Temperature could be such a parameter. The thrust of this approach is to integrate a modified small change sensitivity formula. An alternate implementation via *sparse matrix techniques* is discussed next.

The second approach describes a means for computing the perturbed transfer function matrix when several parameters undergo radical change simultaneously. The key in this method is a clever use of *Householder's formula*.

To begin, we assume the system transfer function matrix (in addition to its frequency dependence) depends only on a single parameter, r. Suppressing the frequency variable, s, we have

$$S(r) = L_{22} + L_{21}Q(r)L_{12} \qquad (3.1)$$

where $Q(r)$ is the *internal system transfer function matrix*.

$$Q(r) = (I-Z(r)L_{11})^{-1}Z(r) \qquad (3.2)$$

Clearly, knowing the variation of Q with respect to r permits immediate computation of the variation of S relative to r. In order to develop a continuations algorithm it is necessary to find a functional expression for dQ/dr. Equation III.3.40 yields the desired expression:

$$\frac{dQ}{dr} = [(I-Z(r)L_{11})^{-1}]\frac{dZ}{dr}[I+L_{11}(I-Z(r)L_{11})^{-1}Z(r)] \qquad (3.3)$$

Equation 3.2 implies that

$$Q(r)Z(r)^{-1} = (I-Z(r)L_{11})^{-1} \qquad (3.4)$$

Simplifying 3.3 leads to the following differential equation.

$$\frac{dQ}{dr} = [Q(r)Z(r)^{-1}]\frac{dZ}{dr}[I+L_{11}Q(r)] \qquad (3.5)$$

The gist of our scheme, then, is to initialize this differential equation at $Q(r_0)$, the nominal value of Q, and then integrate over the extent of the r-variation. Thus we obtain an estimate of $Q(r_f)$ which produces an estimate of the perturbed transfer function matrix, $S(r_f)$.

Although seemingly roundabout, integration of 3.5 is a computationally viable technique. The only inverse required in its solution is that of the block diagonal composite component matrix, $Z(r)$. This is readily computable and is often known analytically. By hypothesis, dZ/dr is known analytically. Thus, integration of 3.5 is computationally straightforward.

EXAMPLE 3.6: Consider the RC voltage divider circuit of figure 2.6. Recall that this circuit was the subject of a small change sensitivity analysis sketched in the preceding section. In this example we fabricate a means for computing the perturbed transfer function matrix given a large variation in the capacitance, C. The nominal value of C is taken to be 1. Suppose it is known that the variation in C is $\Delta C=1$ resulting in a perturbed value of $C=2$.

Let us fix the complex frequency s=j. Now let us use a two step *Euler integration* scheme to estimate Q and thereby S at C=2 and s=j. Observe that equations 2.7 and 2.10 imply that

$$Q(C) \;=\; (I-Z(C)L_{11})^{-1}Z(C) \;=\; \frac{1}{s+1} \begin{bmatrix} Cs & -1 \\ 1 & 1 \end{bmatrix} \qquad (3.7)$$

Setting C=1 and s=j we have

$$Q(1) \;=\; \frac{1}{j+1} \begin{bmatrix} j & -1 \\ 1 & 1 \end{bmatrix} \qquad (3.8)$$

The symbolic equation for an Euler integration of 3.5 is

$$Q(C+ \triangle C) \approx Q(C) + \triangle C\; f(Q(C)) \qquad (3.9)$$

where

$$\frac{dQ}{dC} \;=\; Q(C)Z(C)^{-1} \frac{dZ}{dC} (I+L_{11}Q(C)) \triangleq f(Q(C)) \qquad (3.10)$$

Consequently, at s=j the estimate of Q(1.5) is

$$Q(1.5) \;=\; Q(1) + \triangle C\; f(Q(1))$$

$$= \begin{bmatrix} .75+j.5 & -.25+j.5 \\ .25-j.5 & .25-j.5 \end{bmatrix} \qquad (3.11)$$

Finally, the estimate of Q(2) is

$$Q(2) \;=\; Q(1.5) + \triangle C\; f(Q(1.5))$$

$$= \begin{bmatrix} .764+j.416 & -.104+j.416 \\ .104-j.416 & .104-j.416 \end{bmatrix} \qquad (3.12)$$

Taking into account the crudity of the Euler scheme, this compares favorably with the correct value of Q at C=2 and s=j, which is

$$Q(2) = \begin{bmatrix} .8+j.4 & -.2+j.4 \\ .2-j.4 & .2-j.4 \end{bmatrix} \tag{3.13}$$

The actual and estimated transfer functions respectively are

$$S(2) = .2-j.4 \tag{3.14}$$

and

$$S(2) = .104-j.416 \tag{3.15}$$

Of course, a more accurate estimate would have resulted by using a smaller Δ C and several *corrector iterations* after each *Euler predictor* step. For complete frequency response information it is necessary to do the procedure for all s=jω. Practically speaking,one would do the analysis only over the interesting bandwidth of the system.

For single-input single-output systems, a numerical implementation of the above theory means that one's goal is to construct magnitude and phase plots of the perturbed transfer function. For multi-input multi-output systems, one manufactures a *set* of magnitude and phase plots. Each plot describes the relationship between a particular output and a particular input.

Equation 3.5 requires *no non-trivial matrix inversions*. Thus, compared to a complete re-analysis of a system at a new parameter value, this technique for large change sensitivity analysis requires far fewer computations. Moreover, it supplies complete data on intermediate parameter values. This data is often as valuable as the desired information.

Although the above continuations approach is computationally efficient for large systems, the Q-matrix is generally non-sparse. Thus, since $Q=(I-ZL_{11})^{-1}Z$, it may be numerically preferable to work with the sparse inverses of the *LU-factors* of $(I-ZL_{11})$. A continuations algorithm exploiting such a sparse decomposition follows readily from theorem III.4.34. Specifically, suppose r_0 is the nominal parameter value and that

$$(I-Z(r_0)L_{11})^{-1} = U(r_0)^{-1}L(r_0)^{-1} \tag{3.16}$$

is known in its LU-factored form. Theorem III.4.34 implies that

$$\frac{dU^{-1}}{dr}(r) = U(r)^{-1} \,^{u}\!\left[L(r)^{-1} \frac{dZ}{dr}(r) \, L_{11}U(r)^{-1}\right]$$

$$\frac{dL^{-1}}{dr}(r) = \,^{\ell}\!\left[L(r)^{-1} \frac{dZ}{dr}(r) \, L_{11}U(r)^{-1}\right]L(r)^{-1}$$

(3.17)

where $^{u}[.]$ and $^{\ell}[.]$ denote the operations of picking out the strictly upper and the usual lower triangular parts respectively. Recall that if one integrates the coupled differential equations of 3.17 with the initial conditions given by 3.16, then the resultant product $U(r)^{-1}L(r)^{-1}$ will specify

$$(I-Z(r)L_{11})^{-1} = U(r)^{-1}L(r)^{-1}$$

(3.18)

Clearly $S(r)$ is given as

$$S(r) = L_{22} + L_{21}U(r)^{-1}L(r)^{-1}Z(r)L_{12}$$

(3.19)

These notions combine to give us another avenue for computing perturbed transfer function matrices due to a single radical parameter change. Note that all matrices in 3.19 are sparse.

EXAMPLE 3.20: Consider the negative feedback system shown in figure 3.21.

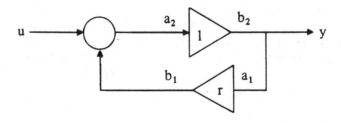

Figure 3.21. Negative feedback system.

with $r \neq -1$. The component connection equations for this system are given by

$$\begin{bmatrix} b_1 \\ b_2 \end{bmatrix} = \begin{bmatrix} r & 0 \\ 0 & 1 \end{bmatrix} \begin{bmatrix} a_1 \\ a_2 \end{bmatrix} \tag{3.22}$$

and

$$\begin{bmatrix} a_1 \\ a_2 \\ \hline y \end{bmatrix} = \begin{bmatrix} 0 & 1 & \vline & 0 \\ -1 & 0 & \vline & 1 \\ \hline 0 & 1 & \vline & 0 \end{bmatrix} \begin{bmatrix} b_1 \\ b_2 \\ \hline u \end{bmatrix} \tag{3.23}$$

Moreover

$$(I-Z(r)L_{11}) = \begin{bmatrix} 1 & -r \\ 1 & 1 \end{bmatrix} \tag{3.24}$$

This is the same matrix whose inverse was computed by a continuations approach in example III.4.42 using an initial value of $r_0=0$. Thus we can use that example to illustrate the computation of an estimate of $S(.1)$, the perturbed transfer function. In other words, suppose r is perturbed from 0.0 to 0.1, then compute $S(.1)$. Using an Euler integration scheme, example III.4.42 showed that

$$U(.1)^{-1} \approx \begin{bmatrix} 1 & -.1 \\ 0 & 1 \end{bmatrix} \tag{3.25}$$

and

$$L(.1)^{-1} \approx \begin{bmatrix} 1 & 0 \\ .9 & .9 \end{bmatrix} \tag{3.26}$$

yielding

$$S(.1) \ = \ [0] + [1 \ \ 0]\begin{bmatrix} 1 & -.1 \\ 0 & 1 \end{bmatrix}\begin{bmatrix} 1 & 0 \\ .9 & .9 \end{bmatrix}\begin{bmatrix} 1 & 0 \\ 0 & .1 \end{bmatrix}\begin{bmatrix} 1 \\ 0 \end{bmatrix} = [.91] \quad (3.27)$$

as the estimate of $S(.1)$. This compares favorably with the exact solution of $10/11 = .9090909$.

Recall the above approaches can handle only single parameter variations though the parameter may affect several components simultaneously. Fortunately, computational requirements are independent of the number of components affected by the underlying parameter. On the other hand, if several independent parameters undergo simultaneous large variations, the algorithm breaks down. Any large change sensitivity analysis scheme grappling with this problem will necessitate some matrix inversions. Application of the *Householder formula* will keep such numerics and difficulties to a minimum.

Consider the following general frequency domain system description.

$$b \ = \ Za \qquad\qquad (3.28)$$

$$\begin{aligned} a \ &= \ L_{11}b + L_{12}u \\ y \ &= \ L_{21}b + L_{22}u \end{aligned} \qquad\qquad (3.29)$$

and

$$S \ = \ L_{22} + L_{21}(I-ZL_{11})^{-1}ZL_{12} \qquad\qquad (3.30)$$

Assume one can model the perturbed *composite component transfer function matrix* as the sum of the nominal transfer function matrix and a "perturbation" matrix as

$$Z_p \ = \ Z + \begin{bmatrix} Z_m & 0 \\ 0 & 0 \end{bmatrix} \qquad\qquad (3.31)$$

where Z_m is an mxm perturbation. The size of Z_m is limited only by the size of Z. The assumption that the perturbation occurs in the upper left hand corner of Z is a notational convenience only. Observe that

$$(I-Z_pL_{11}) \quad = \quad (I-ZL_{11}) + \begin{bmatrix} -1 \\ 0 \end{bmatrix} Z_m \quad L_{11}^m \qquad (3.32)$$

where L_{11}^m denotes the upper m rows of the L_{11} matrix. Equation 3.32 is in a form amenable to application of Householder's formula to compute $(I-Z_pL_{11})^{-1}$. In particular,

$$(D+REC)^{-1} \quad = \quad [I + D^{-1}RE(I+CD^{-1}RE)^{-1}C] \, D^{-1} \quad (3.33)$$

(as per III.6.15) where $D = (I-ZL_{11})$, $R = col(-I,0)$, $E = Z_m$, and $C = L_{11}^m$. The explicit formula is

$$(I-Z_pL_{11})^{-1} \quad = $$

$$\left[I - (I-ZL_{11})^{-1} \begin{bmatrix} Z_m \\ 0 \end{bmatrix} \left(I - L_{11}^m (I-ZL_{11})^{-1} \begin{bmatrix} Z_m \\ 0 \end{bmatrix} \right)^{-1} L_{11}^m \right] (I-ZL_{11})^{-1} \quad (3.34)$$

whenever $(I-ZL_{11})^{-1}$ exists. Once $(I-ZL_{11})^{-1}$ is known, $(I-Z_pL_{11})^{-1}$ is computed simply by inverting an mxm matrix. Thus if only an mxm portion of the *composite component transfer function matrix* is perturbed, a large change sensitivity analysis would proceed by a straightforward application of equation 3.34. The procedure is both accurate and numerically clean. The perturbed transfer function matrix is given by

$$S_p \quad = \quad L_{22} + (I-Z_pL_{11})^{-1}Z_pL_{12} \qquad (3.35)$$

as expected. Of course, if only a single component is perturbed the analysis requires only a scalar inversion. If the goal is to compute the composite system transfer function matrix for an entire array of perturbed parameter values, significant computational savings can be achieved by a clever reordering of the parameters in the array. The intent is to reorder so that only one parameter varies per computation. Such a *systematic exploration* may permit an entire analysis without any matrix inversions. Calculating S_p for a fixed array of perturbed parameters occurs quite often in system simulation, as in, for instance, a *Monte Carlo analysis* of component tolerances.[6] Thus, it behooves the engineer to find efficient algorithms for large change sensitivity analysis.

EXAMPLE 3.36: Again consider the feedback system of example 3.20. Taking r=0 as the nominal value, we have

$$Z = \begin{bmatrix} 0 & 0 \\ 0 & 1 \end{bmatrix} \tag{3.37}$$

Suppose the perturbation is $\Delta r=.1$, then

$$Z_p = Z + \begin{bmatrix} .1 & 0 \\ 0 & 0 \end{bmatrix} \tag{3.38}$$

Thus

$$(I-Z_p L_{11}) = (I-ZL_{11}) + \begin{bmatrix} -1 \\ 0 \end{bmatrix} [.1] \begin{bmatrix} 0 & 1 \end{bmatrix} \tag{3.39}$$

Inverting we have

$$(I-ZL_{11})^{-1} = \begin{bmatrix} 1 & 0 \\ 1 & 1 \end{bmatrix}^{-1} = \begin{bmatrix} 1 & 0 \\ -1 & 1 \end{bmatrix} \tag{3.40}$$

The overall task is to compute $(I-Z_p L_{11})^{-1}$ via Householder's formula. Since the perturbation on Z is of rank one, it suffices to perform only a scalar inversion.

$$\left([1] - \begin{bmatrix} 0 & 1 \end{bmatrix} \begin{bmatrix} 1 & 0 \\ -1 & 1 \end{bmatrix} \begin{bmatrix} .1 \\ 0 \end{bmatrix} \right)^{-1} = [1.1]^{-1} = 1/1.1 \tag{3.41}$$

Substituting this into equation 3.34 yields

$$(I-Z_p L_{11})^{-1} = \frac{1}{1.1} \begin{bmatrix} 1 & .1 \\ -1 & 1 \end{bmatrix} \tag{3.42}$$

Finally calculating the perturbed transfer function matrix, $S(.1)$, we have

$$S(.1) = \frac{.1}{1.1} = \frac{10}{11} \tag{3.43}$$

This solution is exact as opposed to the estimates obtained from the continuations approaches. This is because Householder's formula yields an exact solution no matter how large the perturbation.

4. PERFORMANCE MEASURE SENSITIVITY

Nonlinear systems have no well-defined frequency response. Moreover, from a time domain perspective, given an input, a (t), and a component described by a nonlinear state model, there is, in general, no closed form expression for the output, b(t). Such theoretical barricades preclude the use of traditional forms of sensitivity analysis. How then does one even talk about the sensitivity of a nonlinear system relative to some parameter? As motivation, let us scrutinize the relationship between the time and frequency domain representations of a system.

Intrinsically, the *Laplace transform* is a prism which picks out the frequency dependent behavior of a linear system. From an abstract viewpoint, the Laplace transform is an *integral performance measure*. The *quadratic performance measure* used in *optimal control theory* is another integral transform, or window, picking out energy dependent behavior of a control system. Intuitively, then, an integral performance measure is a means for extracting one or more aspects of the behavior of a system. The sensitivity analysis discussed in this section revolves around the use of an integral performance measure which characterizes the important behavior of the system. In point, we focus on the sensitivity of a given performance measure with respect to a parameter variation. Fortunately, the central calculations needed for the analysis duplicate those needed for the *sparse tableau simulation*.

The form of the integral performance measure will be:

$$P = \int_0^T h(x,a,b,r)dt \tag{4.1}$$

where h is termed the *objective function*. The objective function is a smooth nonlinear function of the composite system state vector, the composite component input and output vectors, and a vector of possibly variable parameters, denoted as r. Physically speaking, h characterizes some pertinent system attribute. Theoretically, the smoothness assumption on h is somewhat restrictive. However, it is surely general enough for the study of a wide spectrum of system attributes such as frequency sensitivity, transient behavior, potential instability, etc. Of course, the *measure of sensitivity* will be the derivative of P with respect to the parameter vector – i.e., dP/dr.

The system model we will be using is a *nonlinear state model* as per equations 4.2 and 4.3.

$$
\begin{aligned}
\dot{x} &= f(x,a,r) \\
b &= g(x,a,r)
\end{aligned}
\tag{4.2}
$$

$$
\begin{aligned}
a &= L_{11}b + L_{12}u \\
y &= L_{21}b + L_{22}u
\end{aligned}
\tag{4.3}
$$

We assume the input, u, and the initial state, $x(0)$, have been fixed a-priori.

Recall from section IV.2 that a sparse tableau simulation of such a system begins by first *discretizing the derivative* of x, as follows: the derivative at the k-th time instant has the approximation:

$$
\dot{x}_k \approx \sum_{i=0}^{m} d_i x_{k-i}
\tag{4.4}
$$

where the subscript denotes the particular time instant. In other words, the derivative of the state at the k-th time instant is a weighted sum of the states at the present and past time instants. The simulation proceeds by solving the sparse tableau equation.

$$
T_r(x_k, a_k, b_k) \triangleq \begin{bmatrix} f(x_k, a_k, r) - d_0 x_k \\ g(x_k, a_k, r) - b_k \\ -a_k + L_{11} b_k \end{bmatrix}
$$

$$
= \begin{bmatrix} \sum_{i=1}^{m} d_i x_{k-i} \\ 0 \\ -L_{12} u_k \end{bmatrix} \triangleq v
\tag{4.5}
$$

The potentially variable parameter vector, r, is independent of time. Basically one picks an initial guess, $\mathrm{col}(x_k^0, a_k^0, b_k^0)$, and updates the guess via the formula

$$
\begin{bmatrix} x_k^{i+1} \\ a_k^{i+1} \\ b_k^{i+1} \end{bmatrix} = \begin{bmatrix} x_k^{i} \\ a_k^{i} \\ b_k^{i} \end{bmatrix} - J_{T_{r_i}}^{-1} [T_{r_i} - v] \tag{4.6}
$$

where

$$
J_{T_r} = \begin{bmatrix} \dfrac{df}{dx} - d_0I & \dfrac{df}{da} & 0 \\ \hline \dfrac{dg}{dx} & \dfrac{dg}{da} & -I \\ \hline 0 & -I & L_{11} \end{bmatrix} \tag{4.7}
$$

With this background completed, we may formally state the *nonlinear sensitivity problem* in the following way:[4,5]
Compute dP/dr under the constraints

$$
P = \int_0^T h(x,a,b,r,)dt \tag{4.8}
$$

$$
\dot{x} = f(x,a,r) \tag{4.9}
$$
$$
b = g(x,a,r)
$$

$$
a = L_{11}b + L_{12}u \tag{4.10}
$$
$$
y = L_{21}b + L_{22}u
$$

and

$$
\frac{dr}{dt} = 0 \tag{4.11}
$$

The overall goal is more difficult than merely finding an expression for dP/dr. The sensitivity dP/dr must be numerically computable so that no significant additional computations are required over and above those required for a system simulation. The first step, in a rather elegant solution to this problem, is to replace the objective function h by a *Lagrangian*, L, as follows.

$$L(x,a,b,r,\lambda,\theta,\phi,\gamma) = h(x,a,b,r) + \lambda^t[\dot{x} - f(x,a,r)]$$
$$+ \theta^t[b - g(x,a,r)] + \phi^t[a - L_{11}b + L_{12}u] - \gamma^t[\dot{r}] \qquad (4.12)$$

where λ,θ,ϕ, and γ are termed *adjoint variables*. They are in general vector valued functions of time. Observe that all *multiplier terms* in the Lagrangian are zero if the problem constraints, 4.8 through 4.11, are satisfied. But any accurate simulation guarantees the validity of 4.8 through 4.10. Moreover, the physical nature of the problem forces 4.11 to be satisfied. Thus, under the given constraints, the Lagrangian reduces to the original objective function, h. Secondly, this fact permits us to choose the adjoint variables "arbitrarily." The advantage of working with the Lagrangian is that λ, θ, ϕ, and γ may be judiciously chosen so as to simplify the expression for dP/dr.

Rewriting the performance measure produces

$$P = \int_0^T L(x,a,b,r,\lambda,\theta,\phi,\gamma)dt \qquad (4.13)$$

Using equation 4.13, let us compute the variation of P with respect to the vectors z and $\dot{z}$ where $z = col(x,a,b,r\lambda,\theta,\phi,y)$ and $\dot{z} = col(\dot{x},\dot{a},\dot{b},\dot{r},\dot{\lambda},\dot{\theta},\dot{\phi},\dot{\gamma})$.

LEMMA 4.14: The variation in P, δP, with respect to z and $\dot{z}$ is given by:

$$\delta P = \int_0^T \left[\frac{dL}{dz} - \frac{d}{dt}\frac{dL}{d\dot{z}}\right](\delta z)dt + \frac{dL}{d\dot{z}}(\delta z)\Big|_0^T \qquad (4.15)$$

PROOF: By definition, the desired variation is given by:

$$\delta P = \int_0^T \left[\frac{dL}{dz}(\delta z) + \frac{dL}{d\dot{z}}(\delta\dot{z})\right]dt \qquad (4.16)$$

Distributing the integration over the integrand results in 4.17.

$$\delta P = \int_0^T \frac{dL}{dz}(\delta z)dt + \int_0^T \frac{dL}{d\dot{z}}(\delta\dot{z})dt \qquad (4.17)$$

Integrating the second integral by parts produces:

$$\int_0^T \frac{dL}{d\dot{z}} [(\delta\dot{z})dt] = \frac{dL}{d\dot{z}} (\delta z)\Big|_0^T - \int_0^T \frac{d}{dt} \frac{dL}{d\dot{z}} (\delta z)dt \qquad (4.18)$$

The proof now follows by substitution of 4.18 into 4.17.

Since it is permissible to choose the adjoint variables "arbitrarily," let us pick them so that

$$\lambda(T) = 0 \qquad (4.19)$$

and the *Euler equation*

$$\left[\frac{dL}{dz} - \frac{d}{dt} \left(\frac{dL}{d\dot{z}} \right) \right] \equiv 0 \qquad (4.20)$$

is satisfied. Momentarily delaying a discussion of how to so pick the adjoint variables, observe that if the Euler equation is zero as per 4.20, then the variation in P becomes:

$$\delta P = \frac{dL}{d\dot{z}} (\delta z)\Big|_0^T \qquad (4.21)$$

The following **lemma** proffers a specific formula for computing the sensitivity, dP/dr.

LEMMA 4.22: Under the problem constraints and the satisfaction of 4.19 and 4.20, we have the following expression for the desired sensitivity:

$$\frac{dP}{dr} = \lim_{\delta r \to 0} \left[\frac{\delta P}{\delta r} \right] = \int_0^T \dot{\gamma}(t)dt \qquad (4.23)$$

PROOF: The crux of the proof is simply a straightforward evaluation of equation 4.21. Differentiating the Lagrangian with respect to $\dot{z}$ yields

$$\frac{dL}{d\dot{z}} = [\lambda'(t), 0,0,\gamma'(t),0,0,0,0] \qquad (4.24)$$

Therefore:

$$\frac{dL}{d\dot{z}}(\delta z) = \lambda'(t) (\delta x) + \gamma'(t) (\delta r) \qquad (4.25)$$

It now follows that:

$$\delta P \;=\; \frac{dL}{d\dot{z}}\,(\delta z)\bigg|_0^T \;=\; \lambda^t(T)\,[\delta x(T)] + \gamma^t(T)\,[\delta r(T)]$$

$$- \lambda^t(0)\,[\delta x(0)] - \gamma^t(0)\,[\delta r(0)] \qquad (4.26)$$

By assumption, $\lambda(T) = 0$ and the state x (0) is fixed, implying that $\delta x(0) = 0$. Furthermore, the assumption that $dr/dt = 0$ implies that

$$\delta r(T) \;=\; \delta r(0) \;\triangleq\; \delta r \qquad (4.27)$$

Consequently, equation 4.26 reduces to:

$$\delta P \;=\; [\gamma^t(T) - \gamma^t(0)]\,(\delta r) \qquad (4.28)$$

Obviously then:

$$\frac{dP}{dr} \;=\; \lim_{\delta r \to 0}\left[\frac{\delta P}{\delta r}\right] = \int_0^T \dot{\gamma}^t(t)dt \qquad (4.29)$$

The proof is now complete.

This **lemma** simplifies the task of computing dP/dr. The keys to evaluating the integral in equation 4.23 or equation 4.29 are a set of eight constraints which force the Euler equation to be zero. These constraints provide an expression for γ in terms of λ, θ, and ϕ, peeling away another layer of complexity. Moreover, they will supply the numerical key to the solution of our problem.

Expanding the Euler equation, 4.20, into its component parts, we obtain:

$$0 \;=\; \frac{dL}{dx} - \frac{d}{dt}\frac{dL}{d\dot{x}} \;=\; \frac{dh}{dx} - \lambda^t\frac{df}{dx} - \theta^t\frac{dg}{dx} - \frac{d\lambda^t}{dt} \qquad (4.30)$$

$$0 \;=\; \frac{dL}{da} - \frac{d}{dt}\frac{dL}{d\dot{a}} \;=\; \frac{dh}{da} - \lambda^t\frac{df}{da} - \theta^t\frac{dg}{da} + \phi^t \qquad (4.31)$$

$$0 \;=\; \frac{dL}{db} - \frac{d}{dt}\frac{dL}{d\dot{b}} \;=\; \frac{dh}{db} + \theta^t - \phi^t\,L_{11} \qquad (4.32)$$

$$0 = \frac{dL}{dr} - \frac{d}{dt}\frac{dL}{dr} = \frac{dh}{dr} - \lambda^{\iota}\frac{df}{dr} - \theta^{\iota}\frac{dg}{dr} - \frac{d\gamma^{\iota}}{dt} \qquad (4.33)$$

$$0 = \frac{dL}{d\lambda} - \frac{d}{dt}\frac{dL}{d\lambda} = \dot{x} - f(x,a,r,) \qquad (4.34)$$

$$0 = \frac{dL}{d\theta} - \frac{d}{dt}\frac{dL}{d\dot{\theta}} = b - g(x,a,r) \qquad (4.35)$$

$$0 = \frac{dL}{d\phi} - \frac{d}{dt}\frac{dL}{d\dot{\phi}} = a - L_{11}b - L_{12}u \qquad (4.36)$$

$$0 = \frac{dL}{d\gamma} - \frac{d}{dt}\frac{dL}{d\dot{\gamma}} = \dot{r} \qquad (4.37)$$

Equations 4.34 through 4.37 are the original problem constraint conditions which also define the system. Their satisfaction is assured via a sparse tableau simulation. Rearranging equation 4.33 implies

$$\frac{d\gamma^{\iota}}{dt} = \frac{dh}{dr} - \lambda^{\iota}\frac{df}{dr} - \theta^{\iota}\frac{ag}{dr} \qquad (4.38)$$

Substituting this expression into equation 4.29 yields:

$$\frac{dP}{dr} = \int_0^T \dot{\gamma}^{\iota}(t)dt = \int_0^T [\frac{dh}{dr} - \lambda^{\iota}\frac{df}{dr} - \theta^{\iota}\frac{dg}{dr}]\,dt \qquad (4.39)$$

This equation indicates that the sensitivity measure, dP/dr depends on λ and θ which, in turn, depend on ϕ. Equations 4.30 to 4.32 specify this interdependence. Thus, the sensitivity problem dwindles to first solving these coupled equations simultaneously for λ, θ, and ϕ. Then one integrates equation 4.38 as per 4.39. Note that dh/dr, df/dr, and dg/dr are quantities known a priori. In particular, the functions f and g are composite component representations. Thus, their change relative to r is either known or easily computed. The notion is directly akin to the use of dZ/dr in the previous discussion of transfer function sensitivity.

To numerically implement these notions, first discretize the equations; then *approximate the derivative* of λ at the k-th time instant as:

$$\dot{\lambda}_k \approx \sum_{i=0}^{m} - d_i\,\lambda_{k+i} \qquad (4.40)$$

This approximation runs backwards in time since $\lambda(T) = 0$ is a final value condition. To compute $\lambda, \theta,$ and ϕ, it is necessary to solve the simultaneous set of linear equations.

$$
[\lambda_k^t \; \theta_k^t \; \phi_k^t]
\begin{bmatrix}
\dfrac{df}{dx} - d_0 I & \dfrac{df}{da} & 0 \\
\dfrac{dg}{dx} & \dfrac{dg}{da} & -I \\
0 & -I & L_{11}
\end{bmatrix}
$$

$$
\underset{=}{\triangle} \; [\lambda_k^t \; \theta_k^t \; \phi_k^t] \, J_{T_r} =
\begin{bmatrix}
\displaystyle\sum_{i=1}^{m} d_i \, \lambda_{k+i}^t + \dfrac{dh}{dx} & \dfrac{dh}{da} & \dfrac{dh}{db}
\end{bmatrix}
\tag{4.41}
$$

Now simulate the system by solving the tableau equation as per 4.6. Then solve 4.41 for the adjoint variables with the matrix J_{T_r} evaluated along the solution trajectories computed in the sparse tableau simulation. Finally after calculating the values of the discretized adjoint variables, one calls upon a standard numerical integration routine to evaluate the sensitivity as per 4.39.

Observe that the inverse needed to solve 4.41 is the inverse of the *Jacobian matrix* of the tableau equations. Solving the tableau equation already requires the inversion. Thus, no real additional matrix inversion over and above that for the simulation procedure is necessary for computation of the performance measure sensitivity. However, since the sensitivity is a final value problem, the computation must iterate *backwards in time*. Thus, some minor storage requirements will accrue.

In the linear time invariant case, the Jacobian is independent of time so only a single inversion occurs. For nonlinear systems with a small number of nonlinear components, *Householder's formula* provides an efficient means for computing J_T^{-1}.

This kind of overlapping calculation has the advantage of making procedures such as parameter optimization and system diagnostics highly efficient. Intrinsic to such procedures is the obvious need for system simulation and the "minimization" of performance measure sensitivities. Everything then appears to coalesce into a tightly woven package.

EXAMPLE 4.42: Suppose the nominal value of the parameter, r, in the linear feedback system of figure 4.43 is one. Example IV.2.12 gave a

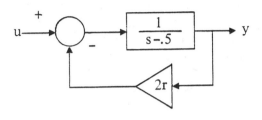

Figure 4.43. Feedback system with potentially variable parameter, r.

simulation of this system with $u(t) = t$ over the time interval $[0,2]$. The time step size was one; the derivative of the state vector was approximated by a first order Euler formula, resulting in the crude estimated solution of:

$$\begin{bmatrix} x \\ a_1 \\ a_2 \\ b_1 \\ b_2 \end{bmatrix}_{t=0} = \begin{bmatrix} 0 \\ 0 \\ 0 \\ 0 \\ 0 \end{bmatrix} \; ; \begin{bmatrix} x \\ a_1 \\ a_2 \\ b_1 \\ b_2 \end{bmatrix}_{t=1} = \begin{bmatrix} .4 \\ .2 \\ .4 \\ .4 \\ .8 \end{bmatrix} \; ; \begin{bmatrix} x \\ a_1 \\ a_2 \\ b_1 \\ b_2 \end{bmatrix}_{t=2} = \begin{bmatrix} .88 \\ .24 \\ .88 \\ .88 \\ 1.76 \end{bmatrix} \tag{4.44}$$

The inverse of the tableau matrix at the nominal value of $r=1$ is

$$T^{-1} = \begin{bmatrix} 1 & 2 & 0 & 1 & -2 \\ 0 & 1 & 0 & .5 & -1 \\ 0 & 0 & 1 & 0 & .5 \\ 0 & 0 & 0 & 1 & -2 \\ 0 & 0 & 0 & 0 & 1 \end{bmatrix} \begin{bmatrix} -2 & 0 & 0 & 0 & 0 \\ 1 & .5 & 0 & 0 & 0 \\ 0 & 0 & .5 & 0 & 0 \\ -2 & -1 & 0 & -2 & 0 \\ -.8 & -.4 & -.2 & -.8 & -.4 \end{bmatrix} \tag{4.45}$$

The goal is to use this data to compute dP/dr where

$$P = \int_0^2 b_1(t)dt \tag{4.46}$$

First compute the adjoint vectors using T^{-1} in lieu of the inverse Jacobian. Observe that

$$\left[\sum_{i=1}^{m} d_i\, \lambda_{k+2}^t + \frac{dh}{dx} \;\middle|\; \frac{dh}{da} \;\middle|\; \frac{dh}{db}\right] = \left[-\lambda_{k+1}^t \;\middle|\; 0\ 0 \;\middle|\; 1\ 0\right] \quad (4.47)$$

Assuming that all adjoint variables are zero at t=2, we may successively calculate the variables obtaining

$$\begin{bmatrix} \lambda_1 \\ \theta_1 \\ \theta_2 \\ \phi_1 \\ \phi_2 \end{bmatrix}_{t=2} = \begin{bmatrix} 0 \\ 0 \\ 0 \\ 0 \\ 0 \end{bmatrix}\ ;\ \begin{bmatrix} \lambda \\ \theta_1 \\ \theta_2 \\ \phi_1 \\ \phi_2 \end{bmatrix}_{t=1} = \begin{bmatrix} -.4 \\ -.2 \\ .4 \\ -.4 \\ .8 \end{bmatrix}\ ;\ \begin{bmatrix} \lambda \\ \theta_1 \\ \theta_2 \\ \phi_1 \\ \phi_2 \end{bmatrix}_{t=0} = \begin{bmatrix} -.504 \\ -.12 \\ -5.6 \\ -5.6 \\ 1.12 \end{bmatrix} \quad (4.48)$$

Plugging into equation 4.39, we have

$$\frac{dP}{dr} = \int_0^T \left[\frac{dh}{dr} - \lambda^t \frac{df}{dr} - \theta^t \frac{dg}{dr}\right] dt = \int_0^T - [\theta_1\ \theta_2] \begin{bmatrix} 0 \\ 2a \end{bmatrix} dt \quad (4.49)$$

Using *Simpson's rule* to carry out the integration results in

$$\frac{dP}{dr} = \int_0^2 - 2\theta_2 a_2 dt = \frac{1}{3} [0 + 4(-.32) + 0] = -.43 \quad (4.50)$$

This number is the numerically calculated sensitivity measure. Let us compare this with the exact number. First, observe that for the given input

$$b_1(t) = \left[\frac{e^{-\beta t}}{\beta^2} + \frac{t}{\beta} - \frac{1}{\beta^2}\right] U(t) \quad (4.51)$$

where $\beta = (2r - .5)$. Therefore

$$P = \frac{e^{-2\beta}-1}{-\beta^3} + \frac{2}{\beta} - \frac{2}{\beta^2} \quad (4.52)$$

Taking the derivative with respect to r yields

$$\frac{dP}{dr} = \frac{4e^{-2\beta}}{\beta^3} + \frac{6[e^{-2\beta}-1]}{\beta^4} - \frac{4}{\beta^2} + \frac{8}{\beta^3} \tag{4.53}$$

Evaluating this expression at the nominal value of r = 1 results in

$$\frac{dP}{dr} = -.47 \tag{4.54}$$

Despite our crude approximation both in the system simulation and the integration for P, we obtained good correspondence.

EXAMPLE 4.55: As a final example, let us compute a performance measure sensitivity for the nonlinear circuit simulated in example IV.2.33. The circuit diagram is redrawn below. Recall that the

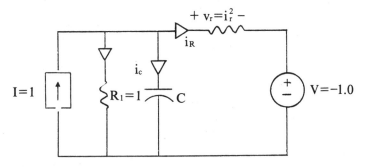

Figure 4.56. Nonlinear RC network with $V_c(0) = 1$.

solution vectors for t = 0,1, and 2 respectively were given by:

$$\begin{bmatrix} v_c \\ i_c \\ v_R \\ v_c \\ i_R \end{bmatrix} = \begin{bmatrix} 1.0 \\ \sqrt{2} \\ 2.0 \\ 1.0 \\ \sqrt{2} \end{bmatrix}_{t=0} ; \begin{bmatrix} 0.4069 \\ -.5931 \\ 1.4069 \\ 0.4069 \\ 1.1860 \end{bmatrix}_{t=1} ; \begin{bmatrix} 0.1637 \\ -.2432 \\ 1.1652 \\ 0.1637 \\ 1.0795 \end{bmatrix}_{t=2} \tag{4.57}$$

Specifically, let us compute the sensitivity of the performance measure

$$P = \int_0^2 [v_c(t)]^3 \, dt \tag{4.58}$$

with respect to the parameter C. The nominal value of C is taken as one.

The first step is to compute the adjoint variables via the following formula given that the objective function $h = v_c^3$ and that $\lambda \approx -\lambda_k + \lambda_{k+1}$.

$$[\lambda_k{}^t \ \theta_k{}^t \ \phi_k{}^t] \ = \ [-\lambda_{k+1}{}^t + \frac{dh}{dx} \Big| \frac{dh}{da} \Big| \frac{dh}{db}] \ J_T{}^{-1}(x_k, a_k, b_k) \tag{4.59}$$

Clearly we have that:

$$\frac{dh}{dx} \ = \ 3v_c{}^2$$

$$\frac{dh}{da} \ = \ [0 \ \ 0] \tag{4.60}$$

$$\frac{dh}{db} \ = \ [0 \ \ 0]$$

Moreover, by assumption we have all adjoint variables equal to zero at $t = 2$. To compute the adjoint variables at $t = 1$, we must solve

$$[\lambda \ \ \theta_1 \ \theta_2 \ \ \phi_1 \ \phi_2] \ = \ [.4967 \ \ 0 \ 0 \ \ 0 \ 0] \ J_T{}^{-1} \tag{4.61}$$

where $J_T{}^{-1}$ can be computed from the formulas given in IV.2.52 and IV.2.53. As such, we have that

$$[\lambda \ \ \theta_1 \ \theta_2 \ \ \phi_1 \ \phi_2]_{t=1} \ = \ [-.205 \ \ .2899 \ .205 \ \ -.205 \ .0864] \tag{4.62}$$

Similarly,

$$[\lambda \ \ \theta_1 \ \theta_2 \ \ \phi_1 \ \phi_2]_{t=0} \ = \ [-1.362 \ \ 1.843 \ 1.362 \ \ -1.362 \ .4808] \tag{4.63}$$

To compute the sensitivity, we must evaluate

$$\frac{dp}{dr} \ = \ \int_0^2 [\frac{dh}{dr} - \lambda^t \frac{df}{dr} - \theta^t \frac{dg}{dr}] \ dt \tag{4.64}$$

For this circuit we may point out the following relationships:

$$\frac{dh}{dC} \ = \ \frac{dg}{dC} \ = \ 0 \tag{4.65}$$

and

$$f(x,a) = (\frac{1}{C}) i_c \tag{4.66}$$

which implies that

$$\frac{df}{dC} = -\frac{1}{C^2} i_c \tag{4.67}$$

Evaluating 4.66 at the nominal value of C=1, the sensitivity measure simplifies to

$$\frac{dP}{dC} = \int_0^T [-\lambda^t \frac{df}{dC}] dt \tag{4.68}$$

The pertinent numerical values are given in table 4.69 below.

time	$\frac{df}{dC}$
0	2
1	.592
2	.243

Table 4.69. Table of values

Thus, using Simpson's rule to evaluate 4.68, we obtain:

$$\frac{dP}{dC} = \frac{1}{3} [2 (1.362) + 4(.592)(.205) + 0] = 1.6098 \tag{4.70}$$

Similarly, one may compute the sensitivity of P with respect to R whose nominal value is taken to be one also. As such:

$$\frac{dP}{dR} = \int_0^2 -\theta^t [\frac{dg}{dR}] dt \tag{4.71}$$

Using Simpson's rule to evaluate this integral, we have:

$$\frac{dP}{dR} = \frac{1}{3} [1.843 + 4(.29)(.407) + 0] = .7717 \tag{4.72}$$

This essentially completes our discussion of performance measure sensitivity. One area needing intensive study is the fabrication of a criterion to choose suitable performance measures for the modeling of interesting system characteristics. The performance measures in the above examples were arbitrarily chosen without regard to physical meaning. Within this context, it is then necessary to properly interpret the sensitivity measure, dP/dr.

5.   COMPUTER-AIDED DESIGN

Often the system design process is heuristic. During the past half century, there have been numerous attempts to automate the design process. There now exist specialized design algorithms (computerized) for *telephone filters*, *logic networks*, etc. The purpose of this section is to formulate a widely applicable class of *parameter optimization* algorithms. As usual, we assume the system is represented by the component connection model.

In the linear case we assume a frequency dependent *composite component model* of the form

$$b \;=\; Z(s,r)a \tag{5.1}$$

where r is a vector of variable parameters. For the nonlinear case we assume a *nonlinear state model* for the components.

$$\dot{x} \;=\; f(x,a,r)$$
$$b \;=\; g(x,a,r) \tag{5.2}$$

In both cases the connection equations remain

$$a \;=\; L_{11}b + L_{12}u$$
$$y \;=\; L_{21}b + L_{22}u \tag{5.3}$$

The objective is to choose r so that the system behaves optimally. Of course, an optimal design may mean anything from a least cost strategy to a performance spec-fastest switching time, to the approximation of some idealized system behavior.

Parameter optimization schemes depend on the choice of performance measure which mathematically defines the *designer's notion of optimal*. Several classes of performance measures commonly encountered in system design are worthy of special attention.

Integral performance measures of the form

$$P = \int_0^T h(x,a,b,r)dt \qquad (5.4)$$

as discussed in the previous section are important in CAD problems. For instance, suppose it is necessary to minimize the *switching time* of a circuit whose output signal $x(t)$ changes from 0 to 5 volts. A useful integral performance measure is

$$P = \int_0^T [\, |x(t,r)-2.5| - 2.5\,]\, dt \qquad (5.5)$$

This measure penalizes the circuit when $x(t,r)$ is between 0 and 5 but gives no penalty when $x(t,r)$ takes on either extremal value. Clearly there are many performance measures characterizing circuit switching time. Some use *time constants*, slopes, etc. However, the one in 5.5 belongs to the class studied in section 4 and is thus amenable to the analytic techniques studied there.

For linear systems it is frequently convenient to formulate a *frequency dependent performance measure*. For example, suppose the task is to choose a parameter vector, r, so that the composite system transfer function matrix approximates some prespecified $T(s)$. One may choose to minimize the norm (with respect to r) of the difference at one or more frequencies via

$$P_1 = \sum_{i=1}^{n} ||\, S(s_i) - T(s_i)\,|| \qquad (5.6)$$

or

$$P_2 = \int_0^{\infty} ||\, S(j\omega) - T(j\omega)\,||\, d\omega \qquad (5.7)$$

EXAMPLE 5.8: Choose a parameter vector, r, so that $S(s)$ approximates the frequency response of an *ideal low-pass filter* with linear phase. The ideal frequency response is

$$T(j\omega) = \begin{cases} \exp(-j\omega) & |\omega| \leqslant \omega_0 \\ \\ 0 & |\omega| > \omega_0 \end{cases} \qquad (5.9)$$

For this problem, $P(r)$ could take the form of 5.6 or 5.7. If one relaxes the phase constraint and only requires that the magnitudes of $S(j\omega)$ and $T(j\omega)$

match, then the following performance measure would be adequate.

$$P = \int_{-\infty}^{-\omega_0} |S(j\omega)|\, d\omega + \int_{-\omega_0}^{\omega_0} \left[|S(j\omega)| - 1\right]\, d\omega + \int_{\omega_0}^{\infty} |S(j\omega)|\, d\omega$$

$$(5.10)$$

Great flexibility exists in choosing performance measures which characterize some optimal behavior. The two classes above are worthwhile because of their adaptability to efficient analysis via the *sensitivity theory* developed in the preceding sections. Remember that optimality will occur at a point in the set of points which force the derivative of P to be zero.

EXAMPLE 5.11: Maximize the power dissipated in the resistor of the π-network sketched in figure 5.12 at the frequency $\omega = 1$ rad. The optimization is to be done with respect to R.

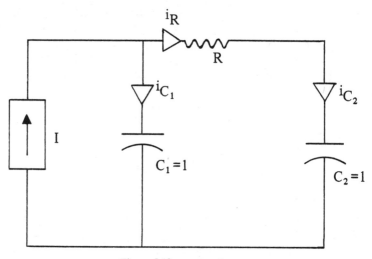

Figure 5.12. π-network.

The component connection equations for this circuit are

$$\begin{bmatrix} v_{c_1} \\ i_{c_2} \\ v_R \end{bmatrix} = \begin{bmatrix} s^{-1} & 0 & 0 \\ 0 & s & 0 \\ 0 & 0 & R \end{bmatrix} \begin{bmatrix} i_{c_1} \\ v_{c_2} \\ i_R \end{bmatrix} \qquad (5.13)$$

and

$$
\begin{bmatrix} i_{c_1} \\ v_{c_2} \\ i_R \\ i_R \end{bmatrix} = \left[\begin{array}{ccc|cc} 0 & -1 & 0 & 1 \\ 1 & 0 & -1 & 0 \\ \hline 0 & 1 & 0 & 0 \\ 0 & 1 & 0 & 0 \end{array}\right] \begin{bmatrix} v_{c_1} \\ i_{c_2} \\ v_R \\ I \end{bmatrix} \qquad (5.14)
$$

To compute the composite system transfer function

$$
(I-Z(R)L_{11}) = \begin{bmatrix} 1 & 1/s & 0 \\ -s & 1 & s \\ 0 & -R & 1 \end{bmatrix} \qquad (5.15)
$$

and

$$
(I-Z(R)L_{11})^{-1} = \frac{1}{Rs+2} \begin{bmatrix} Rs+1 & -1/s & 1 \\ s & 1 & -s \\ Rs & R & 2 \end{bmatrix} \quad (5.16)
$$

which lead to

$$
S(R) = L_{22}+L_{21}(I-Z(s)L_{11})^{-1}Z(s)L_{12} = \frac{1}{Rs+2} \qquad (5.17)
$$

Moreover

$$
\frac{dS}{dR} = [L_{21}(I-Z(s)L_{11})^{-1}]\frac{dZ}{dR}[I+L_{11}(I-Z(s)L_{11})^{-1}Z(s)]L_{12}
$$

$$
= \frac{-s}{(Rs+2)^2} \qquad (5.18)
$$

The transfer function, $S(R)$, maps the network input, I, to the network output, i_R. For our performance measure we take the power dissipated in the resistor at $s = j\omega = j$ which is

$$
P = R|i_R|^2 = R[S(R)I][S(R)I]^* = \frac{R|I|^2}{R^2+4} \qquad (5.19)
$$

where "x" indicates complex conjugate transpose. Since I is independent of R, the right hand side of 5.19 may be differentiated with respect to R. By equating this to zero, we may solve for the optimal R. Alternately, we may use the expression for ds/dR of equation 5.18 in an expression for dP/dR according to

$$(5.20)$$

$$\frac{dP}{dR} = [S(R)S(R)*+R\frac{dS}{dR}S(R)*+RS(R)\frac{dS}{dR}*]\,|\,I\,|^2 = \frac{(4-R^2)|\,I\,|^2}{(4+R^2)^2}$$

Since 5.20 is zero for R=2, the optimal parameter which maximizes $P = R|i_R|^2$ is R=2.

Real world engineers seldom optimize over scalar parameters. *Multiparameter optimization* problems are more germane although they are certainly less clearcut. For these problems one uses the generalized derivative, the *Jacobian*, or, in the present framework, the *gradient* of the performance measure. The gradient denoted by ∇P is

$$\nabla P = \frac{dP}{dr} = [\frac{dP}{dr_1}\,\frac{dP}{dr_2}\,\cdots\,\frac{dP}{dr_k}] \qquad (5.21)$$

The property of the gradient apropos to this discussion is its relationship to the *directional derivative*, i.e., for any k-vector d

$$[\nabla P(r)]d = \lim_{h\searrow 0}\frac{P(r+hd)-P(r)}{h} \qquad (5.22)$$

This property leads to an immediate solution of the *unconstrained nonlinear programming* problem: optimize $P(r)$ for r in $\mathbf{R}^k$. Assuming $P(r)$ has continuous partials we have

THEOREM 5.23: Let r* minimize (maximize) the scalar valued function $P(r)$ over all r in $\mathbf{R}^k$. Then

$$\frac{dP}{dr}(r*) = 0 \qquad (5.24)$$

PROOF: The proof is by contradiction. Suppose 5.24 is not satisfied. Then there exists a nonzero vector d in $\mathbf{R}^k$ (note in particular $d = -[\nabla P]^t$ such that

$$[\nabla P(r*)]d < 0 \qquad (5.25)$$

From 5.22 the directional derivative of P in the d-direction is negative for sufficiently small positive h. Consequently

$$P(r^*+hd) < P(r^*) \tag{5.26}$$

which contradicts the optimality of r^*.

In essence, this theorem substitutes the given optimization problem with the hopefully more tractable problem of simultaneously solving the set of equations

$$\nabla P(r) = 0 \tag{5.27}$$

Unfortunately, 5.27 is only a necessary condition. The global min's and max's, the local min's and max's, and the inflection points of P all satisfy 5.27. Usually in practice the designer has enough knowledge/intuition to properly pick the correct answer from the set induced by solving 5.27.

EXAMPLE 5.28: Minimize

$$P(r_1,r_2) = -\sin(r_1)\exp[-r_1^2+r_2^2] \tag{5.29}$$

By direct differentiation, the gradient of P is

$$\nabla P = \left[\frac{dP}{dr_1} \mid \frac{dP}{dr_2} \right] \tag{5.30}$$

where

$$\frac{dP}{dr_1} = -[\cos(r_1)-2r_1\sin(r_1)]\exp[-(r_1^2+r_2^2)] \tag{5.31}$$

$$\frac{dP}{dr_2} = 2r_2\sin(r_1)\exp[-(r_1^2+r_2^2)] \tag{5.32}$$

Since the exponential terms are never zero, solving 5.30 reduces to solving

$$\left[\begin{array}{c} \cos(r_1)-2r_1\sin(r_1) \\ \\ 2r_2\sin(r_1) \end{array} \right] = \left[\begin{array}{c} 0 \\ \\ 0 \end{array} \right] \tag{5.33}$$

Obviously, there are countably many solution points to 5.33. In all cases

$r_2=0$. The four, in some sense, smallest solutions are

$$r_2 = 0, \quad r_1 = \pm.653272, \pm3.29231 \tag{5.34}$$

To determine which solution minimizes $P(r_1,r_2)$ we may evaluate P at the different points or plot $P(r_1,0)$ as below.

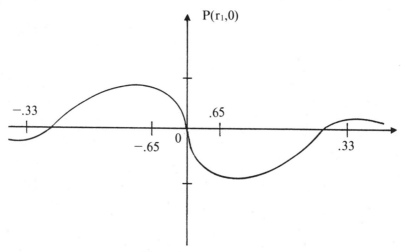

Figure 5.35. Sketch of $P(r_1,0)$.

Clearly P has a global minimum at $(r_1,r_2) = (.65,0)$; a global maximum at $(-.65,0)$. All remaining points satisfying 5.33 are local relative min's or max's.

Few real world optimization problems are unconstrained. In other words most have the form: "optimize $P(r)$ subject to the constraints . . . " For instance, one might have the problem "optimize power transfer of a network under the constraints: all components are passive, all resistors have values exceeding one unit." Frequently physical limitations, such as velocity, acceleration, bandwidth, op-amp gain, torque, etc., upper bound system parameters. Thus, real world optimization problems typically have the form: optimize $P(r)$ subject to the *inequality constraints*

$$\phi_i(r) \geqslant 0 \quad i=1,...,m \tag{5.36}$$

Such inequality constraints include all other constraint forms in that $\phi_i(r)=0$ is equivalent to "$\phi_i(r) \geqslant 0$ and $-\phi_i(r) \geqslant 0$." Similarly $\phi_i(r) \geqslant c$ is equivalent to $\phi_i(r)-c \geqslant 0$. This then is the fundamental *nonlinear programming problem*.[9]

EXAMPLE 5.37: Minimize the performance measure

$$P(r_1, r_2) = -\sin(r_1)\exp[-(r_1^2 + r_2^2)] \qquad (5.38)$$

subject to the constraint $r_1 \leqslant 0$ (equivalently $-r_1 \geqslant 0$). Clearly, the minimum lies on the line $r_2 = 0$. Inspection of figure 5.35 indicates that the *constrained* global minimum occurs at $r^* = (-3.29231, 0)$ where $P(r^*) = -2.94 \times 10^{-6}$. If the constraint were changed to $r_1 \leqslant .5$, r^* would be $(.5, 0)$ and $P(r^*)$ would equal $-.3734$. Note that for $r_1 \leqslant .5$, the solution r^* lies on the *boundary of the feasible region*.

As illustrated by this example, the solution, r^*, to a constrained optimization problem may be a global or local optimum satisfying 5.24 or r^* may lie on the *boundary* of the feasible region. The *feasible region* is the set of all vectors satisfying the inequality constraints. One of the two situations described above always occurs.

Again, the solutions to 5.24 may not yield a valid optimum point. Fortunately, by adding certain terms to 5.24, the situation may be set right. Naturally, these terms are outgrowths of the constraints. The result is an "equality" which the solution of the constrained optimization problem must satisfy. The derivation requires some preliminaries.

DEFINITION 5.39: A region $\Omega \subset \mathbf{R}^k$ is *convex* if and only if for any two points w_1 and w_2 in Ω and all ξ in $[0,1]$, then

$$(1-\xi)w_1 + \xi w_2 \; \epsilon \; \Omega \qquad (5.40)$$

Hyperplanes, linear subspaces, half spaces, etc. are all examples of a convex region.

DEFINITION 5.41: A region $\Omega \subset \mathbf{R}^k$ is a *cone* if and only if for all scalars, $\xi \geqslant 0$, and all w in Ω, ξw is in Ω.

Loosely speaking, a cone consists of half lines pointing from the origin. Obviously a convex cone is convex. Moreover, Ω is a *convex cone* if and only if Ω is a cone and for any w_1 and w_2 in Ω we have that $w_1 + w_2$ is in Ω. An example of a convex cone is

$$\Omega = \{\, r \,|\, Mr \leqslant 0 \,\} \qquad (5.42)$$

for a given matrix M with r in $\mathbf{R}^k$. In this case Ω is precisely the set of all vectors satisfying a set of homogeneous linear inequalities.

A technique for synthesizing a convex cone in $\mathbf{R}^k$ is to take a finite set of vectors $\{w_i \mid i=1,...,m\}$ and define Ω according to

$$\Omega = \left\{ w \mid w = \sum_{i=1}^{m} \xi_i w_i \; : \; \xi_i \geqslant 0 \right\} \qquad (5.43)$$

Figure 5.44 illustrates two regions in $\mathbf{R}^2$ which are convex cones. Figure 5.45 shows two regions which are not convex cones.

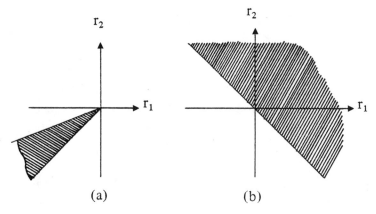

(a)   (b)

Figure 5.44. Regions forming convex cones.

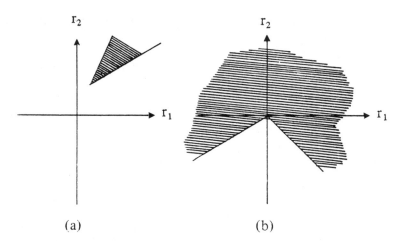

(a)   (b)

Figure 5.45. Regions which do not form convex cones.

Figure 5.45(a) fails to be a cone since the vertex of the region is not the origin. Part (b) of the figure fails because the region is not closed under addition.

An interesting characterization of a convex cone constructed and/or determined by 5.43 is *Farkas' lemma*.

FARKAS' LEMMA 5.46: Let $\Omega \subset \mathbf{R}^k$, then $w \in \Omega$ if and only if for any vector $r \in \mathbf{R}^k$ satisfying

$$r^t w_i \geqslant 0 \tag{5.47}$$

for $i = 1, \ldots m$, then

$$r^t w \geqslant 0 \tag{5.48}$$

This is actually a special case of the following more generalized *Farkas' lemma*.

LEMMA 5.49: Suppose $\{w_i \mid i=1,\ldots,m\}$ and $\{v_i \mid i=1,\ldots,n\}$ are two sets of k-vectors. For any r in $\mathbf{R}^k$, if the satisfaction of the equalities

$$r^t w_i \geqslant 0 \tag{5.50}$$

for $i = 1, \ldots, m$ implies that at least one of inequalities

$$r^t v_i \geqslant 0 \tag{5.51}$$

is satisfied for some i, $1 \leqslant i \leqslant n$, there exist non-negative scalars $\{\xi_i \mid i = 1, \ldots, m\}$ and $\{\zeta_i \mid i = 1, \ldots, n\}$ such that

$$\sum_{i=1}^{m} \xi_i w_i = \sum_{i=1}^{n} \zeta_i v_i \tag{5.52}$$

where

$$\sum_{i=1}^{n} \zeta_i = 1 \tag{5.53}$$

The proof of this highly functional lemma is prohibitively long and provides little insight into the current optimization problem. It is thus relegated to Appendix B. However, the statement of the lemma proves to be a key tool in deriving the solution to the *constrained optimization problem*.

$$\text{Minimize } P(r) \tag{5.54}$$

subject to

$$\phi_i(r) \geqslant 0 \tag{5.55}$$

for $i = 1, \ldots, m$, r in $\mathbf{R}^k$, where P and ϕ_i are assumed to be continuously differentiable. Denote by Φ the *feasible set* of 5.55. Explicitly

$$\Phi = \{ r \, \epsilon \, \mathbf{R}^k \mid \phi_i(r) \geqslant 0 \} \tag{5.56}$$

for $i = 1, \ldots, m$. In words Φ is the set of all vectors r satisfying problem constraints, 5.55. Define $I(r)$ to be the set of indices where $\phi_i(r) = 0$. This definition is r-dependent

$$I(r) = \{ i \mid \phi_i(r) = 0 \} \tag{5.57}$$

The set $\{ \phi_p(r) \mid p \epsilon I(R) \}$ are called *active constraints*, the other being *inactive*. The point being, with regard to a "local optimization criterion" $\phi_j(r) > 0$ has no effect on the solution, hence it may be neglected.

One further concept germane to our goal is the notion of *feasible directions*. Suppose r is in Φ; a vector w in $\mathbf{R}^k$ is a *feasible direction* for Φ at r if there exists a sequence of vectors $\{ r_i \}_0^\infty$ contained in Φ and a corresponding sequence of positive scalars $\{ \xi_i \}_0^\infty$ such that

$$\lim_{i \to \infty} r_i = r \tag{5.58}$$

and

$$\lim_{i \to \infty} \frac{(r_i - r)}{\xi_i} = w \tag{5.59}$$

Denote the set of all feasible directions for Φ at r by $\Omega(\Phi, r)$. Figure 5.60 helps clarify the notion of feasible direction.

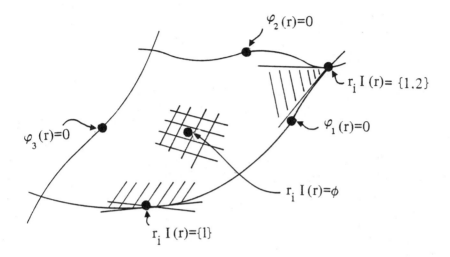

Figure 5.60. Feasible set defined by three inequality constraints and possible I(r) and the corresponding $\Omega(\Phi,r)$ (shown translated to the point r) for three distinct choices of r in Φ.

Intuitively, $\Omega(\Phi,r)$ represents a set contained in Φ of possible directed approaches to the point, r, within the feasible set.

LEMMA 5.61: $\Omega(\Phi,r)$ is a convex cone.

PROOF: We first show Ω is a cone. Let w be in Ω; we must show αw is in Ω. From 5.58 and 5.59

$$\alpha w = \lim_{i \to \infty} \frac{r_i - r}{(\xi_i/\alpha)} \tag{5.62}$$

Hence αw is in Ω.

   Showing that $\Omega(\Phi,r)$ is convex is equivalent to showing $\Omega(\Phi,r)$ is closed under addition, i.e., w_1 and w_2 in Ω implies $w_1 + w_2$ in Ω. Suppose w_1 is contructed as per 5.58 and 5.59. Let v_i be an appropriate sequence of vectors in Φ and μ_i an appropriate sequence of positive scalars such that

$$\lim_{i \to \infty} v_i = r \tag{5.63}$$

and

$$\lim_{i \to \infty} \frac{v_i - r}{\mu_i} = w_2 \tag{5.64}$$

The task is to suitably define a sequence of vectors and scalars so that "$w_1 + w_2$" may be constructed in the form of 5.58 and 5.59.

Choose positive scalars α_i and β_i so that

$$\frac{\alpha_i + \beta_i}{\alpha_i} = \xi_i \tag{5.65}$$

and

$$\frac{\alpha_i + \beta_i}{\beta_i} = \mu_i \tag{5.66}$$

Define a sequence of vectors $\{q_i\}_0^\infty$ according to

$$q_i = \frac{\alpha_i r_i + \beta_i v_i}{(\alpha_i + \beta_i)} \tag{5.67}$$

Since $r_i \to r$ and $v_i \to r$, it follows that

$$\lim_{i \to \infty} q_i = \lim_{i \to \infty} \frac{\alpha_i r_i + \beta_i v_i}{(\alpha_i + \beta_i)} = r \tag{5.68}$$

On the other hand, setting $\zeta_i = 1$ for all i we have

$$\lim_{i \to \infty} \frac{q_i - r}{\zeta_i} = \lim_{i \to \infty} \frac{\alpha_i r_i + \beta_i v_i}{\alpha_i + \beta_i} - \frac{(\alpha_i + \beta_i) r}{(\alpha_i + \beta_i)}$$

$$= \lim_{i \to \infty} \frac{\alpha_i}{\alpha_i + \beta_i} (r_i - r) + \lim_{i \to \infty} \frac{\beta_i}{\alpha_i + \beta_i} (v_i - r)$$

$$= \lim_{i \to \infty} \frac{r_i - r}{\xi_i} + \lim_{i \to \infty} \frac{v_i - r}{\mu_i} = w_1 + w_2 \tag{5.69}$$

Therefore, $\Omega(\Phi,r)$ is convex. The proof is complete.

Another interpretation is that $\Omega(\Phi,r)$ consists of the directions along which r may be perturbed so that the resultant perturbed vector remains in the feasible set Φ. Conceptually for w in $\Omega(\Phi,r)$, r in Φ, and a sufficiently small positive scalar ξ, then $r+\xi w$ is in Φ. This intuition sometimes fails if w happens to lie on the boundary of $\Omega(\Phi,r)$. In particular, $r+\xi w$ may be arbitrarily close to Φ but not in Φ for all $\xi > 0$.

The coming arguments pertaining to the constrained optimization problem are manifestly simplified if the above anomaly is prohibited from occurring. As such we assume $r+\xi w$ lives inside Φ for all sufficiently small ξ. With this assumption, if r^* is optimal (say a minimum), then

$$P(r^*) \leq P(r^*+\xi w) \tag{5.70}$$

for w in $\Omega(\Phi,r^*)$ and sufficiently small ξ - i.e., since $r^*+\xi w$ is feasible, the opposite inequality would contradict the optimality of r^*.

Expanding $P(r^*+\xi w)$ in a *Taylor series* about $\xi=0$ leads to

$$P(r^*+\xi w) = P(r^*) + \xi\left[\frac{dP}{dr}(r^*)\right]w + R_2(\xi^2) \tag{5.71}$$

where $R_2(\xi^2)$ represents the remainder term for the truncated taylor series and is proportional in magnitude to ξ^2. Since the remainder term is negligible for sufficiently small ξ and since $\xi > 0$, the only valid way 5.71 can be consistent with 5.70 is for

$$\left[\frac{dP}{dr}(r^*)\right]w \geq 0 \tag{5.72}$$

for all w in $\Omega(\Phi,r^*)$. Intuitively, 5.72 is the natural extension of 5.24 for constrained optimization problems. Indeed in the unconstrained case or a case where there are no active constraints $\Omega(\Phi,r^*) = \mathbf{R}^k$ implying the validity of 5.72 for both $\pm w$. In this circumstance 5.72 reduces to

$$\left[\frac{dP}{dr}(r^*)\right] = 0 \tag{5.73}$$

which is precisely 5.24. In the constrained case, however, "-w" may not live in $\Omega(\Phi,r^*)$. Formalizing the above discussion leads to the following lemma.

LEMMA 5.74: Suppose r* minimizes P(r) subject to the constraints $\phi_i(r) \geq 0$ for i = 1,2,...,m. Assuming P(r) has continuous first order partial derivatives at r*, then

$$\left[\frac{dP}{dr}(r^*)\right] w \geq 0 \tag{5.75}$$

for all w in $\Omega(\Phi, r^*)$.

The impending task is to use *Farkas' lemma* to reformulate 5.75 as an equality by the insertion of *"multiplier" terms*. Specifically, under minor assumptions we will characterize $\Omega(\Phi, r)$ in terms of the derivatives of the constraints. By inspecting figure 5.60, $\Omega(\Phi, r)$ seems determined by vectors tangent to the active constraints at r. This intuition fails in certain pathological cases. For most cases of practical significance, it yields a viable characterization of $\Omega(\Phi, r^*)$.

LEMMA 5.76: Let r be in Φ and suppose $\phi_i(r)$ has continuous first order partials. Then

$$\Omega(\Phi, r) \subset \left\{ w | \left[\frac{d\phi_i}{dr}(r)\right] w \geq 0 \quad \text{for } i \in I(r) \right\} \tag{5.77}$$

PROOF: The proof uses essentially the same taylor series argument of lemma 5.74. As declared earlier, we assume if w is in $\Omega(\Phi, r)$, then for sufficiently small positive ξ, "r+ξw" is in Φ. By definition, i in I(r) implies $\phi_i(r) = 0$. As such $\phi_i(r+\xi w) \geq 0$ for all w in Ω and sufficiently small ξ. Expanding $\phi_i(r+\xi w)$ in a Taylor series about $\phi=0$ yields

$$0 \leq \phi_i(r+\xi w) = \phi_i(r) + \xi\left[\frac{d\phi_i}{dr}(r)\right] w + R_2(\xi^2) \tag{5.78}$$

Since ϕ_i is an active constraint and since $R_2(\xi^2)$ is negligible for small ξ, 5.78 reduces to

$$\left[\frac{d\phi_i}{dr}(r)\right] w \geq 0 \tag{5.79}$$

for each active constraints and all w in Ω. The proof is complete.

In most instances

$$\Omega(\phi, r) = \left\{ w | \left[\frac{d\phi_i}{dr}(r)\right] w \geq 0 \; i \in I(r) \right\} \tag{5.80}$$

Counter examples to this equality are rare in practice, occurring in such pathological situations as *cusps on the boundary* of Φ or at *isolated points* of Φ. Hence we neglect these rare cases and assume 5.77 holds with equality — i.e., 5.80 is valid. An example of such a rare occurrence is:

EXAMPLE 5.81: Consider the constraint equation

$$-r^2 \geqslant 0 \tag{5.81}$$

for real scalars r, in which case $\phi(r) = -r^2$. Clearly $\Phi = \{0\}$ implying $\Omega(\Phi,r) = \{0\}$. On the other hand

$$\left[\frac{d\phi}{dr}(0)\right] = 0 \tag{5.82}$$

implying

$$0 = \Omega(\Phi,0) \neq \{w \mid \left[-\frac{d\phi}{dr}(0)\right] w \geqslant 0\} = \mathbf{R} \tag{5.83}$$

Whenever 5.80 holds, the parameter optimization problem is said to satisfy a *constraint qualification at r*.

THEOREM 5.84: Let r* minimize P(r) subject to $\phi_i(r) \geqslant 0$, i=1,...,m, where P and ϕ_i have continuous first order partials. Suppose the ϕ_i satisfy a constraint qualification at r*. Then there exist non-negative scalars, α_i, iϵI(r*), such that

$$-\left[\frac{dP}{dr}(r^*)\right] + \sum_{i\epsilon I(r^*)} \alpha_i \left[\frac{d\phi_i}{dr}(r^*)\right] = 0 \tag{5.85}$$

PROOF: Given the assumption of a constraint qualification at r*, whenever w satisfies the set of inequalities

$$\left[\frac{d\phi_i}{dr}(r^*)\right] w \geqslant 0 \tag{5.86}$$

for iϵI(r*), then w is in $\Omega(\Phi,r^*)$ hence by lemma 5.74

$$\left[\frac{dP}{dr}(r^*)\right] w \geqslant 0 \tag{5.87}$$

In other words, 5.86 implies 5.87 which is precisely the hypothesis of Farkas' lemma. Therefore, there exist non-negative scalars α_i such that

$$\left[\frac{dP}{dr}(r^*)\right] = \sum_{i \in I(r^*)} \alpha_i \left[\frac{d\phi_i}{dr}(r^*)\right] \tag{5.88}$$

The conditions of this theorem are known as the *Kuhn-Tucker conditions*. The crux of the theorem is the replacement of the originally specified *inequality* constraints with a set of *equality* constraints. The Kuhn-Tucker conditions simplify to the classical *Lagrange multiplier theorem* whenever the problem constraints are specified as equalities. In other words, all constraints are active and the sum in 5.85 is taken over all i.

Since the above used perturbation arguments are valid only in a neighborhood of r*, the similarity between the two theorems is not surprising. Perturbation arguments make the *inactive constraints* "invisible." On the other hand, the active constraints could have originally been specified as equalities, *had we known they would be equalities at r**. Consequently, in a post-factum sense, the optimization criterion for *inequality constraints* coincides with that for *equality constraints,* i.e., when the active constraints are transformed into equalities. Finally note that max may be substituted for min, in which case one removes the minus sign from dP/dr in 5.85.

Since one does not know a priori which constraints will be active, it behooves one to reformulate the Kuhn-Tucker conditions. Specifically, it is necessary to eliminate the I(r*)-dependence. This is done by letting the sum in 5.05 run over all i and adding another set of equations which essentially forces the additional scalars, α_j, to be zero. The modified equations are

$$-\left[\frac{dP}{dr}(r^*)\right] + \sum_{i=1}^{m} \alpha_i \left[\frac{d\phi_i}{dr}(r^*)\right] = 0 \tag{5.89}$$

and

$$\alpha_i \phi_i(r^*) = 0 \tag{5.90}$$

To verify that 5.89 and 5.90 are equivalent to 5.85, observe that if ϕ_i is active then $\phi_i(r^*) = 0$, rendering 5.90 redundant. If ϕ_i is inactive, then $\phi_i(r^*) > 0$ in which case 5.90 forces $\alpha_i = 0$. When actually solving an optimization problem it becomes necessary to use 5.89 and 5.90 and solve them simultaneously. The result will be a set of vectors r in $\mathbf{R}^k$. The solution r* will be a member of this set.

EXAMPLE 5.91: Minimize

$$P(r_1, r_2) = -\sin(r_1)\exp[-(r_1^2 + r_2^2)] \tag{5.92}$$

subject to

$$\phi(r_1, r_2) = r_1 \geqslant 0 \tag{5.93}$$

The goal is to find the solution set of the Kuhn-Tucker equations of 5.89 and 5.90 and choose r* from this set. The Kuhn-Tucker equations are

$$-\left[[2r_1\sin(r_1)-\cos(r_1)]\exp[-(r_1^2+r_2^2)] \;\middle|\; 2r_2\sin(r_1)\exp[-(r_1^2+r_2^2)] \right]$$
$$+\left[\alpha 1 \;\middle|\; 0\right] = \left[0 \;\middle|\; 0\right] \tag{5.94}$$

and

$$\alpha r_1 = 0 \tag{5.95}$$

Solving these two equations is equivalent to solving the following three equations simultaneous

$$-2r_2\sin(r_1)\exp[-r_1^2+r_2^2)] = 0 \tag{5.96}$$

$$\alpha r_1 = 0 \tag{5.97}$$

and

$$[2r_1\sin(r_1)-\cos(r_1)]\exp[-(r_1^2+r_2^2)] = \alpha \tag{5.98}$$

Clearly, the solutions to 5.96 are $r_2 = 0$ or $r_1 = \pm n\pi$ for n=0,1,2,3... This potential solution set is drastically reduced by observing that 5.97 and 5.98 are inconsistent for $r_1 = \pm n\pi$, n > 0. Hence the potential solution set for 5.94-5.95 reduces to $(0, r_2)$ or $(r_1, 0)$.

Now if $r_1 = 0$, then 5.97 is satisfied for all α and 5.98 implies $\alpha = -\exp(-r_2^2)$. Hence part of the desired solution space is

$$(0, r_2) \text{ and } \alpha = -\exp(-r_2^2) \tag{5.99}$$

Finally, if $r_2 = 0$, we have $\alpha = 0$, and example 5.28 showed that $r_1 = \pm.653272, \pm3.29231, \ldots$ The negative values of r_1 fail to satisfy the constraints; hence, $r_1 = .653272, 3.29231, \ldots$ This shows that the family of solutions of the Kuhn-Tucker equations is countably infinite. At any rate the remaining solution set is

$$r_1 = .653, 3.292, \ldots ; r_2 = 0; \alpha = 0 \qquad (5.100)$$

Inspection of the performance measure indicates that r_1^2 should be small for the optimal solution. Thus, we may surely neglect values of r_1 larger than 3.3. This then leaves us with three possible candidates for r*. Plugging into the performance measure verifies r* $= (.653,0)$, as intuitively expected from figure 5.35. The solution $(3.292,0)$ represents a relative max whereas 5.99 depicts a local min on the boundary of the set.

EXAMPLE 5.101: Repeat the above example under the different constraint

$$\phi(r_1,r_2) = .5-r_1 \geqslant 0 \qquad (5.102)$$

For this case the Kuhn-Tucker equations are

$$-2r_2\sin(r_1)\exp[-(r_1^2+r_2^2)] = 0 \qquad (5.103)$$

$$\alpha(.5-r_1) = 0 \qquad (5.104)$$

and

$$\alpha = -[2r_1\sin(r_1)-\cos(r_1)]\exp[-(r_1^2+r_2^2)] \qquad (5.105)$$

Of course, the solutions to 5.103 are the same as those of 5.96 and $r_1 = \pm n\pi, n > 0$ remain inconsistent with 5.104 and 5.105. For the case of $r_1 = 0$, 5.104 forces $\alpha = 0$. From 5.105, this in turn requires that $\exp[-r^2] = 0$, which is impossible. As such, $r_1 = \pm n\pi, n = 0,1,2,\ldots$, are inconsistent with the Kuhn-Tucker equations.

If $r_2 = 0$, then 5.104 forces $\alpha = 0$, or $r_1 = .5$. The case of $\alpha = 0$ results in a family of solutions

$$r_1 = -.653, -3.292, \ldots; r_2 = 0; \alpha = 0 \qquad (5.106)$$

On the other hand, if $r_1 = .5$, we have

$$\alpha = [\cos(.5)-\sin(.5)]\exp(-.25) = .310085 \qquad (5.107)$$

As such, the remaining solution to the Kuhn-Tucker equations is

$$r_1 = .5; \quad r_2 = 0; \quad \alpha = .31 \qquad\qquad (5.108)$$

In this case the global min is given by 5.108 which is a solution living on the boundary of the feasible region. The remaining solutions are either relative min's or relative max's.

Clearly, solving the Kuhn-Tucker equations does not automatically produce an r*. However, the optimal solution will be among the solutions. Since the number of such solutions is small, picking the actual optimum is straightforward. Simply substitute the solutions into the performance measure and/or invoke one's previous experience with similar problems, engineering intuition, or some other empirical reason.

EXAMPLE 5.109: As a final example, consider a *pulse detector* circuit as sketched in figure 5.110.[2] Here, a pulse normalized to unit height and width and corrupted by *white noise* is passed through an RC filter so as to improve the signal to noise ratio at the output. Since an RC filtered pulse has the general shape sketched in figure 5.111, it is natural to look for

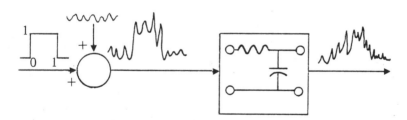

Figure 5.110. Pulse detector.

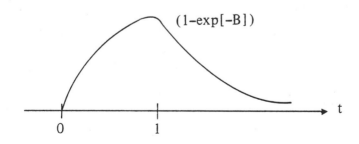

Figure 5.111. RC filtered pulse.

evidence of a pulse at time $t = 1$ where the filtered pulse assumes its maximum value of [1-exp(-B]. Here $B = 1/RC$ is taken as a definition of the filtered *bandwidth* (indeed, it represents the 3db. point of the filter's frequency response). Moreover, since the magnitude spectrum of *white noise* is uniform for all frequency, the mean squared amplitude of the noise, passed through an RC filter, is proportional to the filter bandwidth, B. Thus, for this problem we define the *signal to noise ratio* as the ratio of the square of the maximum signal amplitude to the mean of the noise amplitude.

The optimization problem requires the maximization of

$$P(B) = \frac{[1-\exp(-B)]^2}{B} \tag{5.112}$$

Of course, as in most practical CAD problems, reality imposes a set of constraints. System stability requires a positive bandwidth bounded away from zero, say $B \geqslant .1$. The use of real rather than idealized components imposes an upper bound on bandwidth which, relative to the normalization of the time scale, we take as $B \leqslant 1$. The design problem reduces to maximizing 5.112 subject to the constraints

$$\phi_1(B) = (B-.1) \geqslant 0 \tag{5.113}$$

and

$$\phi_2(B) = (1-B) \geqslant 0 \tag{5.114}$$

Differentiating 5.112 through 5.114 sets up the following Kuhn-Tucker equations.

$$[1-\exp(-B)] [2B\exp(-B) \textbf{ V } (1-\exp(-B))] + \alpha_1 - \alpha_2 = 0 \tag{5.115}$$

$$\alpha_1(B-.1) = 0 \tag{5.116}$$

and

$$\alpha_2(1-B) = 0 \tag{5.117}$$

There are three solutions to these equations as given in table 5.118.

	α_1	α_2	B
1	0	0	1.26
2	0	.07	1
3	-.82	0	.1

Table 5.118. Solutions to Kuhn-Tucker equations.

Solution 1 can be excluded since B fails to satisfy the constraints. In fact, solution 1 is optimal for the unconstrained problem. Solution 2 is optimal, i.e., $B^* = 1$, for this constrained problem. The third solution is invalid since $\alpha_1 < 0$. In fact, solution 3 represents a local min at the boundary of the feasible set.

As in the preceding examples, the Kuhn-Tucker equations for most real world problems are constructed through ad hoc techniques. However, with regard to the component connection model, it is possible to take advantage of the analytical expressions for the derivative of a *composite system transfer function matrix* and the derivative of an *integral performance measure* so as to express the Kuhn-Tucker equations *analytically* for a large class of parameter optimization problems.

Suppose for instance, it is necessary to minimize a function of the system frequency response at ω_0: minimize $P[S(j\omega_0)]$ subject to $\phi_i(r) \geqslant 0$ for $i = 1,2,...,m$. With the aid of equation 2.5 the Kuhn-Tucker equations may be written according to

$$-\frac{dP}{dS}\left[[L_{21}(I-Z(\omega_0,r)L_{11})^{-1}]\frac{dZ}{dr}(\omega_0,r)\ [I+L_{11}(I-Z(\omega_0,r)L_{11})^{-1}Z(\omega_0,r)]L_{12}\right]$$

$$+ \sum_{i=1}^{m} \alpha_i \frac{d\phi_i}{dr}(r) = 0 \qquad (5.119)$$

and

$$\alpha_i\phi_i(r) = 0 \qquad (5.120)$$

for $i = 1, \ldots, m$.

Of course, performance measures of the type $\sum P_j[S\omega_j)]$ or $\int P[S(\omega)]d\omega$ are equally adaptable. Indeed, using the *Lagrange multiplier* approach to compute the derivative of an integral performance measure with respect to a component parameter, it is possible to analytically specify the Kuhn-Tucker equations for surprisingly complex classes of CAD problems even for nonlinear systems. In the nonlinear case we use the formula for dP/dr derived in 4.39 together with the side equations, 4.30 through 4.37, and the Kuhn-Tucker equations to obtain a set of design equations for a given nonlinear parameter optimization problem. Conceptually, this is all well and good. Unfortunately, the resultant equations are unusually extensive since they include all the original system variables, plus the Lagrange multipliers used to compute dP/dr and the multipliers (the α_i's) from the Kuhn-Tucker equations. Thus, it may happen that the resultant design package is numerically unfeasible. To circumvent this difficulty it may be necessary to resort to direct numerical optimization algorithms ignoring such design equations.

6. NUMERICAL OPTIMIZATION

The previous section outlined the *computer aided design* problem and its solution for nonlinear interconnected dynamical systems. The solution of the dynamic design equations reduced to solving a set of nonlinear algebraic equations. The cost for this exact analytical solution was the large increase in dimensionality of the design equations through augmentation by *Lagrange multipliers.*

In practice, numerical optimization procedures show better efficiency. No prior on-line computation or augmentation by Lagrange multipliers is required. In reality, the procedures are *"educated searches."* Specifically, it is frequently faster to optimize $P(r)$ through a search technique than through solving dP/dr and usually the additional evaluation of d^2P/dr^2.

One dimensional search algorithms are commonly called *line searches.* There are many line search schemes. Most multi-dimensional optimization routines adopt some type of line search. Such programs choose an "optimal direction" in a multi-dimensional space and implement a line search in this optimal direction. In this light, our study of numerical optimization begins with a brief noncomprehensive review of some of the more common line search algorithms. The section concludes with the discussion of several multi-dimensional schemes. Since $\max[P(r)]$ is equivalent to $\min[-P(r)]$, we restrict the discussion to minimization problems.

The first scheme discussed is the *Powell line search.* This procedure uses a quadratic curve fit through three suitably chosen points to estimate the minimum of a given objective function. It has the advantage of requiring no derivative information.

To fit a quadratic, $q(r)$, through three points, say $P_1=P(r_1)$, $P_2=P(r_2)$, and $P_3=P(r_3)$, we use the formula

$$q(r) \ = \ \sum_{i=1}^{3} \ P_i \ \frac{\prod_{j\neq i} (r-r_j)}{\prod_{j\neq i} (r_i-r_j)} \tag{6.1}$$

The value of r minimizing $q(r)$ is r given by

$$r \ = \ .5 \ \frac{b_{23}P_1+b_{31}P_2+b_{12}P_3}{a_{23}P_1+a_{31}P_2+a_{12}P_3} \tag{6.2}$$

where $b_{ij} = r_i^2 - r_j^2$ and $a_{ij} = r_i - r_j$.

Briefly, given a *good* initial guess, r_0, and an initial *search increment*, d, one generates a sequence $[P_n]$ in one of two possible ways. The form of $[P_n]$ depends on the relative values of $P(r_0)$ and $P(r_0+d)$. If $P(r_0)>P(r_0+d)$, define $r_n = r_0 + nd$ and $P_n = P(r_n)$. Terminate the sequence after obtaining three consecutive points, P_{n-2}, P_{n-1}, and P_n such that

$$\begin{aligned} P_{n-2} &> P_{n-1} \\ P_{n-1} &< P_n \end{aligned} \tag{6.3}$$

Fit a quadratic through these points. Compute r and $P(r)$.

On the other hand, if $P(r_0) < P(r_0+d)$, define $r_n = r_0-nd$ and $P_n = P(r_n)$. Generate $[P_n]$ until one finds three consecutive points P_{n-2}, P_{n-1}, and P_n, again satisfying 6.3. Do a quadratic curve fit; compute r and $P(r)$. If three points satisfying 6.3 cannot be found, decrease d and start again.

After finding r and $P(r)$, one reinitializes the algorithm by setting $r_0 = r$ and by appropriately decreasing d. Figure 6.6 captures the basic idea.

EXAMPLE 6.4: Numerically maximize $\sin(r)\exp(-r^2)$. Equivalently, minimize the performance measure

$$P(r) \ = \ -\sin(r)\exp(-r^2) \tag{6.5}$$

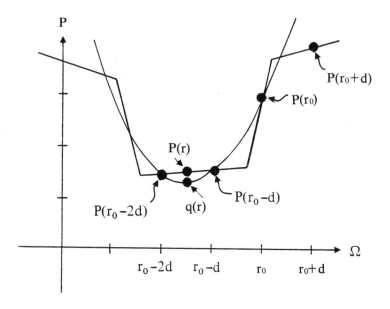

Figure 6.6. Sketch illustrating Powell search.

Take d = .5 and an initial guess of $r_0 = -2$. In this case $P_0 = P(-2) = .0167$
and $P(r_0+d) = P(-1.5) = .1051$. Since $P_0 < P(-1.5)$, define

$$r_n = r_0 - nd$$
$$P_n = P(r_n)$$ (6.7)

This permits us to construct the following table

n	r_n	P_n
0	–2.0	.0167
1	–2.5	1.15×10^{-3}
2	–3.0	1.74×10^{-5}
3	–3.5	-1.68×10^{-6}
4	–4.0	-8.517×10^{-8}

Table 6.8. Optimization values for example.

Since P_2, P_3, P_4 satisfy 6.3, we may compute r as

$$r = -3.6728 \qquad\qquad (6.9)$$

Although this value of r does not produce a relative min of $P(r)$, it is converging toward the value $r^* = -3.29231$ which does. Note that as shown in example 5.28, $P(-3.29231)$ is a relative min, not a global min. This resulted from the initial guess of -2 which was very close to -3.29231. The global min is given by $r = .653272$. For the remaining discussion let r^* denote the minimizing parameter value in the region of concern.

Implicit in the above line search and in the remaining search procedures is the assumption that in the region of concern, the performance measure $P(r)$ is *unimodal* – i.e., $P(r)$ has a single relative minimum. Moreover, the Powell search implicitly assumes $P(r)$ satisfies some kind of smoothness condition whereas the Fibonacci and Golden section line searches do not.

In a *Fibonacci line search*, one is given a unimodal scalar valued function of a scalar, say $P(r)$, on a known interval $[c_1,c_2]$. For the purposes of explanation, let $[c_1,c_2] = [0,1]$. The Fibonacci search minimizes the uncertainty in the location of r^* given exactly N evaluations of $P(r)$. If N=2 – i.e., we are permitted two evaluations of $P(r)$ – then we divide $[0,1]$ into two intervals, $[0,.5]$ and $[.5,1]$ as illustrated in figure 6.10.

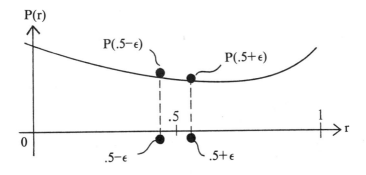

Figure 6.10. Fibonacci line search with N=2.

Let ϵ be an arbitrarily small number and evaluate $P(.5-\epsilon)$ and $P(.5+\epsilon)$. If $P(.5-\epsilon) > P(.5+\epsilon)$ as in figure 6.9, then the unimodality assumption implies that r^* lies in $[.5,1]$. If $P(.5-\epsilon) < P(.5+\epsilon)$, r^* lies in $[0,.5]$.

Now let N=3. Divide $[0,1]$ into three intervals $I_1 = [0,1/3]$, $I_2 = [1/3,2/3]$, $I_3 = [2/3,1]$. Evaluate $P(1/3)$ and $P(2/3)$. If

$P(1/3) > P(2/3)$, r^* lies in I_2 or I_3, otherwise r^* lies in I_1 or I_2. Suppose as in figure 6.11, $P(1/3) > P(2/3)$. We must determine with exactly one more evaluation of $P(r)$ whether r^* is in I_2 or I_3. Again let ϵ be arbitrarily small and evaluate $P(2/3+\epsilon)$. Figure 6.11 indicates that $P(2/3) > P(2/3+\epsilon)$; therefore, r^* is in I_3.

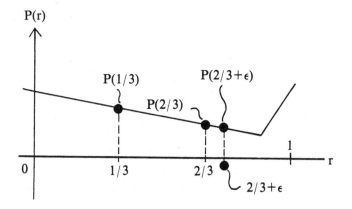

Figure 6.11. Fibonacci search with N=3.

To describe the general approach, take the initial length of uncertainty to be $d_1 = c_2 - c_1$ where the minimization is to take place over $[c_1, c_2]$. Let d_k be the length of uncertainty after k measurements of $P(r)$. The lengths d_1 and d_k satisfy

$$d_k = \frac{F_{N-K+1}}{F_N} d_1 \qquad (6.12)$$

where it is assumed that we are permitted exactly N measurements of $P(r)$ in $[c_1, c_2]$. The numbers F_j belong to the *Fibonacci sequence* generated by the difference equation

$$F_N = F_{N-1} + F_{N-2} \qquad F_0 = F_1 = 1 \qquad (6.13)$$

The first several terms in the Fibonacci sequence are $1,1,2,3,5,8,13,21,34,...$ The search procedure starts by making measurements of $P(r)$ symmetrically at a distance $(F_{N-1}/F_n)d_1$ from the respective ends of the initial interval $[c_1, c_2]$. This produces a new interval of uncertainty of length

$$d_2 \;\; = \;\; \frac{F_{N-1}}{F_N} \;\; d_1 \tag{6.14}$$

In this new interval, we place a third measurement point symmetric with respect to the measurement (obtained from the first step) already in this new interval. Figure 6.16 helps clarify the procedure. When viewing this figure assume $P_2 > P_1$. Notice that evaluating P_3 produces a new uncertainty interval of length

$$d_3 \;\; = \;\; \frac{F_{N-2}}{F_N} \;\; d_1 \tag{6.15}$$

The process continues until N evaluations of $P(r)$ are made.

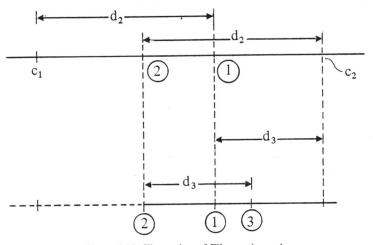

Figure 6.16. Illustration of Fibonacci search.

By letting N approach infinity, the Fibonacci line search reduces to the well-known *Golden Section line search*. It is possible to show that

$$\lim_{N \to \infty} \; \frac{F_{N-1}}{F_N} \;\; = \;\; \frac{1}{\tau} \;\; \approx \;\; .618 \tag{6.17}$$

where

$$\tau \;\; = \;\; \frac{1+\sqrt{5}}{2} \tag{6.18}$$

The width of the uncertainty interval at each step of the process becomes

$$d_k = \left[\frac{1}{\tau}\right]^{k-1} d_1 \qquad (6.19)$$

This concludes the discussion of line searches so we move on to describe some multi-dimensional schemes.

For multi-dimensions, the problem is to minimize $P(r)$ where r is a vector in $\mathbf{R}^n$, given some feasibility region in which the optimization is to take place. The method of *steepest descent* is one of the oldest and most widely used algorithms. It is a gradient method and requires that $P(r)$ have continuous partial derivates on $\mathbf{R}^n$. The idea as illustrated in figure 6.20 is to obtain an initial guess and an optimal direction vector d_0. Using a line search one then finds a non-negative scalar, α_0, such that $P(r_0+\alpha_0 d_0)$ is minimum. Define $r_1 = r_0+\alpha_0 d_0$. Find a new direction vector, d_1, and search for a new non-negative scalar α_1, such that $P(r_1+\alpha d_1)$ is minimum. Repeat the procedure until the minimum is found.

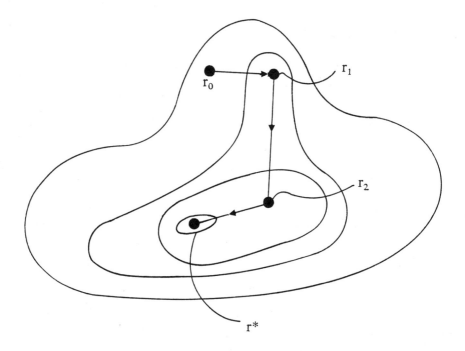

Figure 6.20. Illustration of steepest descent algorithm.

One chooses an optimal direction by making use of derivative information. Define the *gradient* of P(r) as

$$\nabla P(r) = \left[\frac{dP}{dr}(r)\right]$$ (6.21)

The optimal direction of search for minimization problems is

$$-[\nabla P(r_0)]^t$$ (6.22)

Using *Cauchy's inequality*, it can be shown that this vector minimizes the directional derivative of P(r) at r_0.[3] The iterative scheme for choosing successive r_k's is

$$r_{k+1} = r_k - \alpha_k [\nabla P(r_k)]^t$$ (6.23)

where for each k, a line search finds that α_k which minimizes

$$P(r_k + \alpha d_k)$$ (6.24)

where

$$d_k = -[\nabla P(r_k)]^t$$ (6.25)

is the optimal search direction.

EXAMPLE 6.26: Minimize the following performance measure of two variables:

$$P(r_1, r_2) = -\sin(r_1)\exp[-(r_1^2 + r_2^2)]$$ (6.27)

The gradient of P is

$$\nabla P = \left[\frac{dP}{dr_1} \;\middle|\; \frac{dP}{dr_2}\right]$$ (6.28)

where

$$\frac{dP}{dr_1} = [2r_1\sin(r_1) - \cos(r_1)]\exp[-(r_1^2 + r_2^2)]$$ (6.29)

and

$$\frac{dP}{dr_1} = 2r_2\sin(r_1)\exp[-(r_1^2+r_2^2)] \tag{6.30}$$

Given an initial guess of $r_0 = [0,0]^t$, we obtain the initial search direction of

$$-[\nabla P]^t = \begin{bmatrix} 1 \\ 0 \end{bmatrix} \tag{6.31}$$

The next step is to find α such that

$$\begin{bmatrix} 0 \\ 0 \end{bmatrix} + \alpha \begin{bmatrix} 1 \\ 0 \end{bmatrix} \tag{6.32}$$

minimizes $P(\alpha,0)$. Examples 6.4 and 5.28 indicate that $\alpha = .653272$. In fact, $r^* = (.653272, 0)$. Of course, different initial guesses lead to different rates of convergence and for this performance measure different r^*'s. In the above case, the guess of $[0,0]^t$ was highly opportune. The guess $(-2,0)$ would lead to different and erroneous results.

A highly useful optimization routine is the *conjugate gradient method* which is designed specifically for *quadratic performance measures* of the form

$$P(r) = r^tQr + q \tag{6.33}$$

where Q is a symmetric *positive definite matrix* and q is a vector. An extension of the conjugate gradient method to non-quadratic performance measures is the *Fletcher-Reeves method*. This method initializes with a steepest descent step, then switches to an *m-step modified conjugate gradient approach*, repeating the process until convergence. Of course, m is determined beforehand. For notational convenience, define

$$g_k = [\nabla P(r_k)]^t \tag{6.34}$$

The Fletcher-Reeves method is as follows:

STEP 1: Given r_0, compute g_0 and set the initial search direction $d_0 = -g_0$.

STEP 2: For $k = 0,1,...,m-1$
(i) set $r_{k+1} = r_k + \alpha_k d_k$ where α_k is chosen by a line search so as to minimize $P(r_k + \alpha d_k)$.

(ii) compute the updated gradient transpose g_{k+1}.
(iii) For $k \neq m-1$, set the new search direction

$$d_{k+1} = -g_{k+1} + \beta_k \, d_k \qquad (6.35)$$

where

$$\beta_k = \frac{g_{k+1}^t \, g_{k+1}}{g_k^t \, g_k} \qquad (6.36)$$

STEP 3: For $k=m-1$ reinitialize the algorithm by setting $r_0 = r_m$ and return to Step 1.

For all of the above algorithms, the actual optimal, r^*, is seldom achieved. Globally speaking, r_k will converge to r^* so some kind of convergence criterion must always be present so as to properly terminate the program.

Both of the above multi-dimensional schemes employed the gradient to, in some way, pick an optimal search direction. An alternative having poorer convergence but easier programming is the *cyclical search*. In this scheme one sequentially searches for a minimum along each of the coordinate directions in n-space.

Suppose the parameter vector, r, has n entries. For the cyclical search one first minimizes $P(r)$ in the direction $d_0 = [1,0,...,0]^t$ to obtain an updated parameter vector, $r_1 = r_0 + \alpha_0 \, d_0$. Next we find α_1 such that $P(r_1 + \alpha d_1)$ is minimum where $d_1 = [0,1,0,...,0]^t$. One cycles through the n-coordinate directions minimizing and updating r at each step. The process is repeated until convergence occurs.

In any "real world" parameter optimization problem, viable results mandate a close interface between the computer and the design engineer. Often the design engineer must first make a good "educated" initial guess for the algorithm to converge to a true optimum as opposed to a local relative min.

7. MULTIFREQUENCY SYSTEM DIAGNOSIS

To motivate the ideas of this section,[7] we open with a brief problem: Suppose you are given a "black box" known to contain a parallel RC circuit having nominal parameter values $R = C = 1$. On the box are two terminals marked "+" and "−". With the aid of a signal generator and an oscilloscope you are to determine the exact values of R and C using only information obtained at the terminals.

First consider the diagram of figure 7.1 in which I is the output and V

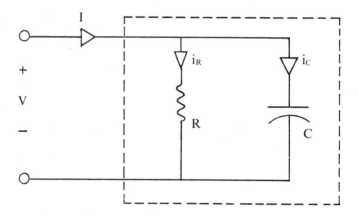

Figure 7.1. Schematic of "black box" with 1 output and V input.

the input. The transfer function matrix for this circuit is easily computed as

$$S(C,R,s) \;=\; Cs + \frac{1}{R} \;=\; [s \;\; 1] \begin{bmatrix} C \\ \frac{1}{R} \end{bmatrix} \qquad (7.2)$$

Suppose by using our signal generator and scope, we physically measure the transfer function at $s_1 = 0$ to be

$$S_1 \;=\; S(R,C,0) \;=\; 1.5 \qquad (7.3)$$

and at $s_2 = -1$ to be

$$S_2 \;=\; S(R,C,-1) \;=\; -1.666 \qquad (7.4)$$

With the aid of 7.2, we may solve for R and C by solving

$$\begin{bmatrix} 1.5 \\ -1.666 \end{bmatrix} \;=\; \begin{bmatrix} 0 & 1 \\ -1 & 1 \end{bmatrix} \begin{bmatrix} C \\ \frac{1}{R} \end{bmatrix} \qquad (7.5)$$

which has the solution $R = .666$ and $C = 1.0$. More generally, one may "diagnose" the parameter values R and C by measuring $S(R,C,s)$ at any two distinct frequencies, s_1 and s_2, by solving

$$\begin{bmatrix} S_1 \\ \\ S_2 \end{bmatrix} = \begin{bmatrix} s_1 & 1 \\ \\ s_2 & 1 \end{bmatrix} \begin{bmatrix} C \\ \\ \dfrac{1}{R} \end{bmatrix} \tag{7.6}$$

Clearly 7.6 has a unique solution if and only if $s_1 \neq s_2$.

This example borders on the naive in that it lacks the nonlinearities inherent in the general diagnostic equations soon to be derived. However, it illustrates the concept of *multifrequency testing* by generating a set of independent equations from the same transfer function matrix measured at distinct frequencies.

Recall that in section III.3 we previewed the question of *system diagnosability* by solving the *connection function equation.*

$$S = f(Z) = L_{22} + L_{21}(I-ZL_{11})^{-1}ZL_{12} \tag{7.7}$$

Unfortunately, the single frequency solution is impractical for solving the "real world" system diagnosis problem since the appropriate inverse matrices typcially fail to exist. As in the motivating example, these problems may be circumvented by assuming (i) the essential dynamical character of Z is constant and (ii) response variation from nominal is due entirely to changes in a given parameter vector, r. Physically this might correspond to a resistor failing by changing its resistance but staying a resistor or a capacitor failing by developing a parallel conductance, etc. The goal of this section is to reopen the investigation of *system diagnosability* in light of such assumptions on Z.

In particular, suppose the *composite component transfer function matrix* is $Z(r,s)$ where r is a given k-vector of parameter values and s is the usual complex variable. In the opening example, r was the vector (R,C). The r-s dependent system transfer function matrix becomes

$$S(r,s) = L_{22} + L_{21}(I-Z(r,s)L_{11})^{-1}Z(r,s)L_{12} \tag{7.8}$$

For fixed s theorem III.3.30 again will characterize the solution of 7.8 if the the appropriate inverses exist, i.e., we *measure* $S(r,s)$ at fixed s and attempt to solve 7.8 for r. Using the intuition gained by studying the parallel RC circuit, it is clear that measuring S at several distinct frequencies may help to alleviate the difficulty in solving 7.8 when L_{12} and L_{21} fail to be invertible. As notation, take $S_i(r) = S(r,s_i)$; $Z_i(r) = Z(r,s_i)$; and

$$f[Z_i(r)] \stackrel{\triangle}{=} L_{22} + L_{21}(I-Z_i(r)L_{11})^{-1}Z_i(r)L_{12} \tag{7.9}$$

for $i = 1,...,n$ where s_i is an arbitrarily specified set of complex frequencies and f is the *connection function*. As such, we may expand 7.8 into a set of n equations in the hope that they will form a set of independent equations whose solution will be possible.

$$S^m \stackrel{\triangle}{=} \begin{bmatrix} S_1(r) \\ S_2(r) \\ \cdot \\ \cdot \\ \cdot \\ S_n(r) \end{bmatrix} = \begin{bmatrix} f[Z_1(r)] \\ f[Z_2(r)] \\ \cdot \\ \cdot \\ \cdot \\ f[Z_n(r)] \end{bmatrix} \tag{7.10}$$

where the superscript "m" on S^m indicates "measured." Observe that S^m is a matrix of known (measured) complex numbers. In general, S^m is rectangular. The right side of 7.10 is a matrix of functions in r which we assume to be rational. By using 7.10, we increase the number of equations without changing the number of unknowns, i.e., the number of entries in the vector, r. In the memoryless case (Z independent of s) the equations of 7.10 form a redundant set though, in general, the number of independent equations of 7.10 may exceed those of 7.9, as in our example. Equation 7.10 is called the *fault diagnosis equation*.

The viability of the fault diagnosis equation begs the answer to two fundamental questions: "What choice of test frequencies maximizes the solvability of 7.10?" and "How solvable is 7.10, given an optimal choice of test frequencies?" The answers to both questions hinge on developing some measure of the *solvability* of a *set* of *nonlinear algebraic equations*. Before proceeding, let us convert the matrix equation 7.10 to a vector equation. Clearly solving 7.10 is equivalent to solving

$$\begin{bmatrix} vec[S_2(r)] \\ vec[S_2(r)] \\ \cdot \\ \cdot \\ \cdot \\ vec[S_n(r)] \end{bmatrix} = \begin{bmatrix} vec[f[Z_1(r)]] \\ vec[f[Z_2(r)]] \\ \cdot \\ \cdot \\ \cdot \\ vec[f[Z_n(r)]] \end{bmatrix} \tag{7.11}$$

which is a vector equation. The left side of 7.11 is a vector of fixed complex numbers whereas the right side is a vector of rational functions, each a function of the k-entries of r. To further compactify the notation, define the right side of 7.11 according to

$$
\text{vec}[S] \quad = \quad
\begin{bmatrix}
\text{vec}[f[Z_1]] \\
\text{vec}[f[Z_2]] \\
\cdot \\
\cdot \\
\cdot \\
\text{vec}[f[Z_n]]
\end{bmatrix}
\tag{7.12}
$$

and the left hand side of 7.11 as

$$
\text{vec}[S^m] \quad = \quad
\begin{bmatrix}
\text{vec}[S_1] \\
\text{vec}[S_2] \\
\cdot \\
\cdot \\
\cdot \\
\text{vec}[S_n]
\end{bmatrix}
\tag{7.13}
$$

taking note that we have deliberately abused the "vec" notation in equations 7.12 and 7.13. Thus, the solution of the fault diagnosis equation as per 7.10 is equivalent to solving

$$
\text{vec}[S^m] \quad = \quad \text{vec}[S(r)]
\tag{7.14}
$$

for the unknown parameter vector, r.

Motivated by classical linear algebra we develop a measure of the solvability of 7.14 by first linearizing 7.14 about the nominal parameter vector, r_0. The resultant linearization is

$$
\text{vec}[S^m] \approx \text{vec}[S(r_0)] + \left[\frac{d\,\text{vec}[S]}{dr}\ (r_0) \right] [r - r_0]
\tag{7.15}
$$

where $d\,\text{vec}[S]/dr$ is the *Jacobian matrix* for $\text{vec}[S]$. The *implicit function theorem* now guarantees that the fault diagnosis equation, 7.14, is solvable in a neighborhood of r_0 up to a manifold of dimension $\delta(r_0)$ where

$$\delta(r_0) \;=\; k \;-\; \mathrm{rank}\left[\frac{\mathrm{dvec}[S]}{\mathrm{dr}}\right](r_0) \qquad (7.16)$$

Plainly speaking, the solution of the fault diagnosis equation will have at most $\delta(r_0)$ arbitrary entries in a neighborhood of r_0. It turns out that this *local measure of solvability*, $\delta(r_0)$, is happily also a *global measure*. Before proving this characteristic of $\delta(r_0)$ let us first pin down the structure of the above Jacobian matrix.

It is clear that

$$\left[\frac{\mathrm{dvec}[S]}{\mathrm{dr}}\right] = \begin{bmatrix} \dfrac{\mathrm{dvec}[f(Z_1)]}{\mathrm{dr}} \\ \hline \dfrac{\mathrm{dvec}[f(Z_2)]}{\mathrm{dr}} \\ \hline \cdot \\ \cdot \\ \cdot \\ \hline \dfrac{\mathrm{dvec}[f(Z_n)]}{\mathrm{dr}} \end{bmatrix} \qquad (7.17)$$

Therefore, computing the Jacobian reduces to finding a convenient expression for the derivative of $\mathrm{vec}[f(Z_i)]$. Equation 2.20 provides the key as follows

$$\left[\frac{\mathrm{dvec}[f(Z_i)]}{\mathrm{dr}}\right] = \left[\,[I+L_{11}(I-ZL_{11})^{-1}Z]\; L_{12}\right]^t$$
$$\otimes [L_{21}(I-ZL_{11})^{-1}]\left[\frac{\mathrm{dvec}(Z)}{\mathrm{dr}}\right] \qquad (7.18)$$

Evaluating this expression at the nominal value, r_0, and substituting into 7.17 produces the desired expression.

Let us now demonstrate that the local solvability measure, $\delta(r_0)$, is in fact a *global measure of solvability*. This property is intrinsic to the rational (in r) nature of the matrix given in 7.17. In the following we will use the term *"for almost all values of r"* to mean for all vectors, r, except possibly those lying in a lower dimensional *algebraic variety* in the parameter space, $\mathbf{R}^k$. In addition, we impose the minor restriction that $Z(r,s)$ be rational in r. Such a restriction still admits the usual RLC components, op-amps, filters, and other devices whose potentially variable parameters take the form of gains, pole or zero locations, etc.

THEOREM 7.19: Let $Z(r,s)$ be rational in r as well as s. Then

$$\text{rank} \left[\frac{\text{dvec}[S]}{dr} (r) \right] \tag{7.20}$$

is constant *for almost all values* of r.

PROOF: For convenience define

$$\Phi = \left[\frac{\text{dvec}[S]}{dr} \right] \tag{7.21}$$

Since $Z(r,s)$ is rational in r, 7.18 guarantees $\Phi(r)$ is also rational in r. Moreover, since we have a priori fixed the set of test frequencies, s_i, Φ is no longer a function of s. It is rational *only* in the k potentially variable entries of r.

The least common denominator of a rational matrix (in particular Φ) is nonzero for almost all r, hence multiplying Φ by its least common denominator leaves its *generic rank* unchanged. Therefore, instead of dealing with Φ, we may equivalently deal with the polynomial matrix

$$P(r) = \Phi \; \text{LCD}[\Phi] \tag{7.22}$$

in k-variables, the potentially variable entries of r.

For any fixed value of r, say r_1, rank $[P(r_1)]$ equals the dimension of the largest square submatrix of $P(r)$ having nonzero determinant. Let r_m be a value of the parameter vector, r, maximizing rank $[P(r)]$. Let $M(r)$ denote an appropriate square submatrix whose dimension equals rank $[P(r)]$ such that

$$\det[M(r_m)] \neq 0 \tag{7.23}$$

Since there are only a finite number of values rank $[\Phi]$ can assume, such an r_m and such a submatrix, $M(r)$, must exist.

Since 7.23 implies the polynomial $\det[M(r)]$ is *not identically equal to zero*, the zero set of $\det[M(r)]$ must lie in an algebraic variety of dimension strictly less than the dimension of $\mathbf{R}^k$. Hence

$$\det[M(r)] \neq 0 \quad \text{a.e.} \tag{7.24}$$

where a.e. denotes "almost everywhere." In consequence

$$(7.25)$$

$$\text{rank } [P(r_m)] \geqslant \text{rank}[P(r)] \geqslant \dim[M(r)] = \text{rank}[P(r_m)] \quad \text{a.e.}$$

In particular

$$\text{rank}[P(r)] = \text{rank}[P(r_m)] \quad \text{a.e.} \qquad (7.26)$$

as was to be shown.

In the sequel rank $[\Phi]$ will *always* mean the *generic rank* of Φ. The significance of the above theorem ought to be clear. The locally defined *measure of solvability* as per 7.16 is now a *global measure of solvability*. As such our measure of solvability

$$\delta = k - \text{rank} \left[\frac{\text{dvec}[S]}{dr} \right] \qquad (7.27)$$

of the fault diagnosis equations is essentially constant.

To obtain an intuitive feel for the meaning of δ, suppose $k=3$ and $\delta=1$. Specifically, r is a 3-vector. The measure of solvability, $\delta=1$, means that we may solve the fault diagnosis equations up to a one-dimensional perturbation on r. In other words, if r is any solution of the fault diagnosis equations, then any other solution in a neighborhood of r will be of the form

$$r + f(\alpha) \qquad (7.28)$$

where α is a real scalar and f is a function taking values in 3-space.

EXAMPLE 7.29: Consider the system shown in figure 7.32 having amplifiers whose gains r_1 and r_2 are the only potentially faulty parameters. Computing the transfer function and invoking the sensitivity theory of section 2 yields

$$\frac{dS(s)}{dr_1} = \frac{-r_2}{s(s+r_1)^2} \qquad (7.30)$$

and

$$\frac{dS(s)}{dr_2} = \frac{1}{s(s+r_1)} \qquad (7.31)$$

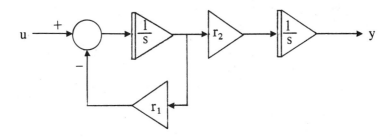

Figure 7.32. System having two potentially faulty gains.

Using two real test frequencies of $s_1 = 1$ and $s_2 = 2$ allows us to compute Φ as

$$
\Phi = \left[\frac{d\text{vec}[S]}{dr}\right](r) =
\begin{bmatrix}
\dfrac{-r_2}{(1+r_1)^2} & \dfrac{1}{(1+r_1)} \\[3mm]
\dfrac{-r_2}{2(2+r_1)^2} & \dfrac{1}{2(2+r_1)}
\end{bmatrix}
\tag{7.33}
$$

Multiplying by the least common denominator, $2(1+r_1)^2(2+r_1)^2$ leaves the polynomial matrix

$$
P(r) =
\begin{bmatrix}
-2r_2(2+r_1)^2 & 2(1+r_1)\,(2+r_1)^2 \\[3mm]
-r_2(1+r_1)^2 & (2+r_1)\,(1+r_1)^2
\end{bmatrix}
\tag{7.34}
$$

which has determinant

$$
\det[P(r)] = -2r_2(2+r_1)^2(1+r_1)^2
\tag{7.35}
$$

Since this determinant is not identically equal to zero, the generic rank of the Jacobian matrix, 7.33, is 2 and $\delta=0$. However, the rank of the Jacobian matrix is not identically a constant since $\det[P(r)] = 0$ whenever $r_2 = 0$, $r_1 = -1$, or $r_1 = -2$. Moreover, since we multiplied 7.33 by its LCD two of the degeneracies corresponding to $r_1 = -1$ or $r_2 = -2$ are points where the Jacobian is not well defined (poles) rather than true rank degeneracies.

Let us now return to the problem of choosing *test frequencies* and optimizing the solvability of the fault diagnosis equations. For this define δ_{min} to be the minimum value of δ achievable over any choice of test frequencies, $s_1,...,s_n$. Here both n and the value of s_i vary over all integers and complex numbers respectively.

THEOREM 7.36: δ_{min} is given by the formula

$$\delta_{min} \; = \; k \; - \; \text{col–rank} \left[\frac{d\,\text{vec}[S(s)]}{dr} \right] \tag{7.37}$$

where k equals the number of entries in r and "col-rank" denotes the number of linearly independent columns of the rational matrix, $d\,\text{vec}[S(s)]/dr$, over the field of complex numbers. Moreover, δ_{min} is achieved for almost any choice of "$k-\delta_{min}$" distinct complex frequencies.

PROOF: For brevity's sake we restrict the proof to the special case in which $S(s)$ is a scalar transfer function. The essential arguments generalize modulo some notational complexities. Since the rank of the Jocobian matrix is constant, a.e., we may fix the parameter vector, r, at any generic point, say r_q. The Jacobian matrix then reduces to a row vector of rational functions in s:

$$\left[\frac{dS(s)}{dr} \right]\Bigg|_{r_q} \; \triangleq \; R(s) \; = \; [R_1(s) \; . \; . \; . \; R_k(s)] \tag{7.38}$$

where

$$R_i(s) \; \triangleq \; \frac{dS(s)}{dr_i}\Bigg|_{r_q} \tag{7.39}$$

For this case, the theorem statement is equivalent to the following: Verify that the number of linearly independent columns of $R(s)$ over the field of complex scalars equals the maximum possible rank of the complex matrix

$$
\operatorname*{col}_{i}[R(S_i)] \;=\; \begin{bmatrix} R(s_1) \\ \hline R(s_2) \\ \cdot \\ \cdot \\ \cdot \\ \hline R(s_n) \end{bmatrix} \;=\; \left[\begin{array}{c|c|c|c} R_1(s_1) & R_2(s_1) & \cdots & R_k(s_1) \\ \hline R_1(s_2) & R_2(s_2) & \cdots & R_k(s_2) \\ \cdot & \cdot & & \cdot \\ \cdot & \cdot & & \cdot \\ \hline R_1(s_n) & R_2(s_n) & \cdots & R_k(s_n) \end{array}\right] \qquad (7.40)
$$

where n ranges over all positive integers and the distinct complex numbers s_i range over all possible values.

The method of proof is as follows: We first show that the rank of col $[R(s_i)]$ is always less than or equal to the number of linearly independent columns of $R(s)$. Then using an inductive argument we show that equality can be achieved for almost any choice of "$k-\delta_{min}$" distinct complex frequencies.

Now if the k-th column of $R(s)$ is dependent on the remaining columns, then

$$
R_k(s) \;=\; \sum_{j=1}^{k-1} \xi_j R_j(s) \qquad (7.41)
$$

for all complex frequencies, s, and for some set of complex scalars, ξ_j, not all zero. Consequently, for any set of n complex frequencies, $s_1,...,s_n$

$$
\operatorname*{col}_{i}[R_k(s_i)] \;=\; \sum_{j=1}^{k-1} \xi_j \operatorname*{col}_{i}[R_j(s_i)] \qquad (7.42)
$$

As such, since 7.41 implies 7.42 and

$$
\operatorname{rank}\left[\operatorname*{col}_{i}[R(s_i)]\right] \;\leqslant\; \text{col-rank}[R(s)] \qquad (7.43)
$$

where col-rank $[R(s)]$ means the number of linearly independent columns of $R(s)$.

Now it is necessary to prove that equality can be achieved by an appropriate number and choice of complex frequencies $s_1,...,s_n$. Suppose col-rank$[R(s)] = q$. Without loss of generality assume $R_1(s)$ through $R_q(s)$ are linearly independent rational functions (scalars are taken as complex numbers). The point is to show there exist complex frequencies, $s_1,...,s_q$ which make the first q columns of col$[R(s_i)]$ linearly independent. Again the argument will be inductive.

If q=1, there must exist an s_1 such that $R_1(s_1) \neq 0$ for otherwise $R_1(s) = 0$, contradicting the basic hypothesis. As such we may assume R^p (as defined below) has p linearly independent columns for some $p < q$.

$$
R^p \triangleq
\begin{bmatrix}
R_1(s_1) & R_2(s_1) & \cdots & R_p(s_1) \\
R_1(s_2) & R_2(s_2) & \cdots & \cdot \\
\cdot & \cdot & \cdot & \cdot \\
\cdot & \cdot & \cdot & \cdot \\
\cdot & \cdot & \cdot & \cdot \\
R_1(s_p) & R_2(s_p) & \cdots & R_p(s_p)
\end{bmatrix}
\tag{7.44}
$$

To conclude our argument we must show that there exists an s_{p+1} such that

$$
R^{p+1} \triangleq
\begin{bmatrix}
R_1(s_1) & R_2(s_1) & \cdots & R_p(s_1) \\
R_1(s_2) & R_2(s_2) & \cdots & \cdot \\
\cdot & \cdot & \cdot & \cdot \\
\cdot & \cdot & \cdot & \cdot \\
R_1(s_{p+1}) & R_2(s_{p+1}) & \cdots & R_p(s_{p+1})
\end{bmatrix}
\tag{7.45}
$$

has p+1 linearly independent columns.

The assumption that $S(s)$ is scalar valued implies R^p and R^{p+1} are square. Hence we may check that R^{p+1} has p+1 linearly independent columns by verifying that $\det[R^{p+1}] \neq 0$. Denote by $R^{p+1}(s)$ the matrix R^{p+1} in which s_{p+1} is replaced by the *variable*, s. In other words, $R^{p+1}(s)$ is a matrix whose botton row is a set of rational functions in s. Now we may expand $\det[R^{p+1}(s)]$ in terms of cofactors along the bottom row according to

$$
\det[R^{p+1}(s)] = \sum_{j=1}^{p+1} (-1)^{p+j+1} \, \Delta_{p+1,j} \, R_j(s)
\tag{7.46}
$$

where $\Delta_{p+1,j}$ is the cofactor associated with $R_j(s)$. Since R^p has linearly independent columns, $\Delta_{p+1,p+1} \neq 0$. Moreover, since $p < q$ and since $R_1(s)$ through $R_q(s)$ are linearly independent by hypothesis, there exists s_{p+1} such that $\det[R^{p+1}] \neq 0$. The proof is now complete.

As with the proofs of most deep theorems, the above generates further insight into the multifrequency fault diagnosis question. In addition to constructing a simple scheme for computing δ_{min}, it generates a natural criterion for choosing new test frequencies. Choose new test frequencies so as to maximize the number of linearly independent columns of the Jacobian matrix. For the scalar case the proof specifically shows the number of required test frequencies exactly equals $k-\delta_{min}$. In the general case where $S(s)$ is not scalar valued, the number of required test frequencies is less than or equal to $k-\delta_{min}$.

Again δ_{min} is the natural *measure of testability* of the system. It measures the diagnosability of a component failure relative to measurements taken at the accessible terminals of the system (i.e., u and y) with an optimal choice of test frequencies.

EXAMPLE 7.47: Consider the problem of fault diagnosis for the circuit of figure 7.48 where v_i is the only admissible input but both the input current I and the output voltage V_0 are admissible test outputs.

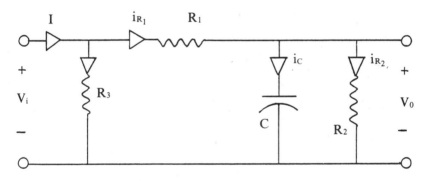

Figure 7.48. Circuit for fault diagnosis.

The transfer function matrix for this circuit, evaluated at nominal parameter values $R_1 = R_2 = R_3 = C = 1$, is

$$S(s) = \frac{1}{s+2} \begin{bmatrix} 2s+3 \\ 1 \end{bmatrix} \tag{7.49}$$

The Jacobian matrix evaluated at the nominal parameter vector is

$$
\frac{d\,\text{vec}[S]}{dr} =
\left[
\begin{array}{c|c|c|c}
\dfrac{-(s+1)^2}{(2s+3)^2} & \dfrac{-(s+1)\,(2s+1)}{(2s+3)^2} & \dfrac{-(s+1)^2}{(2s+3)^2} & \dfrac{-2s^2}{(2s+3)^2} \\[4mm]
\dfrac{-s}{(s+2)^2} & \dfrac{1}{(s+2)^2} & 0 & \dfrac{-s}{(s+2)^2}
\end{array}
\right]
\tag{7.50}
$$

This matrix has four linearly independent columns. However, if the first row of this matrix is considered alone, then only three columns are linearly independent. This situation would reflect having only I as an admissible test output. Thus, $\delta_{min} = 1$. If the only admissible output were V_0, then the corresponding Jacobian would be the second row of the above matrix. This row has only two independent columns making $\delta_{min} = 2$.

Using both test outputs and test frequencies $s_1 = 0$ and $s_2 = 1$, the Jacobian for the fault diagnosis equation is

$$
\left[\frac{d\,\text{vec}[S]}{dr} \right] =
\begin{bmatrix}
-.111 & -.111 & -.111 & 0 \\
0 & .25 & 0 & 0 \\
-.16 & -.24 & -.16 & -.08 \\
-.111 & .111 & 0 & -.111
\end{bmatrix}
\tag{7.51}
$$

The Jacobian matrix for the fault diagnosis equation under the same test frequencies relative to the test output V_0 is

$$
\begin{bmatrix}
0 & .25 & 0 & 0 \\
-.111 & .111 & 0 & -.111
\end{bmatrix}
\tag{7.52}
$$

To compute the Jacobian matrix corresponding to the test output I necessitates a third test frequency taken as $s_3 = -1$. In this case the Jacobian is

$$
\begin{bmatrix}
-.111 & -.111 & -.111 & 0 \\
-.16 & -.24 & -.16 & -.08 \\
0 & 0 & 0 & -2
\end{bmatrix}
\tag{7.53}
$$

The ranks of the above Jacobians matrices are four, two, and three respectively.

In the above example, we randomly chose test frequencies which, consistent with our theorem, resulted in maximal rank matrices with "probability one." In randomly choosing test frequencies one should exercise caution so as to avoid exceptional points. Unfortunately, at the time of this writing no methodology exists for optimally choosing test frequencies.

The final problem addressed in this section is the extension of multi-frequency diagnosis to nonlinear systems. Since nonlinear systems lack any sort of viable frequency domain theory, any direct extensions of the linear theory are out of the question.

Intuitively speaking, a transfer function matrix is an *integral performance measure*—the Laplace transform integral. In some sense we may conceptually generalize the linear multi-frequency approach to a nonlinear *multi-integral performance measure* approach. Specifically, we will consider a set of distinct integral performance measures in somewhat the same manner that we used distinct frequencies in the linear case.

As expected, the system model will be the nonlinear state form of the component connection model where

$$\dot{x} = f(x,a,r)$$
$$b = g(x,a,r)$$
(7.54)

is the component model and the interconnections with test inputs, u, and test outputs, y, are

$$a = L_{11}b + L_{12}u$$
$$y = L_{21}b + L_{22}u$$
(7.55)

The test inputs and outputs do not necessarily coincide with those of the system. Hopefully the test points are a much larger set. As per section 4 of this chapter, our integral performance measures take the form

$$P_i(r) = \int_0^T h_i(y,u,r)dt$$
(7.56)

for $i=1,2,...,n$ subject to the constraint

$$\frac{dr}{dt} \equiv 0$$
(7.57)

Intuitively, evaluation of the various performance measures corresponds to evaluating a transfer function matrix at multiple frequencies. We now attack the fault diagnosis problem by measuring the actual values for the integral performance measures, P_i^m of the system under test and solving equations 7.54, 7.55, 7.56, and 7.57 for r. Clearly there is little hope of finding an analytical solution for the set of equations 7.54, 7.55, 7.56 and 7.57. On the other hand one can still formulate a set of fault diagnosis equations which can be evaluated on a *system simulator* and solved numerically.

EXAMPLE 7.58: Consider the following nonlinear network for which the parameters ξ and k are to be diagnosed.

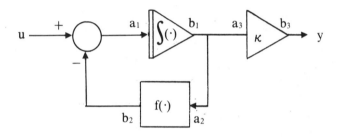

Figure 7.59. Nonlinear network for fault diagnosis.

For component 2, let $f(\cdot)$ be as follows

$$b_2 \; = \; f(a_2) \; = \; 1 - \exp(-\xi a_2) \tag{7.60}$$

for a nonzero scalar ξ. Define the test input u as

$$u \; = \; \cup(t) \; = \; \begin{cases} 1 & t \geqslant 0 \\ 0 & t < 0 \end{cases} \tag{7.61}$$

This inputs admits an exact analytical expression for y given zero initial conditions according to

$$y = \frac{k}{\xi} \ln [\xi t + 1] \tag{7.62}$$

In deriving 7.62 it is important to define $a_1(0) = 1$. In order to identify the parameters k and ξ, we use the following two performance measures,

$$P_1(k,\xi) \;=\; \int_0^1 y(t)dt \qquad\qquad (7.63)$$

and

$$P_2(k,\xi) \;=\; \int_0^1 y(t)\exp[ty(t)]dt \qquad\qquad (7.64)$$

Analytical expressions for both performance measures are given according to

$$P_1(k,\xi) \;=\; \frac{k}{\xi^2}\,[(\xi+1)\ln(\xi+1) - \xi] \qquad\qquad (7.65)$$

and

$$P_2(k,\xi) \;=\; [\xi+1]^{k/\xi} - 1 \qquad\qquad (7.66)$$

As such, equations 7.66 and 7.67 make up our desired set of *nonlinear fault diagnosis equations*.

For this simple set of fault diagnosis equations one can compute the *Jacobian matrix* analytically and thus use a *Newton-Raphson algorithm* for the solution of the fault diagnosis equations.

The required partial derivatives are

$$\frac{dP_1}{dk} \;=\; \frac{P_1}{k} \qquad\qquad (7.67)$$

$$\frac{dP_1}{d\xi} \;=\; \frac{-2k}{\xi^3}\,[(\xi+1)\ln(\xi+1) - \xi] + \frac{k}{\xi^2}\ln(\xi+1) \qquad (7.68)$$

$$\frac{dP_2}{dk} \;=\; \frac{1}{\xi}\,[\xi+1]^{k/\xi}\ln(\xi+1) \qquad\qquad (7.69)$$

and

$$\frac{dP_2}{d\xi} = \frac{k}{\xi}[\xi+1]^{k-\xi/\xi} - \frac{k}{\xi^2}(\xi+1)\ln(\xi+1) \qquad (7.70)$$

The Jacobian matrix J_p is

$$J_p = \begin{bmatrix} \dfrac{dP_1}{dk} & \dfrac{dP_1}{d\xi} \\[2ex] \dfrac{dP_2}{dk} & \dfrac{dP_2}{d\xi} \end{bmatrix} \qquad (7.71)$$

Taking an initial guess, $r_0 = [1\ 1]^t$, for our solution we obtain.

$$P(r_0) = \text{vec}[P_i(r_0)] = \begin{bmatrix} .386 \\ 1 \end{bmatrix} \qquad (7.72)$$

$$J_P(r_0) = \begin{bmatrix} .386 & -.079 \\ 1.386 & -.386 \end{bmatrix} \qquad (7.73)$$

and

$$J_p^{-1}(r_0) = \begin{bmatrix} 9.77 & -2 \\ 35 & -9.77 \end{bmatrix} \qquad (7.74)$$

Then one iteration of the Newton-Raphson algorithm will yield

$$(7.75)$$

$$r_1 = r_0 - J_p^{-1}(r_0)[P(r_0)-P^m] = \begin{bmatrix} 1 \\ 1 \end{bmatrix} - \begin{bmatrix} 9.77 & -2 \\ 35 & -9.77 \end{bmatrix} \left(\begin{bmatrix} .386 \\ 1 \end{bmatrix} - \begin{bmatrix} P_1^m \\ P_2^m \end{bmatrix} \right)$$

The following table summarizes the results of our nonlinear fault diagnosis problem in several cases where $r \approx r_0$ (in which case a single iteration of the Newton-Raphson algorithm can be expected to yield satisfactory results).

Measurements		Estimate (r_1)		Actual (r)	
$P_1{}^m$	$P_2{}^m$	k	ξ	k	ξ
.425	1.144	1.093	.958	1.1	1
.379	.963	1.001	1.101	1	1.1
.324	.732	.93	2.45	1	2

Table 7.76. Results of a fault diagnosis test.

At the time of this writing, almost no theory exists for fault diagnosis of nonlinear systems. However, in this particular example, the Jacobian matrix has rank 2 for almost all values of k and ξ. Consequently, the implicit function theorem implies, as in the linear case, that the fault diagnosis equation will be uniquely solvable in a neighborhood of almost any failure - i.e., $\delta_{min} = 0$. The extent to which such a result is generally applicable is, however, still unknown.

8. SOLUTION OF THE FAULT DIAGNOSIS EQUATIONS

At this point we discuss the solution of the *fault diagnosis equations*, 7.11, written in compact form as

$$\text{vec}[S^m] \;=\; \text{vec}[S(r)] \qquad (8.1)$$

Since 8.1 generally represents a large nonsparse set of nonlinear equations, some sophistication must underlie any efficient solution scheme. As a working hypothesis, let us assume 8.1 is uniquely solvable, i.e., $\delta_{min} = 0$.

Intuitively, a *Newton-Raphson* approach (as taken in example 7.58) may seem appropriate. The efficacy of such an approach depends on a good initial guess. The presence of potential short and open circuit phenomena precludes such a possibility. More precisely, let r_0 and r_f be the nominal and actual values of r respectively. For faulty systems, the *Euclidean norm*, $\| r_f - r_0 \|$, is generally large. Thus, r_0 does not serve as a good initial guess. The nonsparseness of 8.1 is another drawback. Finally, experience indicates the approach is not viable.

Let r^i denote the i-th entry of the parameter vector r. Although $\| r_f - r_0 \|$ is generally very large, $r_f{}^i - r_0{}^i \approx 0$ for most entries, typically all but 1,2,or 3, since it is reasonable to assume that only 1,2,or 3 of the system components have failed. A good solution algorithm should exploit this property. The first such algorithm falls under the general category of *fault simulation*.

The idea here is to construct a discretized set of parameter vectors which accounts, approximately, for all possible faulty parameter combinations. Assuming the number of faulty parameters, p, is much less than k, the number of entries in r will hopefully keep the size of this set within tolerable bounds. Note that the set must also include the nominal value of r. Label the elements of this set as r_j's. For each r_j simulate the system to obtain $vec[S_j]$. Essentially we have constructed a fault table. For each r_j we associate a complex vector, $vec[S_j]$.

To actually test a system, one directly measures $vec[S^m]$ and then searches the fault table to find that $vec[S_j]$ which best matches $vec[S^m]$. Since the match will not be exact, some type of decision process will be required.[8]

EXAMPLE 8.2: Consider the simple parallel RC circuit of figure 8.3 previously discussed in example 7.1.

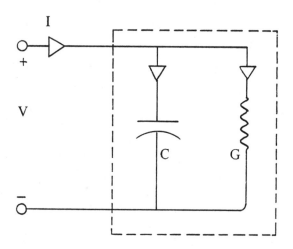

Figure 8.3. Parallel RC circuit with l output and v input.

The fault diagnosis equation 7.6 is rewritten below as

$$
\begin{bmatrix} S_1^m \\ S_2^m \end{bmatrix} = \begin{bmatrix} s_1 & 1 \\ s_2 & 1 \end{bmatrix} \begin{bmatrix} C \\ G \end{bmatrix}
\tag{8.4}
$$

Taking *test frequencies* $s_1 = 0$ and $s_2 = 1$ reduces 8.4 to the following linear algebraic equation.

$$\begin{bmatrix} S_1^m \\ S_2^m \end{bmatrix} = \begin{bmatrix} 0 & 1 \\ 1 & 1 \end{bmatrix} \begin{bmatrix} C \\ G \end{bmatrix} \tag{8.5}$$

Although uniquely solvable via matrix techniques, 8.5 is useful in illustrating the *fault simulation algorithm*. Let us assume, at most, one parameter is faulty. As possible faulty parameter values, we take shorts, opens, $\pm20\%$ off nominal and, of course, we also include the nominal, $(C=G=1)$. As such, the fault table includes 9 pairs of entries including one for the nominal system. Letting r be the column vector $[C,G]^t$, the *fault table* becomes

j	$r_j = r^t$	$vec[S_j] = [S_1,S_2]^t$
0	[1,1]	[1,2]
1	[0,1]	[1,1]
2	[.8,1]	[1,1.8]
3	[1.2,1]	[1,2.2]
4	[∞,1]	[1,∞]
5	[1,0]	[0,1]
6	[1,.8]	[.8,1.8]
7	[1,1.2]	[1.2,2.2]
8	[1,∞]	[∞,∞]

Table 8.6. Fault table for example 8.2.

Suppose in a particular instance a technician physically measured $vec[S^m] = [1.2,2.2]^t$. A search of the fault table reveals $r_f = r_7 = [1,1.2]^6$. In another case, suppose $vec[S^m] = [1.3,2.3]^6$. This vector fails to coincide with any vector in our fault table. However, it most closely approximates $vec[S_7]$. Hence we would choose $r_f \approx r_7 = [1,1.2]^t$. Solving 8.5 shows that $r_f = [1,1.3]^t$. Thus, r_7, the approximate r_f, would serve to correctly identify the faulty parameter. Finally suppose $vec[S^m] = [1.1,2.1]^t$. With respect to the usual Euclidean norm, this measured quantity equally matches $vec[S_0]$, $vec[S_3]$ and $vec[S_7]$. It is now necessary to choose between r_0, r_3 or r_7 as the best approximation to r_f. Clearly a more sophisticated decision process is warranted.[8]

If more than the assumed number of parameters are either faulty or off nominal, the required decision process is further complicated. Thus, the possibility of "new" fault modes and the ambiguities caused by the discretized set of possible parameter vectors is a major drawback to the fault simulation concept.

On the other hand, there are definite advantages to the approach. Fault simulation effectively exploits the fact that $p \ll k$. Most computation is done prior to the fault. The only on-line computation is a search of the fault table. The notion is well suited to a maintenance philosophy in which the manufacturer generates a fault table at the time of system fabrication. The manufacturer stores the data on mag tape and supplies the tape to the user at time of sale. The user then employs a mini-computer measure $vec[S^m]$ and searches the fault table.

A second approach for solving the fault diagnosis equation makes use of the *large change sensitivity analysis* algorithms of section 3. Again the approach is rendered feasible by $p \ll k$. Suppose one knows a priori that only p parameters have failed. Denote the appropriate entries in r by $r^{i(1)},\ldots,r^{i(p)}$. Derive the function $S_{i(1),\ldots,i(p)}(r^{i(1)},\ldots r^{i(p)})$ from S by setting all unknown parameters to their nominal value excepting of course $r^{i(1)},\ldots,r^{i(p)}$. The fault diagnosis equation reduces to solving

$$vec[S^m] = vec[S_{i(1)\ldots\ldots i(p)}(r^{i(1)},\ldots,r^{i(p)})] \qquad (8.7)$$

for p unknowns. Unfortunately, we do not know a priori which p parameters have failed. With the aid of the previously developed large change sensitivity formulas, the solution of 8.7 is sufficiently straightforward to attempt to solve the entire family of equations (the i-dependence of 8.7) for every combination of p or less faulty parameters. An equation of this family will then have a solution if and only if the parameters which have actually failed are among the $r^{i(1)},\ldots,r^{i(p)}$. Essentially, this idea amounts to a *search algorithm*. One implements the idea through *Householder's formula* and/or a *continuations scheme*. For small p, say ≤ 3, it may require less computation to solve $\binom{k}{p}$ nonlinear equations with p unknowns, than a single nonlinear equation with k unknowns.

A third scheme useful in solving the fault diagnosis equation is a variation of the above. This approach deals with the case in which there is no unique solution and provides a natural means for dealing with systems where the in-tolerance parameters are not nominal. Basically we replace the solution of the family of equations 8.7, with the unconstrained minimization of the family of *performance measures*

$$J_{i(1)}, \ldots, i(p) [r^{i^{(1)}}, \ldots, r^{i^{(p)}}] = \tag{8.8}$$
$$|| vec[S^m] - vec[S_{i(1)}, \ldots, i(p) \ r^{i^{(1)}}, \ldots, r^{i^{(p)}})||^2$$

The minimum in 8.8 is zero if and only if 8.7 has a solution. Therefore, they hold the same information for our purposes. Since the family of performance measures, 8.8, always has a minimum, fewer numerical difficulties are encountered over solving 8.7. Given that real world parameters are seldom nominal but within tolerance, none of the performance measures in 8.8 will actually attain a zero minimum. However, a decision as to the faulty parameter vector is possible by taking the parameter set yielding the smallest minimum for its associated performance measure.

The minimization of the performances 8.8 can be carried out by any of the *numerical optimization algorithms* discussed in section 6. In particular, for the very common case where p=1, the *one-dimensional line search* may be employed. For p>1 a *cyclical search* is ideal, since with the aid of our large change sensitivity theory all of the computations involved in the minimization process will be scalar, i.e., vary only one parameter at a time even though the search is in a p-dimensional space.

EXAMPLE 8.9: Again consider the parallel RC circuit of figure 8.3. Using the test frequencies $s_1 = 0$ and $s_2 = 1$, assuming at most one failure, and nominal values C=G=1, we may define

$$vec[S_C] = \begin{bmatrix} 1 \\ 1+C \end{bmatrix} \tag{8.10}$$

and

$$vec[S_G] = \begin{bmatrix} G \\ 1+G \end{bmatrix} \tag{8.11}$$

Natural performance measures then become

$$J_C = [1-S_1]^2 + [1+C-S_2]^2 \tag{8.12}$$

and

$$J_G = [G-S_1]^2 + [1+G-S_2]^2 \tag{8.13}$$

where S_1 and S_2 are defined via 8.5. Suppose one has measured $\text{vec}[S^m] = [1.2, 2.2]^t$. It is necessary to minimize

$$J_C = (1-1.2)^2 + (1+C-2.2)^2 \tag{8.14}$$

and

$$J_G = (G-1.2)^2 + (1+G-2.2)^2 \tag{8.15}$$

Now J_C is minimized by $C = 1.2$ with $J_C(1.2) = .04$, whereas J_G is minimized by $G = 1.2$ with $J_G(1.2) = 0$. As such, G is the faulty component and its value is 1.2 which is consistent with the result of the fault simulation algorithm.

For a second measured vector, $\text{vec}[S^m] = [1.3, 2.3]$ we minimize

$$J_C = (1-1.3)^2 + (1+C-2.3)^2 \tag{8.16}$$

and

$$J_G = (G-1.3)^2 + (1+G-1.3)^2 \tag{8.17}$$

Here, the former is minimized by $C = 1.3$ with $J_C(1.3) = .09$, whereas the latter is minimized by $G = 1.3$ with $J_G(1.3) = 0$. The faulty parameter is G and its value is 1.3. For this case, as with the fault simulation algorithm, we have correctly identified the faulty parameter. Moreover, we have correctly computed its value.

Finally, suppose the measured data vector is $\text{vec}[S^m] = [1.2, 2.1]^t$. The performance measures to be minimized are

$$J_C = (1-1.2)^2 + (1+C-2.1)^2 \tag{8.18}$$

and

$$J_G = (G-1.2)^2 + (1+G-2.1)^2 \tag{8.19}$$

In this case J_C has minimum $J_C(1.1) = .04$ when $C=1.1$, and J_G has minimum $J_G(1.23) = .0001125$ when $G=1.28$. Neither performance measure reduces to zero. Since $J_G(1.28)$ is the smaller of the two, we take G as the faulty parameter. The estimated value for G is 1.28. In fact, G is the faulty parameter with value 1.2. The anomaly here arises because $C = .9$

rather than its nominal value of 1. This in turn corrupts the measured data. In any event, this approach does yield a natural criterion for determining the faulty parameter and for estimating its value.[1]

As with the fault simulation algorithm, the search algorithm exploits the fact that $p \ll k$. This holds down the number of performance measures to be minimized and with the aid of the large change sensitivity formulae, limits the computation required for the minimization process. The algorithm will yield an exact solution whenever the actual number of faulty parameters is less than or equal to p. Moreover, the algorithm can cope with in-tolerance but non-nominal parameter values. The weakness of the algorithm is that it requires on-line computation and is thus feasible only for small values of p.

As should be apparent from the above examples, the solution of the fault diagnosis equations requires sophisticated numerical techniques even for relatively simple examples (hence our restriction to a two component example with at most one failure). The three techniques, however, are quite capable of coping with the nonlinearity of the fault diagnosis equations; they exploit the fact that $p \ll k$, and are amenable to computer implementation.

9. PROBLEMS

1. For the circuit of example 2.6 assume that both the circuit resistance and capacitance are controlled by a single integrated circuit process parameter, r, in the form $R(r) = r^2$ and $C(r) = \ln(r)$. Compute the circuit sensitivity to r.
2. For the circuit of example 5.11 compute the sensitivity with respect to C_1 and C_2 using nominal component values $C_1=1$, $C_2=1$, and $R=2$.
3. For the circuit of example 7.47 with input V_i and output $\mathrm{col}(I,V_o)$ compute the composite system transfer function and the sensitivity with respect to each of the four component values using nominal component values of $R_1=R_2=R_3=C=2$.
4. Derive equation 2.7.
5. Assume that the composite component transfer function matrix for an interconnected dynamical system is of the form $Z(r) = Zr$. Re-derive the large-change sensitivity formula of equation 3.5 for this special case. Repeat for $Z(r) = Z/r$.
6. Repeat example 3.6 using an *Euler-trapezoidal predictor-corrector* integration algorithm. Does the result converge to the actual solution?
7. If one is only interested in a *large-change sensitivity* analysis for the purpose of determining the effect of a change of the composite

component transfer function matrix from Z_0 to Z_f with no parameterization of $Z(r)$ between Z_0 and Z_f, then the continuation algorithm of equation 3.5 may be applied using an arbitrary parameterization, $Z(r)$, so long as $Z(0) = Z_0$ and $Z(1) = Z_f$. Formulate a parameterization satisfying these conditions which will have the effect of simplifying the differential equation 3.5.

8. Derive equation 3.34 from equation 3.32 and the *Householder formula*.

9. Repeat example 3.36 with

$$Z_p = Z + \begin{bmatrix} 100 & 0 \\ 0 & 0 \end{bmatrix}$$

$$Z_p = Z + \begin{bmatrix} 0 & 0 \\ 0 & 100 \end{bmatrix}$$

10. Derive equation 4.16 from equation 4.12.

11. Derive a generalization of the sensitivity theory of section 4, in which the *objective function* is allowed to be a function of time, i.e.

$$P = \int_0^T h(X,a,b,r,t)dt$$

12. For the interconnected dynamical system of example 4.42 compute the sensitivity of the *performance measure*

$$P = \int_0^T b_2(t)dt$$

with respect to variations in r.

13. Consider a system characterized by a scalar transfer function, $S(r,s)$ in which we desire to choose the parameter vector, r, to make $S(j\omega,r)$ approximate an ideal *low-pass filter* with cut-off frequency $\omega = 1$ preferable by using an r vector of small magnitude. Formulate a performance index for the optimization of r which will tend to achieve these ends.

14. For the circuit of example 5.11 choose a positive value for the resistance which will maximize the power dissipated in the resistor at

frequency $\omega = 5$. Choose a negative value of R which will maximize the power dissipated in the resistor at $\omega = 1$.

15. Sketch on a piece of paper the region of the plane in which the parameter vector, $r = (r_1, r_2)$, is allowed by the following simultaneous set of inequality constraints.

$$r_1 + r_2 - 4 \geqslant 0$$

$$r_1^2 + r_2^2 \geqslant 2$$

and

$$e^{r_1} - 1 \geqslant 0$$

16. Derive equation 5.22.

17. Show that if a vector, r_0, minimizes a performance measure $P(r)$ without constraints then the gradient of P evaluated at r_0 is zero.

18. Compute all values of the vector, $r = (r_1, r_2)$, which make the gradient of the performance measure zero.

$$P(r) = \cosh(\sinh(r_1^2 + r_2^2 + 1)$$

Identify the minima and maxima.

19. Use the *steepest descent* method for the optimization problem of example 6.24 with starting points $[1,1]^t$, $[3,0t$, and $[0,1]^t]$. Repeat using the *Fletcher-Reeves method*.

20. Indicate the modifications which would be required in the method of steepest descent to use it to determine the maximum of a performance measure. In this case it is called the method of *steepest ascent*.

21. Write the *Kuhn-Tucker equations* for the constrained optimization problem, maximize $P(r) = -r^2$ subject to the constraint $r \geqslant 1$. Are these equations valid?

22. Let the Jacobian matrix of a scalar frequency response with respect to five specified internal system parameters be:

$$[s^2 + 2s \mid s^3 + s + 1 \mid 3s \mid -s^2 + 2s \mid -1]$$

What is the measure of diagnosibility of the system? Find k-δ complex frequencies which achieve this value.

10. REFERENCES

1. Chen, H. S. M., and R. Saeks, "A Search Algorithm for Solution of the Fault Diagnosis Equations", Unpublished Notes, Texas Tech Univ., 1978.
2. Cooper, G. R., and C. D. McGillen, *Probabilistic Methods of Signal and System Analysis*, Holt, Rinehart, and Winston, New York, 1971.
3. Fleming, W., *Functions of Several Variables*, Addison-Wesley, Reading, 1965.
4. Hachtel, G., Brayton, R. K., and F. Gustavson, "The Sparse Tableau Approach to Network Analysis and Design", *IEEE Trans. on Circuit Theory*, Vol. CT-18, pp. 101-113, (1971).
5. Hachtel, G., and R. A. Rorher, "Techniques for Optimal Design and Synthesis of Switching Circuits", *IEEE Proc.*, Vol. 55, pp. 1864-1877. (1967).
6. Leung, K. H., and R. Spence, "Multiparameter Large-Change Sensitivity Analysis and Systematic Exploration", *IEEE Trans. on Circuits and Systems*, Vol. CAS-22, pp. 796-804, (1975).
7. Saeks, R., and S. R. Liberty, *Rational Fault Analysis*, Marcel Dekker, New York, 1977.
8. Sriynanda, H., Towill, D. R., and J. H. Williams, "Voting Techniques for Fault Diagnosis from Frequency Domain Test Date." *IEEE Trans. on Reliability*, Vol. R-24, pp. 260-267, (1975).
9. Varaiya, P.P., *Notes on Optimization*, Van Nostrand, Reinhold, New York, 1972.

VI. STABILITY

1. INTRODUCTION

This chapter concerns itself with lumped linear components and composite system stability. Our motivation for focusing on the linear theory is due to the fact that the stability theory of interconnected nonlinear systems is so vast and jumbled that suitable exposition requires a book in itself. Indeed, several adequately cover the topic.[4,6]

Sections 2 and 3 of the chapter treat the classical state model and frequency domain approaches to *BIBO and BIBS stability*.[3] They provide the pertinent background intrinsic to an in-depth understanding of the interconnected theory covered in the last half of the chapter.

Fundamental to the whole question are the notions of *eigenvalue/ eigenvector* which are also introduced and developed in sections 2 and 3. Sections 4 and 5 cover *numerical calculation* of these quantities. Section 4 quickly overviews several algorithms whereas the 5th section develops a *continuations approach* for calculating the eigenvalues/eigenvectors of a matrix function, F(r), of a real scalar parameter, r. This section, entitled *Eigenvalue Dynamics,* forms the basis for numerically implementing the *root locus* and *Nyquist theories* of sections 6 and 8 respectively.

A root locus technique takes shape in section 6 where the root locus is a plot of the eigenvalue loci as a function of coupling. Section 7 defines and justifies notions of *internal* and *external* stability from a frequency domain perspective. This leads smoothly into the *Nyquist theory* of section 8. Section 9 describes the concept of *sensitivity of stability* in terms of the root locus and Nyquist plots. This concept is basically a first order approximation to potential instability arising from parameter perturbations. The equations developed in section 5 again play a prominent role in this development.

2. COMPONENT STABILITY—TIME DOMAIN

The assumptions governing the *input-output model* of a linear component, as outlined in chapter I, guarantee that the component input, a, and the component output, b, are related through a *convolution integral*

$$b(t) \;=\; \int_{-\infty}^{\infty} Z(t-q)a(q)dq \tag{2.1}$$

where $Z(t)$ is the impulse matrix of the particular component. For *causal linear* components, $Z(t) = 0$, $t < 0$, which allows one to replace the upper limit in 2.1 by t. A change of variable permits an alternate form of 2.1 as given in 2.2.

$$b(t) \;=\; \int_{-\infty}^{\infty} Z(q)a(t-q)dq \tag{2.2}$$

In the context of this convolutional input-output representation, how may component stability be characterized in the sense that "bounded" inputs produce "bounded" outputs, i.e., *BIBO stability*?

First we pin down the precise meaning of *bounded* for scalar, vector, and matrix valued functions. Intuitively non-causal systems are unstable so our definition of BIBO stability must reflect this intuition. To define the notion of bounded we use the *sup-norm*.

DEFINITION 2.3: The sup-norm or L^{∞} norm of a scalar function $a(t)$ is

$$||a(\cdot)||_{\infty} \;=\; \sup_{t} |a(t)| \tag{2.4}$$

Generalizations to vector and matrix valued functions are straightforward as follows

$$||a(\cdot)||_{\infty} \;=\; \max_{i} ||a_i(\cdot)||_{\infty} \tag{2.5}$$

where $a(t) = \text{col}(a_1(t),...,a_n(t))$; clearly then for a matrix, say $Z(t) = [Z_{ij}(t)]$, $Z(t) = [Z_{ij}(t)]$, we have

$$\|Z(\cdot)\|_\infty = \max_{(i,j)} \|Z_{ij}(\cdot)\|_\infty \tag{2.6}$$

DEFINITION 2.7: A function, whether it be scalar, vector, or matrix valued, is *bounded* if and only if the L^∞ (sup-norm) is finite.

In particular, $a(\cdot)$ is bounded if and only if there exists a finite real number, K, such that

$$\|a(\cdot)\|_\infty < K \tag{2.8}$$

A norm is a mapping from the *space* of admissible functions to the real numbers satisfying properties outlined in proposition I.4.13. For each function, say $a(\cdot)$, whether it be scalar, vector, or matrix valued, we associate a single real number called its norm. Observe that much confusion surrounds the common use of $a(t)$ to mean both the function $a(\cdot)$ and the function evaluated at the point, t. The context typically indicates the precise use.

Intrinsic to the stability question is the concept of causality. A component is *causal* if whenever $a_1(t) = a_2(t)$ for $t \leqslant q$ then $b_1(t) = b_2(t)$ for $t \leqslant q$. For linear time invariant components we may characterize causality via the impulse response matrix.

DEFINITION 2.9: A causal component represented as 2.1 or 2.2 is BIBO stable if and only if for each bounded input, $a(\cdot)$, there exists a finite constant, K, such that

$$\|b\|_\infty \leqslant K\|a\|_\infty \tag{2.10}$$

Equivalently, a component is BIBO stable if and only if it is causal and bounded inputs map to bounded outputs. We may say directly that a component (not necessarily causal) is BIBO stable if and only if

$$\|E^q b(t)\|_\infty \leqslant K\|E^q a(t)\|_\infty \tag{2.11}$$

for all real numbers q where the *projection operator*, E^q, is defined as

$$E^q a(t) = \begin{cases} a(t) & t \leqslant q \\ 0 & t > q \end{cases} \tag{2.12}$$

Before characterizing BIBO stability in terms of the impulse response matrix, $Z(t)$, we need one further notion of norm. Let $Z(t)$ be a scalar valued admissible function. The L_1 *norm* of $Z(\cdot)$ is

$$\|Z\|_1 = \int_{-\infty}^{\infty} |Z(t)| \, dt \tag{2.13}$$

Clearly then for matrix valued functions, $Z(t)$, we have

$$\|Z\|_1 = \max_{(i,j)} \|Z_{ij}\|_1 \tag{2.14}$$

and similarly for vector valued functions.

THEOREM 2.15: A causal component modeled via 2.1 or 2.2 is BIBO stable if and only if there exists a finite constant, K, such that

$$\|Z\|_1 \leqslant K \tag{2.16}$$

PROOF: The proof will only consider the single input single output case. Generalization to the case in which $Z(t)$ is matrix valued is straightforward. First we prove the reverse direction. Let $a(\cdot)$ be a bounded input. By hypothesis we have that

$$\|Z\|_1 = \int_{0}^{\infty} |Z(t)| \, dt \leqslant K \tag{2.17}$$

Consider now the following string of inequalities

$$\|b\|_{\infty} = \sup_t \left| \int_{-\infty}^{t} Z(q) a(t-q) \, dq \right|$$

$$\leqslant \sup_t \int_{-\infty}^{\infty} |Z(q) a(t-q)| \, dq \tag{2.18}$$

$$\leqslant \|a\|_{\infty} \int_{-\infty}^{\infty} |Z(q)| \, dq$$

Thus from the causality of the component and the fact that $a(\cdot)$ is bounded, there exists a finite constant, K_1, such that

$$|| b ||_\infty \leq || a ||_\infty \int_0^{+\infty} | Z(q) | \, dq \leq K_1 || a ||_\infty \tag{2.19}$$

To prove the forward direction we use contradiction. Assume that bounded inputs map to bounded outputs, but that for each finite constant, K, there exists t_K such that

$$\int_{-\infty}^{t_K} | Z(q) | \, dq > k \tag{2.20}$$

In other words $||Z||_1$ is *not* finite. Define an input $a_K(q)$ as follows

$$a_K(q) \; = \; \text{sgn}[Z(t_K - q)] \tag{2.21}$$

where the sgn-function is defined as

$$\text{sgn}(\alpha) \; = \; \begin{cases} 1 & \alpha > 0 \\ 0 & \alpha = 0 \\ -1 & \alpha < 0 \end{cases} \tag{2.22}$$

Observe that

$$|| a_K(q) ||_\infty \; = \; 1 \tag{2.23}$$

This implies that

$$b_K(t) \; = \; \int_{-\infty}^{t_K} Z(t-q) a_K(q) \, dq > k \tag{2.24}$$

Thus we have that

$$|| E^{t_K} b_K(t) ||_\infty > k \; = \; k \, || a_K(t) ||_\infty \; = \; k \, || E^{t_K} a_K(t) ||_\infty \tag{2.25}$$

Since k can be chosen arbitrarily large there is no finite number k satisfying equation 2.10. Thus if 2.20 is true, definition 2.9 is not satisfied, i.e. not all bounded inputs map to bounded outputs. This is the desired contradiction. The theorem is established.

This theorem provides a very elegant characterization of bounded input–bounded output stability. Recall that for a time invariant state model, the impulse response matrix is given by

$$Z(t) = C[\exp(At)]B + D\delta(t) \tag{2.26}$$

Unfortunately, BIBO stability is insufficient for characterizing the stability of a component. Account must also be made for the internal behavior of the component. Intuitively the state vector characterizing the internal structure of the component must also remain bounded for all bounded inputs. Bounded input bounded state *(BIBS) stability* is the name given to internal stability.

Since the state vector $x(t)$ for causal components is given by

$$x(t) = \int_{-\infty}^{t} \exp[A(t-q)]Ba(q)dq \tag{2.27}$$

theorem 2.15 gives an immediate characterization of BIBS stability.

COROLLARY 2.28: A component is BIBS stable if and only if

$$\int_{0}^{+\infty} \|\exp[A(t-q)]B\|_1 < K \tag{2.29}$$

for some finite number K.

The central entity in these characterizations is $\exp(At)$ which is in turn completely characterized by the *eigenvalues* and *eigenvectors* of A. The remainder of the section gives alternate characterizations of BIBS and BIBO stability in terms of the locations of the eigenvalues of A. Recall that λ_i is an *eigenvalue* of A if and only if λ_i is a root of the equation

$$\det[A - \lambda I] = 0 \tag{2.30}$$

Now e_i is an *eigenvector* associated with the eigenvalue λ_i if and only if e_i is not equal to zero and

$$Ae_i = \lambda_i e_i \tag{2.31}$$

Clearly if $\xi \neq 0$ is any complex number and e_i any eigenvector of A, then ξe_i is also an eigenvector of A associated with the eigenvalue λ_i. For the remainder of the section we assume that A has n-distinct eigenvalues, $(\lambda_1,...,\lambda_n)$.

The *natural frequencies* or *modes* of a component are nicely characterized in terms of the eigenvalues and eigenvectors of A. Basically a mode is a complex sinusoid, $\exp(\lambda_i t)$, oscillating in a particular direction, the direction of the eigenvector. The plan then is to pin down the nature of these motions and their relationship to the stability question.

THEOREM 2.32: Let A be nxn and have n-distinct eigenvalues, $(\lambda_1,...,\lambda_n)$ with corresponding eigenvectors, $(e_i,...,e_n)$. Then $(e_1,...,e_n)$ are a *basis* for C^n with scalars in the field C. Since R^n is a subset of C^n, they are also a basis for R^n over C.

PROOF: Since C^n is an n-dimensional space over C and since $(e_1,...,e_n)$ is a set of n vectors, it suffices to show that the set is *linearly independent*. Recall that a set of vectors, $(e_1,...,e_n)$, is linearly independent if and only if

$$\sum_{i=1}^{n} \xi_i e_i = 0 \tag{2.33}$$

implies that all complex scalars $\xi_i = 0$.

If the set is not independent then there exists a set of n scalars, (ξ_i), not all zero, such that

$$\sum_{i=1}^{n} \xi_i e_i = 0 \tag{2.34}$$

where 0 is the zero vector. Assume without loss of generality that $\xi_1 \neq 0$. Multiply both sides of 2.34 by $(A-\lambda_2 I)(A-\lambda_3 I)... (A-\lambda_n I)$. After simplification we obtain

$$(A-\lambda_2 I)(A-\lambda_3 I)...(A-\lambda_n I)\xi_1 e_1 = 0 \tag{2.35}$$

However by moving ξ_1 to the left and successively operating on e_1 by $(A-\lambda_i I)$ we obtain

$$\xi_1(\gamma_1-\gamma_2)(\gamma_1-\gamma_3)...(\gamma_1-\gamma_n) = 0 \tag{2.36}$$

But since $\xi_1 \neq 0$ and all eigenvalues are distinct, equation 2.36 cannot be valid. This is the contradiction. Therefore $(e_1,...,e_n)$ is a basis for $\mathbf{C}^n$ over $\mathbf{C}$.

This is an important theorem for it allows us to completely characterize the "directions of motion" of any linear, lumped, time invariant component. The ultimate goal is to break $\exp(At)$ up into a sum of *directed modes*.

LEMMA 2.37: Let A have eigenvalues and eigenvectors as above, then

$$e^{At} e_i = e^{\lambda_i t} e_i \tag{2.38}$$

The proof to this lemma is straightforward upon considering the Taylor series expression for $\exp(At)$ of definition I.4.9.

In view of the above theorem and lemma let us analyze the zero-input state-response of a component. In other words let us consider the solution of

$$\dot{x}(t) = Ax(t) ; \quad x(0) = x_0 \tag{2.39}$$

From chapter I

$$x(t) = e^{At} x_0 \tag{2.40}$$

Since x_0 is a fixed n-vector, theorem 2.32 implies that there exists n complex scalars, $\xi_i(0)$, $i = 1,2,...,n$, such that

$$x_0 = \sum_{i=1}^{n} \xi_i(0) e_i \tag{2.41}$$

With the aid of lemma 2.37, we may express $x(t)$ (as per 2.40) in the following way

$$x(t) = \sum_{i=1}^{n} \xi_i(0) e^{\lambda_i t} e_i \tag{2.42}$$

Thus we have decomposed the state trajectory as a scaled sum of "natural oscillations" in "canonical directions". From this decomposition it becomes clear that the zero-input state trajectory remains bounded if and only if all eigenvalues of A lie in the *closed left half complex plane*.

Now observe that

$$e^{At} = e^{At}I \tag{2.43}$$

Each column of I can be expressed as a linear combination of eigenvectors via theorem 2.32. Thus, as per the derivation of 2.42, we may write

$$e^{At} = \sum_{i=1}^{n} R_i e^{\lambda_i t} \tag{2.44}$$

where the R_i are complex matrices satisfying

$$\sum_{i=1}^{n} R_i = I \tag{2.45}$$

and each column of R_i is a scalar multiple of e_i. The R_i matrices are called *residue matrices*. We will characterize these residue matrices again from a frequency domain perspective in the following section.

However let us return to the question of BIBS stability by substituting equation 2.44 into equation 2.27. We obtain

$$x(t) = \int_{-\infty}^{t} [\sum_{i=1}^{n} e^{\lambda_i(t-q)} R_i B] a(q) dq \tag{2.46}$$

Clearly, by inspection of 2.46 a sufficient condition for BIBS stability is that all eigenvalues, λ_i, lie in the *open left half plane*. Moreover, if any λ_i is the open right half complex plane, some state trajectory will contain a growing exponential and hence be unbounded. What about eigenvalue on the imaginary axis? By assumption all λ_i's are distinct. Therefore all eigenvalues appearing on $j\omega$-axis are simple. The state trajectory will remain bounded provided we cannot excite the eigenvalues at a frequency coincident with their location. Physically speaking, the B matrix must *decouple* all inputs from any mode on the imaginary axis. Equivalently the *range space* of the B-matrix must be *orthogonal* to any directed mode corresponding to an eigenvalue on the imaginary axis.

A straightforward derivation shows that the rows of each R_i are the eigenvectors of A^t (the transpose of A), denoted $(\underline{e}_1, \underline{e}_2, ..., \underline{e}_n)$, associated with eigenvalues $(\overline{\lambda}_1, \overline{\lambda}_2, ..., \overline{\lambda}_n)$ where (i) the over-bar load indicates complex conjugate and (ii) the eigenvalues of A^t are the complex conjugates of those of A. As will be shown in lemma 5.4 of this chapter, the eigenvectors of A and A^t satisfy

$$<\underline{e}_i, e_j> \ = \ 0 \quad \text{and} \quad <\underline{e}_i, e_i> \ \ne \ 0 \tag{2.47}$$

for $j \ne i$. Thus for B to mathematically decouple an input from any mode on the imaginary axis (for the range space of B to be orthogonal to such modes) no column of B may have a component along the direction of motion associated with an eigenvalue on the imaginary axis—i.e., in the direction of the eigenvector associated with the eigenvalue. In other words, suppose λ_1 is an eigenvalue and e_1 its eigenvector. If we express each column of B as a linear combination of the e_i's, then the coefficient of e_1, in each case, must be zero. This establishes the following theorem.

THEOREM 2.48: Let A be nxn and have distinct eigenvalues. Then the component is BIBS stable if and only if
 (i) all eigenvalues are in the closed left half complex plane,
 (ii) no column of B may have a component in the direction of an eigenvector associated with an eigenvalue on the imaginary axis.

This is much deeper than corollary 2.27 since it describes the interplay between the eigenvalues of A and the B matrix. Using this theorem as motivation, the BIBO case becomes obvious. One can think of the C-matrix in component model (A,B,C,D) as a filter on the internal system motions. If a natural frequency or directed motion of the system lies in the *null space* of the C-matrix, then the mode will be *unobservable* at the component output.

PROPOSITION 2.49: A component may be internally unstable (not BIBS stable) but BIBO stable provided the unstable responses lie in the null space of the C-matrix.

EXAMPLE 2.50: Suppose we have a component whose (A,B,C,D) state model is given by

$$A = \begin{bmatrix} -3 & 2 & 2 \\ 0 & -1 & 0 \\ -6 & 6 & 4 \end{bmatrix}; \ B = \begin{bmatrix} 1 & 1 & 2 \\ 0 & 1 & 1 \\ 2 & 0 & 2 \end{bmatrix}; \ C = \begin{bmatrix} -2 & 1 & 1 \\ 2 & 1 & -1 \\ 0 & 0 & 0 \end{bmatrix}; \ D = [0] \quad (2.51)$$

The eigenvalues of A are

$$\lambda_1 = -1, \lambda_2 = 0, \lambda_3 = 1 \tag{2.52}$$

whereas the corresponding eigenvectors are

$$e_1 = \begin{bmatrix} 1 \\ 1 \\ 0 \end{bmatrix}; \ e_2 = \begin{bmatrix} 2 \\ 0 \\ 3 \end{bmatrix}; \ e_3 = \begin{bmatrix} 1 \\ 0 \\ 2 \end{bmatrix} \tag{2.53}$$

Now since $\lambda_3 = 1$, the component is *not* BIBS stable. However, the component is BIBO stable. To see this, observe that the response corresponding to $\lambda_3 = 1$, lies in the null space of C. In particular e_3 is in the null space of C. Moreover, no column of B has a component in the direction of e_2. Therefore no input will excite the eigenvalue on the imaginary axis.

Practically speaking these refinements are highly superfluous. Clearly real world components are subjected to all types of noise which have infinite frequency spectrums. Thus any imaginary axis eigenvalues will always be excited by noise. Moreover, parameter variations do not permit eigenvalues to sit fixedly on the $j\omega$–axis.

In the following section we formulate analogous ideas in a frequency domain context.

3. COMPONENT STABILITY—FREQUENCY DOMAIN

Two objectives underlie this section. The first is to describe component BIBO stability via the component's transfer function matrix, $Z(s)$. The second is to tie together the frequency domain formulation of this section with the state model description of the previous.

In the first endeavor, note that equation 2.1 of the previous section is equivalent to equation 3.1 below.

$$b(s) = Z(s)a(s) \tag{3.1}$$

The equivalence was established in I. 3.13 and I. 3.14. Recall that $Z(s)$ is the

Laplace transform of the impulse response matrix, $Z(t)$. As with the convolutional representation, this is strictly an input-output model without cognizance of internal behavior. Implicitly we trust that the engineer who modeled the component accounted for all pertinent internal parameters. As per the assumptions in chapter I, $Z(s)$ is a matrix of rational functions. As such, it has an expansion as per 3.2.

$$Z(s) = \sum_{i=1}^{k} \sum_{j=1}^{m_i} \frac{Y_{ij}}{(s-\lambda_i)^j} + P(s) \qquad (3.2)$$

Here, each λ_i is a pole of $Z(s)$; m_i is the multiplicity of the pole; k is the number of poles; Y_{ij} are constant matrices having possibly complex entries; $P(s)$ is a polynomial matrix characterizing the poles of $Z(s)$ in a neighborhood of infinity.

EXAMPLE 3.3: Let us expand the scalar transfer function given in equation 3.4 as indicated in formula a 3.2.

$$Z(s) = \frac{(1 + 2s)^4}{s^2(s + 1)} \qquad (3.4)$$

As such, we must determine constants Y_1, Y_2, Y_3, and Y_4 so that

$$\frac{(1 + 2s)^4}{s^2(s + 1)} = \frac{Y_1}{s} + \frac{Y_2}{s^2} + \frac{Y_3}{s + 1} + Y_4 s + Y_5 \qquad (3.5)$$

Using the classical formulas for *continued fraction expansion*

$$Y_4 = Y_5 = 16 \qquad (3.6)$$

while a *partial fraction expansion* of the remainder yields

$$Y_1 = 7 \qquad (3.7)$$

$$Y_2 = 1 \qquad (3.8)$$

and

$$Y_3 = 1 \qquad (3.9)$$

With these facts in mind, we may state and prove the following theorem.

THEOREM 3.10: Let $Z(s)$ be a rational transfer function matrix characterizing a component. Then the component is BIBO stable if and only if $Z(s)$ has no infinite poles and the real part of each finite pole is negative.

PROOF: Suppose $Z(s)$ has no infinite poles and that $Re(\lambda_i) < 0$ for all poles, λ_i, of $Z(s)$. Considering the equivalence between convolutional and transfer function representation as per I. 3.13 and I. 3.14, we have

$$b(t) = \int_{-\infty}^{t} Z(t-q)a(q)dq \qquad (3.11)$$

where the impulse response matrix, $Z(t)$, has the form

$$Z(t) = \sum_{i=1}^{k} \sum_{j=1}^{m_i} Y_{ij}t^{j-1}\exp(\lambda_i t) + P_0\delta(t) \qquad (3.12)$$

for $t \geqslant 0$ and $Z(t) = 0$ for $t < 0$. The matrix $Z(t)$ is nothing more than the inverse Laplace transform of $Z(s)$ with due regard to regions of convergence and causality.

Observe that P_0 is the zeroth order term of the polynomial $P(s)$. All other coefficients of $P(s)$ are zero since by hypothesis $Z(s)$ has no poles at infinity. Moreover, since $Re[\lambda_i] < 0$ for all i, $Z(t)$ approaches the zero matrix exponentially as $t \rightarrow$. Clearly $Z(t)$ is exponentially decaying with t and consequently,

$$\int_{0}^{\infty} \max_{(i,j)} ||Z_{ij}(t)||_1 \ dt < \infty \qquad (3.13)$$

Thus, by theorem 2.15, the component is BIBO stable.

To prove the converse, we will consider three separate cases. The three cases will represent all conceivable ways for which $Z(s)$ will fail to satisfy the hypotheses of the theorem. To simplify the exposition, let $Z(s)$ be a scalar rational function. The matrix case is nothing more than a tedious manipulation of the scalar one.

First suppose $Z(s)$ has a pole at infinity. Consider the bounded input

$$a(t) \;\;=\;\; \sin{(t^2)} \cup (t) \tag{3.14}$$

Poles at infinity represent differentiators. Consequently, the component output (due to $a(t)$ in 3.14) will have terms of the form $t^n \cos(t^2) \cup (t)$ and $t^n \sin(t^2) \cup (t)$ for $n \geqslant 1$. These terms cannot be cancelled and the output will therefore be unbounded.

As a second case, suppose $Z(s)$ has a finite pole in the right half complex plane; i.e., a pole, λ_q, such that $\mathrm{Re}[\lambda_q] > 0$. Let w be a complex number which is neither a pole nor a zero of $Z(s)$. Exciting the component with an input of the form

$$a(t) \;\;=\;\; \exp[wt] \cup (t) \tag{3.15}$$

yields a response

$$b(s) \;\;=\;\; \frac{Z(s)}{s-w} \tag{3.16}$$

The time domain form of b will be

$$b(t) \;\;=\;\; \sum_{i=1}^{k} \sum_{j=1}^{m_i} y_{ij}\, t^{j-1} \exp(\lambda_i t) \cup (t)$$

$$+\; y\, \exp(wt) \;+\; \sum_{i=0}^{n} p_i w^i \exp(wt) \cup (t) \tag{3.17}$$

Clearly $b(t)$ will be unbounded since it will contain the exploding term

$$\sum_{j=1}^{m_q} y_{qj}\, t^{j-1} \exp(\lambda_q t) \tag{3.18}$$

This follows because the highest order "residue" of the pole λ_q will be non-zero since the input $\exp(wt) \cup (t)$ does not affect the multiplicity of the pole. Since distinct exponentials are linearly independent, the exploding term cannot be cancelled and will always be non-zero.

For poles λ_q with $\mathrm{Re}[\lambda_q] = 0$ having multiplicity strictly greater than one, the above argument carries over. The component output will be unbounded. On the other hand, suppose $\text{Re}[\lambda_q] = 0$ and the multiplicity of this imaginary axis pole is one. By choosing the following bounded input

$$a(t) = \exp(\lambda_q t) \cup (t) \tag{3.19}$$

we produce an unbounded output. This follows by considering a partial fraction expansion of $b(s)$ followed by the inverse Laplace transform to obtain $b(t)$. The proof is now complete.

In the final part of this section, we concretize the relationship between the theory of this section and the previous one. For this discussion, we restrict our attention to the distinct eigenvalue (pole) case. Given a component state model of the form

$$\dot{x}(t) = Ax(t) + Ba(t)$$
$$\tag{3.20}$$
$$b(t) = Cx(t) + Da(t)$$

it becomes a simple matter to derive (I.4.39) the transfer function matrix model

$$Z(s) = C(sI-A)^{-1}B + D \tag{3.21}$$

where the inverse of $(sI-A)$ is the Laplace transform of $\exp(At)$. The only trying step in computing $Z(s)$ then is in computing $(sI-A)^{-1}$. Since $(sI-A)$ is a polynomial matrix, using the classical inversion formula, we have

$$(sI-A)^{-1} = \frac{R(s)}{d(s)} \tag{3.22}$$

where $R(s)$ is a polynomial matrix and $d(s) = \det[sI-A]$ is called the *characteristic polynomial* of the matrix A. Expanding 3.22 into partial fractions we obtain

$$\frac{R(s)}{d(s)} = \frac{R_1}{s-\lambda_1} + \ldots + \frac{R_n}{s-\lambda_n} \tag{3.23}$$

where λ_i is an eigenvalue of A (and consequently a zero of $d(s)$) and R_i is the corresponding *residue matrix*. We have assumed A has n distinct eigenvalues. A little thought justifies the following formula for the R_i.

$$R_i = \lim_{s \to \lambda_i} \frac{(s - \lambda_i) R(s)}{d(s)} \tag{3.24}$$

Simply by taking the inverse Laplace transform of 3.23 we arrive at the expression for exp(At) given in 2.46, i.e.,

$$e^{At} = \sum_{i=1}^{n} R_i e^{\lambda_i t} \tag{3.25}$$

The following theorem provides a characterization of the residue matrices.

THEOREM 3.26: The residue matrices have the following properties:

(i) $\sum_{i=1}^{n} R_i = I$

(ii) $AR_k = \lambda_k R_k$

(iii) $R_k A = \lambda_k R_k$

(iv) $R_i R_k = \delta_{ik} R_i$

where

$$\delta_{ik} = \begin{cases} 0 & i \neq k \\ 1 & i = k \end{cases} \tag{3.27}$$

PROOF: The first three properties are straightforward as per the discussion in the previous section. Note that property (iii) implies that the rows of R_i are eigenvectors of A^t. As such, we will only prove the fourth property. Before proceeding, observe that the matrix product $(\lambda_i I - A) R_i = 0$ does not imply that "$(\lambda_i I - A) = 0$ or $R_i = 0$." Both factors may be non-zero with their product zero as shown in 3.28.

$$\begin{bmatrix} 1 & 0 \\ 0 & 0 \end{bmatrix} \begin{bmatrix} 0 & 0 \\ 1 & 1 \end{bmatrix} = \begin{bmatrix} 0 & 0 \\ 0 & 0 \end{bmatrix} \tag{3.28}$$

To prove (iv), consider that property (ii) implies

$$R_j A R_i = \lambda_i R_j R_i \tag{3.29}$$

Furthermore, property (iii) implies

$$R_j A R_i = \lambda_i R_j R_i \tag{3.30}$$

Subtracting 3.30 from 3.29 yields

$$[0] = (\lambda_i - \lambda_j) R_j R_i \tag{3.31}$$

for all i and j. Since the eigenvalues are distinct, for $i \neq j$, we have that

$$R_j R_i = 0 \tag{3.32}$$

In addition, from property (i) and the equality of 3.41, we have

$$R_j = R_j I = R_j \left[\sum_{i=1}^{n} R_i \right] = R_j R_j \tag{3.33}$$

Therefore

$$R_j R_i = \delta_{ji} R_j \tag{3.34}$$

The theorem is now complete.

Before closing the section with an example, we give a simple scheme for computing $R(s)$ in equation 3.22. Since $R(s)$ is a polynomial matrix, we may express $(sI-A)^{-1}$ as

$$\frac{R(s)}{d(s)} = \frac{N_0 s^{n-1} + N_1 s^{n-2} + \dots + N_{n-1}}{s^n + d_1 s^{n-1} + \dots + d_n} \tag{3.35}$$

THEOREM 3.36: Assuming the polynomial $d(s)$ is known, then the R_k matrices satisfy the following formulas.

    (i) $N_0 = I$
    (ii) $N_k = N_{k-1} A + d_k I$ '; $n-1 \geq k \geq 1$
    (iii) $[0] = N_{n-1} A + d_n I$

PROOF: Multiplying both sides of 3.35 on the right by $(sI-A)d(s)$ simplifies to

$$d(s) = [s^{n-1} N_0 + s^{n-2} N_1 + \dots + N_{n-1}](sI-A) \tag{3.37}$$

As such

$$[s^n + d_1 s^{n-1} + \ldots + d_n]I \tag{3.38}$$

$$= s^n N_0 + s^{n-1}(N_1 - N_0 A) + \ldots + s(N_{n-1} - N_{n-2}A) - N_{n-1}A$$

By equating coefficients in like powers of s, we arrive at the desired formulas.

EXAMPLE 3.39: Let us reformulate example 2.50 in the frequency domain. Recall the (A, B, C, D) component state model is given by

$$A = \begin{bmatrix} -3 & 2 & 2 \\ 0 & -1 & 0 \\ -6 & 6 & 4 \end{bmatrix}; B = \begin{bmatrix} 1 & 1 & 2 \\ 0 & 1 & 1 \\ 2 & 0 & 2 \end{bmatrix}$$

$$C = \begin{bmatrix} -2 & 1 & 1 \\ 2 & 1 & -1 \\ 0 & 0 & 0 \end{bmatrix}; D = [0] \tag{3.40}$$

Initially let us compute $d(s) = \det(sI-A)$.

$$d(s) = s^3 - s \tag{3.41}$$

To compute $(sI-A)^{-1}$ we must find N_0, N_1, and N_2 such that

$$(sI-A)^{-1} = \frac{R(s)}{d(s)} = \frac{N_0 s^2 + N_1 s + N_2}{s^3 - s} \tag{3.42}$$

The formulas in theorem 3.36 yield

$$(sI-A)^{-1} = \frac{Is^2 + \begin{bmatrix} -3 & 2 & 2 \\ 0 & -1 & 0 \\ -6 & 6 & 4 \end{bmatrix} s + \begin{bmatrix} -4 & 4 & 2 \\ 0 & 0 & 0 \\ -6 & 6 & 3 \end{bmatrix}}{s^3 - s} \tag{3.43}$$

Multiplying on the left by C and on the right by B produces Z(s) as

$$Z(s) = \frac{1}{s+1} \begin{bmatrix} 0 & -1 & -1 \\ 0 & 3 & 3 \\ 0 & 0 & 0 \end{bmatrix} \tag{3.44}$$

In a neighborhood of t = $-\infty$, our mathematics requires that all initial conditions be zero. Thus the effects of the "internal" eigenvalues, $\lambda_2 = 0$ and $\lambda_3 = 1$, will not be evident in Z(s) as 3.53 so indicates.

This completes our discussion of component stability. Two purposes directed the discussion. The dominant aim was to provide intuition and motivation for the forthcoming investigation of interconnected system stability. Here we will define internal and external system stability along with numerically implementable tests for determining such stability. The second aim was to review some of the more basic results of classical state space stability theory necessary for a deep understanding of the remainder of the chapter.

4. EIGENVALUE COMPUTATION

Eigenvalue/eigenvector calculation via the *continuations approach* (discussed in section 5) requires *initial condition* information. In turn, algorithms for evaluating eigenvalues/eigenvectors become necessary. This section outlines two approaches to the eigenvalue problem. The eigenvector problem can then be solved in a simultaneous linear equation context. The first numerical technique works for symmetric matrices and rests on the property that *similar matrices* have identical eigenvalues. •

DEFINITION 4.1: Two matrices A and B are said to be similar if there exists a nonsingular transformation T such that

$$B = TAT^{-1} \tag{4.2}$$

LEMMA 4.3: The eigenvalues of similar matrices coincide.

PROOF: A complex number, λ, is an eigenvalue of a matrix, A, if and only if the matrix, $(\lambda I - A)$, is singular. Observe that for any nonsingular transformation, T,

$$T(\lambda I - A)T^{-1} = (\lambda I - TAT^{-1}) = (\lambda I - B) \tag{4.4}$$

The nonsingularity of T implies that $(\lambda I - A)$ is singular if and only if $(\lambda I - B)$ is singular. Therefore, the eigenvalues of A and B coincide.

The scheme for computing the *eigenvalues of a symmetric matrix*, (i.e., having real eigenvalues), A, is an iterative one. The algorithm constructs a sequence of matrices (A_k), each similar to A, converging to a lower triangular matrix, B. The diagonal entries of B are the eigenvalues of A.

Define the first matrix in the sequence as $A_1 = A$. Decompose A_1 into an *LU-factorization* as

$$A_1 = L_1 U_1 \tag{4.5}$$

where U_1 is upper triangular with units on the diagonal. The *Crout Algorithm* (III.4) is a convenient technique for executing the factorization. Define the second matrix in the sequence as

$$A_2 = U_1 L_1 \tag{4.6}$$

The matrix A_2 is just a juxtaposition of the LU-factorization of A, and it is similar to A, since

$$A_2 = U_1(A_1 U_1^{-1}) = U_1 A_1 U_1^{-1} \tag{4.7}$$

Iterating, one factors $A_2 = L_2 U_2$ and defines A_3 by juxtaposing L_2 and U_2.

$$A_3 = U_2 L_2 \tag{4.8}$$

Again A_3 is similar to A as follows

$$A_3 = U_2[U_1 A_1 U_1^{-1}] = [U_2 U_1] A [U_2 U_1]^{-1} \tag{4.9}$$

In general

$$A_{k+1} = U_k L_k = [U_k U_{k-1}...U_1] A_1 [U_k U_{k-1}...U_1]^{-1} \tag{4.10}$$

where $A_k = L_k U_k$ is an LU-factorization of A_k. This then determines a sequence of matrices, (A_k), each of which is similar to A by a unit upper triangular matrix.

THEOREM 4.11: If the sequence $([U_k U_{k-1}...U_1])$ converges, then the sequence (A_k) converges to a lower triangular matrix whose diagonal entries are the eigenvalues of A.

PROOF: Clearly if the sequence $([U_k U_{k-1}...U_1])$ converges, then

$$\lim_{k \to \infty} [U_k] = I \tag{4.12}$$

Consequently equations 4.10 and 4.12 imply

$$\lim_{k \to \infty} A_k = \lim_{k \to \infty} L_{k-1} \tag{4.13}$$

and

$$\lim_{k \to \infty} [U_{k-1}^{-1} A_k] = \lim_{k \to \infty} [(U_{k-2}U_{k-3}...U_1) A_1 (U_{k-1}...U_1)] \tag{4.14}$$

Specifically, 4.14 guarantees that the A_k converge if the $[U_k U_{k-1}...U_1]$ do, and that the limiting A_k is similar to $A_1 = A$ whereas 4.13 guarantees that the limit is lower triangular. Therefore, lemma 4.3 certifies that the eigenvalues of the limiting A_k coincide with those of A. The proof is now complete.

It can be shown that the sequence $([U_k U_{k-1}...U_1])$ converges whenever the eigenvalues are distinct. More generally the sequence (A_k) will converge to an appropriate *block lower triangular* matrix for nondistinct eigenvalues. Each block will correspond to a distinct eigenvalue.[9]

Observe that if A is not symmetric, the theorem is false. As a counter example, let

$$A = \begin{bmatrix} -3 & 2 & 2 \\ 0 & -1 & 0 \\ -6 & 6 & 4 \end{bmatrix} \tag{4.15}$$

which has distinct eigenvalues $\lambda_1 = 1$, $\lambda_2 = 0$, and $\lambda_3 = -1$.

EXAMPLE 4.16: Find the eigenvalues of the matrix A where

$$A = \begin{bmatrix} 1 & 0 & 1 \\ 0 & 1 & 0 \\ 1 & 0 & 1 \end{bmatrix} \tag{4.16}$$

Observe that

$$A = A_1 = L_1 U_1 = \begin{bmatrix} 1 & 0 & 0 \\ 0 & 1 & 0 \\ 1 & 0 & 0 \end{bmatrix} \begin{bmatrix} 1 & 0 & 1 \\ 0 & 1 & 0 \\ 0 & 0 & 1 \end{bmatrix} \tag{4.17}$$

Consequently

$$A_2 = U_1 L_1 = \begin{bmatrix} 2 & 0 & 0 \\ 0 & 1 & 0 \\ 1 & 0 & 0 \end{bmatrix} \qquad (4.18)$$

which is lower triangular. Hence the eigenvalues of A are $\lambda_1 = 2$, $\lambda_2 = 1$, and $\lambda_3 = 0$.

A variation on the above scheme is the well documented *QR-algorithm*. Observe that the derivation of the *LU-algorithm* made no explicit use of the upper triangularity of the U_k matrices. The QR-algorithm is essentially the same as the LU scheme except that U_k is replaced by an arbitrary non-singular *orthogonal matrix*, Q_k. An orthogonal matrix is one whose columns are pairwise orthonormal. In particular,

$$\sum_{i=1}^{n} q_{ij} \, q_{im} = \delta_{jm} \qquad (4.19)$$

for $j = 1,2,...,n$, where $Q_k = [q_{ij}]$ and δ_{jm} is the well-known *Kronecker delta* defined by equation 3.37.

Correspondingly, define $A_1 = A$ and A_{k+1} from A_k as

$$A_{k+1} = Q_k L_k \qquad (4.20)$$

where we have previously factored A_k as

$$A_k = L_k Q_k \qquad (4.21)$$

Of course, L_k is lower triangular and Q_k is orthogonal. Observe that A_{k+1} is *congruent* to $A = A_1$ since

$$A_{k+1} = [Q_k Q_{k-1}...Q_1] A_1 [Q_k Q_{k-1}...Q_1]^{-1} \qquad (4.22)$$

Lemma 4.3 again certifies that the eigenvalues of A_k and A coincide for all k. Under appropriate conditions the sequence (A_k) converges to a *lower triangular matrix* whose diagonal entries are the eigenvalues of A.

To apply the QR-algorithm to eigenvalue calculation of a nonsymmetric A, it is usually convenient to first convert A to a similar matrix B which is in upper *Hessenberg form*. It is necessary to find a nonsingular matrix T such that

$$B = TAT^{-1} \tag{4.23}$$

where B is in upper Hessenberg Form if $b_{ij} = 0$ whenever $i > j-1$. Intuitively B is almost upper triangular in that the non-zero entries lie on or above the *sub-diagonal*. Conversion to Hessenberg form comes easily for *sparse matrices* since few non-zero entries fall below the sub-diagonal. Typically the process is accomplished with *row and column permutations* followed by *elementary row and column operations* only when necessary.

EXAMPLE 4.24: Let A be the following 7x7 matrix.

$$A = \begin{bmatrix} 1 & 0 & 0 & 0 & 0 & 0 & 1 \\ 1 & 0 & 0 & 0 & 1 & 0 & 2 \\ 0 & 0 & 2 & 1 & 0 & 0 & 1 \\ 0 & 1 & 0 & 0 & 0 & 0 & 0 \\ 1 & 0 & 0 & 0 & -1 & -1 & 0 \\ 0 & 0 & 0 & 0 & 0 & 1 & 0 \\ 0 & 0 & 1 & 0 & 0 & 1 & 1 \end{bmatrix} \tag{4.25}$$

Conversion to upper Hessenberg form proceeds with one elementary row operation and three row and column permutations. Specifically, subtract row two from five, interchange rows three and four, five and seven, and six and seven; finally, carry out corresponding column permutations to obtain:

$$B = \begin{bmatrix} 1 & 0 & 0 & 0 & 1 & 0 & 0 \\ 1 & 0 & 0 & 0 & 2 & 1 & 0 \\ 0 & 1 & 0 & 0 & 0 & 2 & 0 \\ 0 & 0 & 1 & 2 & 1 & 0 & 0 \\ 0 & 0 & 0 & 1 & 1 & 0 & 1 \\ 0 & 0 & 0 & 0 & -2 & -1 & -2 \\ 0 & 0 & 0 & 0 & 0 & 0 & 1 \end{bmatrix} \tag{4.26}$$

Upper Hessenberg form makes eigenvalue computation readily achievable. Either evaluate $\det[\lambda I - B]$ or apply the QR-algorithm. Note that the sequence (B_k) resulting from a QR-algorithm will all be in upper Hessenberg form.

Once a particular eigenvalue, λ_i, is known, the eigenvector, e_i, results by solving the linear equation

$$(\lambda_i I - A)e_i = 0 \tag{4.27}$$

The matrix, $(\lambda_i I - A)$, will always be singular hence 4.27 has infinitely many solutions. To obtain a particular solution, fix one of the entries of e_i at a convenient value.

In the following section we will construct a continuations algorithm whose initial conditions require knowledge of eigenvalues/eigenvectors of both A and A* (the complex conjugate transpose of A). The eigenvalues of A and A* are complex conjugates of each other whereas the eigenvectors are not.

Suppose the eigenvalues of A are distinct. Let T be a matrix whose columns are the corresponding eigenvectors of A. It is a simple matter to show that the eigenvectors of A* are the columns of the matrix, $[T^{-1}]^*$.

5. EIGENVALUE DYNAMICS

Computationally speaking, execution of the *Nyquist* and *root locus* theories presented in the following three sections demands knowledge of the *eigenvalues* of a matrix, $F(r)$, as a function of a variable, r. For the Nyquist theory knowledge of the eigenvalue loci of $Z(j\omega)L_{11}$ and $(I-Z(j\omega)L_{11})$ as function of ω is necessary.

Certainly one approach to identifying the eigenvalue loci of $F(r)$ as a function of r is to repeatedly compute the eigenvalues at successive values of r. Greater computational speed and some theoretical elegance arises via use of a suitable *continuations algorithm*. Recall that the central idea of a continuations algorithm is to write a differential equation whose solution identifies the desired loci. For the Nyquist theory, we will use a set of three coupled differential equations whose solution identifies the eigenvalue and eigenvector trajectories of $(I-Z(j\omega)L_{11})$. The equations are initialized at the eigenvectors and eigenvalues of $(I-Z(j\omega_0)L_{11})$ for some ω_0.

To insure the flexibility of our theory, we derive the differential equations with regard to an arbitrary matrix, $F(r)$. The key relationship in the derivation is the interplay between the *eigenvalues/eigenvectors* of $F(r)$ and those of its *adjoint matrix*, $F(r)^*$. To obtain $F(r)^*$, first transpose $F(r)$, then take the complex conjugate of every entry. Mathematically, $F(r)^*$ is the *unique* matrix satisfying the equality

$$<F(r)x,y> \;=\; <x,F(r)^*y> \tag{5.1}$$

for all complex n vectors x and y where $F(r)$ is n x n. Note that $<\cdot,\cdot>$ denotes the classical complex *inner product* defined as

$$<x,y> \;=\; \sum_{i=1}^{n} x_i \bar{y}_i \tag{5.2}$$

where $x = \text{col}(x_i)$, $y = \text{col}(y_i)$ and $\bar{y}_i$ is the complex conjugate of y_i.

To explore the kinship between the eigenvalues and eigenvectors of $F(r)$ and $F(r)^*$, observe that for any complex number λ, $[\lambda I - F(r)]$ is nonsingular if and only if

$$[\lambda I - F(r)]^* = [\bar{\lambda} - F(r)^*] \tag{5.3}$$

is nonsingular. Essentially, $\det[\lambda I - F(r)] = 0$ if and only if $\det[\bar{\lambda} I - F(r)^*] = 0$. Clearly then, the eigenvalues of $F(r)^*$ are the complex conjugates of the eigenvalues of $F(r)$.

Assume $F(r)$ has n-distinct eigenvalues, $(\lambda_1, \lambda_2, ..., \lambda_n)$. Denote those of $F(r)^*$ as $(\bar{\lambda}_1, ..., \bar{\lambda}_n)$. Unfortunately, the *eigenvectors* do not share a complex conjugate relationship. For notational convenience, denote the eigenvalues of $F(r)$ and $F(r)^*$ as $(e_1, ..., e_n)$ and $(\underline{e}_1, ..., \underline{e}_n)$ respectively. Whenever the eigenvalues of $F(r)$ are distinct, the complex conjugate relationship between those of $F(r)$ and $F(r)^*$, defines a one-to-one correspondence between the eigenvectors of $F(r)$ and those of $F(r)^*$, i.e., between the eigenvector e_i associated with λ_i and $\underline{e}_i$ associated with $\bar{\lambda}_i$. It is important, especially in programming, to be aware of the *specific correspondence*. Fortunately, the eigenvectors do satisfy a highly utile relationship.

LEMMA 5.4: Let $F(r)$ and $F(r)^*$ be as described above. In particular, let $F(r)$ be n x n with $n \geqslant 2$. Then

$$\langle e_i, \underline{e}_j \rangle = 0 \text{ and } \langle e_i, \underline{e}_i \rangle \neq 0 \tag{5.5}$$

for $j \neq i$.

PROOF: Since the eigenvalues of $F(r)$ are distinct, at least one is non-zero. Without loss of generality, assume it is λ_i. Whenever $j \neq i$, the linearity of the inner product implies

$$\begin{aligned} \langle e_i, \underline{e}_j \rangle &= \frac{1}{\lambda_i} \langle \lambda_i e_i, \underline{e}_j \rangle = \frac{1}{\lambda_i} \langle F(r) e_i, \underline{e}_j \rangle \\ &= \frac{1}{\lambda_i} \langle e_i, F(r)^* \underline{e}_j \rangle = \frac{1}{\lambda_i} \langle e_i, \bar{\lambda}_j \underline{e}_j \rangle \\ &= \frac{\lambda_j}{\lambda_i} \langle e_i, \underline{e}_j \rangle \end{aligned} \tag{5.6}$$

The distinctness of the eigenvalues forces $\lambda_j / \lambda_i \neq 1$. Therefore, $\langle e_i, \underline{e}_j \rangle = 0$ for $i \neq j$.

To verify that $\langle e_i, \underline{e}_i \rangle \neq 0$ assume the contrary, i.e., assume

$\langle e_i, \underline{e}_i \rangle = 0$. Recall from section 2 of this chapter that $(e_1,...,e_n)$ are a basis for $\mathbf{C}^n$ over $\mathbf{C}$. Thus, $\underline{e}_i$ has the following expansion

$$\underline{e}_i = \sum_{j=1}^{n} c_j e_j \qquad (5.7)$$

Clearly, $\langle \underline{e}_i, \underline{e}_i \rangle \neq 0$. However, by taking the inner product of 5.7 with $\underline{e}_i$ we obtain

$$\langle \underline{e}_i, \underline{e}_i \rangle = \sum_{j=1}^{n} c_j \langle e_j, \underline{e}_i \rangle = 0 \qquad (5.8)$$

under our initial assumption and the relationship established through 5.6. This is the desired contradiction. Therefore, $\langle e_i, \underline{e}_i \rangle \neq 0$.

With the above notation and lemma, the following theorem pins down the salient aspects of our theory.[5]

THEOREM 5.9: Let $F(r)$ and its adjoint, $F(r)^*$, have *eigenvector trajectories*, $e_i(r)$ and $\underline{e}_i(r)$, and *eigenvalue trajectories*, $\lambda_i(r)$ and $\bar{\lambda}_i(r)$, respectively for $i = 1,2,...,n$. Then for any value of r where the eigenvalues of $F(r)$ are *all distinct*

$$\frac{d\lambda_i}{dr} = \frac{\langle \dfrac{dF}{dr} e_i, \underline{e}_i \rangle}{\langle e_i, \underline{e}_i \rangle} \qquad (5.10)$$

$$\frac{de_i}{dr} = \sum_{\substack{j=1 \\ j \neq i}}^{n} \frac{\langle \dfrac{dF}{dr} e_i, \underline{e}_j \rangle}{(\lambda_i - \lambda_j) \langle e_j, \underline{e}_j \rangle} e_j \quad ; \quad i=1,2,...,n \qquad (5.11)$$

$$\frac{d\underline{e}_i}{dr} = \sum_{\substack{j=1 \\ j \neq i}}^{n} \frac{\langle \underline{e}_i, \dfrac{dF}{dr} e_j \rangle}{(\bar{\lambda}_i - \bar{\lambda}_j) \langle \underline{e}_j, e_j \rangle} \underline{e}_j \quad ; \quad i=1,2,...,n \qquad (5.12)$$

Again, note that the eigenvalues of $F(r)$ and $F(r)^*$ are complex conjugates of each other, but that the corresponding eigenvectors, $e_i(r)$ and $\underline{e}_i(r)$, are not complex conjugates.

PROOF: The key to the derivation is the equality

$$\left\langle \frac{dF}{dr} e_i, \underline{e}_j \right\rangle = \frac{d\lambda_i}{dr} \langle e_i, \underline{e}_j \rangle + (\lambda_i - \lambda_j) \left\langle \frac{de_i}{dr}, \underline{e}_j \right\rangle \qquad (5.13)$$

whose validity is shown as follows. Note that we are suppressing the dependent variable which could be frequency, s. Differentiating both sides of $Fe_i = \lambda_i e_i$ produces

$$\frac{dF}{dr} e_i + F \frac{de_i}{dr} = \frac{d\lambda_i}{dr} e_i + \lambda_i \frac{de_i}{dr} \qquad (5.14)$$

Taking the inner product of 5.14 with $\underline{e}_j$ produces

$$\left\langle \frac{dF}{dr} e_i, \underline{e}_j \right\rangle + \left\langle F \frac{de_i}{dr}, \underline{e}_j \right\rangle = \frac{d\lambda_i}{dr} \langle e_i, e_j \rangle + \lambda_i \left\langle \frac{de_i}{dr}, \underline{e}_j \right\rangle \qquad (5.15)$$

Now equations 5.1 and 5.8 imply that

$$\left\langle F \frac{de_i}{dr}, \underline{e}_j \right\rangle = \left\langle \frac{de_i}{dr}, F^* \underline{e}_j \right\rangle = \left\langle \frac{de_i}{dr}, \overline{\lambda}_j \underline{e}_j \right\rangle$$

$$= \lambda_j \left\langle \frac{de_i}{dr}, \underline{e}_j \right\rangle \qquad (5.16)$$

Substituting this expression into 5.15 and rearranging yields the equality of 5.13.

To obtain equation 5.10 first set $i = j$ in 5.13 which forces the rightmost term to drop out. Dividing through by $\langle e_i, \underline{e}_i \rangle$ yields 5.10.

To derive 5.11 observe that

$$\frac{de_i}{dr} = \sum_{\substack{j=1 \\ j \neq i}}^{n} a_{ij} e_j \qquad (5.17)$$

for some appropriate set of scalars, a_{ij}. Since the eigenvalues are distinct, $(e_1, \ldots, e_n)$ form a basis for $\mathbf{C}^n$ over $\mathbf{C}$. Hence, any vector, including de_i/dr, can be written as a linear combination of the e_i's. The justification for omitting the e_i term in the summation of 5.17 follows because the defining equation for the eigenvectors always leaves one parameter arbitrary. This permits us to choose $a_{ii} = 0$ without loss of generality. The goal is to find and expression for a_{ij}.

Taking the inner product of 5.17 with any $\underline{e}_k$, $k \neq i$, and simplifying via lemma 5.4, we obtain

$$\left\langle \frac{de_i}{dr}, \underline{e}_k \right\rangle = a_{ik} \left\langle e_k, \underline{e}_k \right\rangle \qquad (5.18)$$

In addition, put $j = k$ in equation 5.13 to obtain

$$\left\langle \frac{de_i}{dr}, \underline{e}_k \right\rangle = \frac{\left\langle \frac{dF}{dr} e_i, \underline{e}_k \right\rangle}{(\lambda_i - \lambda_k)} \qquad (5.19)$$

Solving 5.18 and 5.19 simultaneously for a_{ik} implies

$$a_{ik} = \frac{\left\langle \frac{dF}{dr} e_i \underline{e}_k \right\rangle}{(\lambda_i - \lambda_j) \left\langle e_k, \underline{e}_k \right\rangle} \qquad (5.20)$$

Substituting this expression for a_{ij} into 5.17 yields equation 5.11 as intended. A dual derivation will produce equation 5.12 so the proof of the theorem is now complete.

This theorem is the crux of the desired algorithm. With the equations of this theorem, we now have three coupled differential equations which will serve the intended purpose of illustrating a continuations algorithm for implementing the Nyquist and root locus tests of this chapter. In addition, they are useful in constructing the Nyquist plots and for computing the sensitivities of stability, to be explained in the final section.

EXAMPLE 5.21: Consider the matrix

$$F(r) = \begin{bmatrix} r-1 & 2r^2 + r \\ 0 & -(2 + r) \end{bmatrix} \qquad (5.22)$$

The derivative of $F(r)$ with respect to r at $r = 0$ is

$$\frac{dF}{dr} = \begin{bmatrix} 1 & 4r + 1 \\ 0 & -1 \end{bmatrix} \qquad (5.23)$$

Observe that $F(0)$ is given as

$$F(0) = \begin{bmatrix} -1 & 0 \\ 0 & -2 \end{bmatrix} \tag{5.24}$$

Hence, the eigenvalues and eigenvectors are trivially computed as

$$\lambda_1(0) = -1; \; \lambda_2(0) = -2 \; ; \; e_1(0) = \begin{bmatrix} 1 \\ 0 \end{bmatrix}$$
$$e_2(0) = \begin{bmatrix} 0 \\ 1 \end{bmatrix}; \; \underline{e}_1(0) = \begin{bmatrix} 1 \\ 0 \end{bmatrix}; \; \underline{e}_2(0) = \begin{bmatrix} 0 \\ 1 \end{bmatrix} \tag{5.25}$$

Let us use these values to initialize the coupled differential equations of 5.10, 5.11, and 5.12. Then let us use an *Euler integration* scheme to obtain the corresponding values at $r = n/10$ where n is the nth integration step. Implicitly then we will use a step size of .1.

Substituting these values into 5.10 through 5.12 we obtain

$$\dot{\lambda}_1(0) = 1 \tag{5.26}$$

$$\dot{\lambda}_2(0) = -1 \tag{5.27}$$

$$\dot{e}_1(0) = \begin{bmatrix} 0 \\ 0 \end{bmatrix} \tag{5.28}$$

$$\dot{e}_2(0) = \begin{bmatrix} -1 \\ 0 \end{bmatrix} \tag{5.29}$$

$$\dot{\underline{e}}_1(0) = \begin{bmatrix} -1 \\ 1 \end{bmatrix} \tag{5.30}$$

$$\dot{\underline{e}}_2(0) = \begin{bmatrix} 0 \\ 0 \end{bmatrix} \tag{5.31}$$

Doing the required integration via the Euler formula yields

$$\lambda_1(.1) = -.9; \quad \lambda_2(.1) = -2.1; \quad e_1(.1) = \begin{bmatrix} 1 \\ 0 \end{bmatrix}$$
$$e_2(.1) = \begin{bmatrix} -.1 \\ 1 \end{bmatrix}; \quad \underline{e}_1(.1) = \begin{bmatrix} 1 \\ .1 \end{bmatrix}; \quad \underline{e}_2(.1) = \begin{bmatrix} 0 \\ 1 \end{bmatrix} \tag{5.32}$$

Repeating the process we obtain

$$\lambda_1(.2) = -.8; \quad \lambda_2(.2) = 2.2; \quad e_1(.2) = \begin{bmatrix} 1 \\ 0 \end{bmatrix}$$

$$e_2(.2) = \begin{bmatrix} -.2 \\ 1 \end{bmatrix}; \quad \underline{e}_1(.2) = \begin{bmatrix} 1 \\ .2 \end{bmatrix}; \quad \underline{e}_3(.2) = \begin{bmatrix} 0 \\ 1 \end{bmatrix} \qquad (5.33)$$

Iterating the process n-times produces

$$\lambda_1(n/10) = \frac{n}{10} - 1; \quad \lambda_2(n/10) = -\frac{n}{10} + 2$$

$$e_1(n/10) = \begin{bmatrix} 1 \\ 0 \end{bmatrix}; \quad e_2(n/10) = \begin{bmatrix} \frac{-n}{10} \\ 1 \end{bmatrix}$$

$$\underline{e}_1(n/10) = \begin{bmatrix} 1 \\ \frac{n}{10} \end{bmatrix}; \quad \underline{e}_2(n/10) = \begin{bmatrix} 0 \\ 1 \end{bmatrix} \qquad (5.34)$$

As such, the eigenvalue loci for $F(r)$ may be plotted as in figure 5.35.

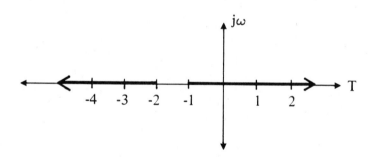

Figure 5.35. Eigenvalue loci for matrix of equation 5.22.

Note that generally speaking, the initial eigenvalues and eigenvectors must be tabulated via one of the standard algorithms. If it happens that during the integration, two or more eigenvalues become identical or very close, it is necessary to perturb r to another value and reinitialize the coupled differential equations before proceeding.

As a final point, these coupled equations provide us with a means for implementing the forthcoming Nyquist, root locus, and sensitivity of stability theories. Unfortunately, the equations are not numerically efficient. What is needed is a *continuations QR-algorithm*. In particular, it would be highly desirable to have a continuations algorithms for eigenvalue computation alone which could make use of sparse matrix techniques when necessary.

6. A ROOT LOCUS TECHNIQUE

Classically, the control engineer's *root locus plot* sketches the pole location of a *closed loop feedback* system as a function of the *open loop gain*.[2] An extension to interconnected systems arises naturally in the component connection model context. This section presents a "root locus" technique where the root locus originates at the *eigenvalues* of the *composite component state model*, traces out eigenvalue trajectories as intercomponent coupling (L_{11} information) is continuously and uniform introduced, and terminates at the *composite system eigenvalues*. In addition to answering the system stability question, the technique is also applicable to the analysis of short/open circuit phenomena as well as investigating the effect on composite system eigenvalues of increasing/decreasing coupling gains between components. Numerical implementation of the method is a straightforward application of equations 5.10, 5.11, and 5.12 of the previous section.

At this level, there is a need to distinguish between *internal and external stability* of a composite system. The ideas are analogous to BIBS and BIBO stability respectively as discussed in sections 2 and 3.

Suppose the composite component state model is given by

$$\begin{aligned}
\dot{x} &= Ax + Ba \\
b &= Cx + Da
\end{aligned} \tag{6.1}$$

and the usual connection equations by

$$\left[\frac{a}{y}\right] = \left[\begin{array}{c|c} L_{11} & L_{12} \\ \hline L_{21} & L_{22} \end{array}\right] \left[\frac{b}{u}\right] \tag{6.2}$$

Under appropriate assumptions (see section III.5), if the *composite system state vector* and the *composite component state vector* coincide, the system admits a composite system state model as

$$\dot{x} = Fx + Gu$$
$$y = Hx + Ju \tag{6.3}$$

where F, G, H and J are defined by III.5.8, III.5.9, III.5.10, and III.5.11 respectively. Since F is the only matrix pertinent to the present discussion, its form is rewritten in equation 6.4.

$$F = A + B(I - L_{11}D)^{-1}L_{11}C \tag{6.4}$$

A composite system is said to be *internally stable* if all the eigenvalues of F lie in the open left half complex plane. Essentially the system as per 6.3 must be BIBS stable. Note that we have disregarded the case where eigenvalues could possibly lie on the imaginary axis. Practically speaking, such mathematical refinements are superfluous since constant parameter variations make fixed eigenvalue locations ridiculous. In addition, the ever present noise phenomena, having an infinite frequency spectrum, would excite such poles irrespective of their decoupling from the input.

Coupling between components is naturally introduced by replacing L_{11} in 6.4 by rL_{11}, for a real scalar, r, to obtain

$$F(r) = A + B[I - (rL_{11})D]^{-1}(rL_{11})C \tag{6.5}$$

Clearly $F(0) = A$, the important decoupled system matrix, and $F(1) = F$, the relevant composite system matrix.

Obviously the "root locus" is a plot of the eigenvalue loci of $F(r)$ as r varies between zero and one. To numerically compute the loci, we adopt a *continuations approach* as per equations 5.10, 5.11 and 5.12. Capsulized, one initializes the coupled differential equations at the eigenvalues/eigenvectors of the decoupled matrix A (in 6.1) and A*. One then integrates with respect to r over the interval [0,1] to obtain the root locus; i.e., the *eigenvalue trajectories* as coupling is introduced. If any of the eigenvalue trajectories cross, it is necessary to perturb r around the point of intersection and reinitialize the algorithm.

Equations 5.10, 5.11 and 5.12 require knowledge of $F(r)$. Using the *matrix identities*

$$[\dot{M^{-1}}] = -M^{-1}\dot{M}M^{-1} \tag{6.6}$$

$$X(I - YX)^{-1} = (I - XY)^{-1}X$$

whenever either inverse exists, and

$$(I - XY)^{-1} = [I + X(I - YX)^{-1}Y] \qquad (6.7)$$

it is a simple matter to show that

$$\dot{F}(r) = B(I - rL_{11}D)^{-2}L_{11}C \qquad (6.8)$$

where the superscript "–2" means inverse squared. Note that equations 6.6 and 6.7 were derived and explained in theorem III.2.8.

When the composite component state model D-matrix is zero, $F(r) = BL_{11}C$, a constant. When $D \neq 0$, it is necessary to compute $(I - (I - rL_{11}D)^{-1}$ at Each step of the integration. Typically rank $[D] \ll \dim[I]$. Viewing $[-rL_{11}D]$ as a low rank perturbation on I, *Householder's formula* (III.6.15) provides convenient and quick means for computing $(I-rL_{11}D)^{-1}$ at each integration step. Indeed, viewing $L_{11}D$ as a low rank perturbation or the identity greatly simplifies Householder's formula.

EXAMPLE 6.9: As an illustration of the above idea, consider the block diagram of a system given by figure 6.10.

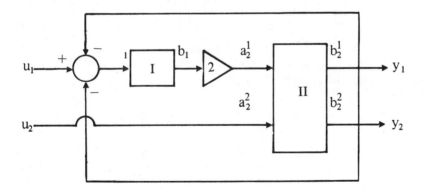

Figure 6.10. Block diagram of system for Example 6.9.

Suppose the state model for component I is

$$\dot{x}_1 = -x_1 + a_1$$

$$b_1 = x_1 \qquad (6.11)$$

and the state model for component II is

$$
\begin{bmatrix} \dot{x}_2^1 \\ \dot{x}_2^2 \end{bmatrix} = \begin{bmatrix} -1 & -2 \\ 2 & -1 \end{bmatrix} \begin{bmatrix} x_2^1 \\ x_2^2 \end{bmatrix} + \begin{bmatrix} a_1^2 \\ a_2^2 \end{bmatrix} ; \quad \begin{bmatrix} b_2^1 \\ b_2^2 \end{bmatrix} = \begin{bmatrix} x_2^1 \\ x_2^2 \end{bmatrix} \tag{6.12}
$$

Defining the appropriate vectors, the composite component state model is

$$
\begin{aligned}
\dot{x} &= Ax + Ba \\
b &= Cx
\end{aligned} \tag{6.13}
$$

where $B = C = I$ (the identity) and

$$
A = \begin{bmatrix} -1 & 0 & 0 \\ 0 & -1 & -2 \\ 0 & 2 & -1 \end{bmatrix} \tag{6.14}
$$

By inspection of figure 6.10, the connection equations are given by

$$
\begin{bmatrix} a_1 \\ a_2^1 \\ a_2^2 \\ \hline y_1 \\ y_2 \end{bmatrix} = \left[\begin{array}{ccc|cc} 0 & -1 & -1 & 1 & 0 \\ 2 & 0 & 0 & 0 & 0 \\ 0 & 0 & 0 & 0 & 1 \\ \hline 0 & 1 & 0 & 0 & 0 \\ 0 & 0 & 1 & 0 & 0 \end{array} \right] \begin{bmatrix} b_1 \\ b_2^1 \\ b_2^2 \\ \hline u_1 \\ u_2 \end{bmatrix} \tag{6.15}
$$

Since $D = 0$, the composite system F-matrix is $F = A + BL_{11}C$. As such we have

$$
F(r) = \begin{bmatrix} -1 & 0 & 0 \\ 0 & -1 & -2 \\ 0 & 2 & -1 \end{bmatrix} + \begin{bmatrix} 0 & -r & -r \\ 2r & 0 & 0 \\ 0 & 0 & 0 \end{bmatrix} \tag{6.16}
$$

which implies

$$
\dot{F}(r) = \begin{bmatrix} 0 & -1 & -1 \\ 2 & 0 & 0 \\ 0 & 0 & 0 \end{bmatrix} \tag{6.17}
$$

Implementing the root locus technique to this example via equations 5.10,

5.11, and 5.12 (executed in an Euler integration scheme) produces good results. Table 6.18 illustrates the results. In this table, $\triangle$ refers to the integration step size whereas "actual" means the actual rounded-off composite system values. Note that the eigenvalues are normalized to unit length and appropriate direction. A plot of the locus follows in figure 6.19.

	Actual	$\triangle$ = .1	$\triangle$ = .05
λ_1	−1.625	−1.54	−1.589
λ_2	−.687+j2.5	−.729+j2.37	−.706+j2.44
λ_3	−.687−j2.5	−.729−j2.37	−.706−j2.44
e_1	$\begin{bmatrix} +.724 \\ -.206 \\ +.658 \end{bmatrix}$	$\begin{bmatrix} +.740 \\ -.154 \\ +.655 \end{bmatrix}$	$\begin{bmatrix} +.731 \\ -.181 \\ +.658 \end{bmatrix}$
e_2	$\begin{bmatrix} .203+j.33 \\ .721-j.039 \\ .04-j.576 \end{bmatrix}$	$\begin{bmatrix} .197+j.321 \\ .712+j.013 \\ .076-j.588 \end{bmatrix}$	$\begin{bmatrix} .2+j.325 \\ .718-j.021 \\ .056-j.579 \end{bmatrix}$
e_3	$\begin{bmatrix} .203-j.33 \\ .721+j.039 \\ .04+j.576 \end{bmatrix}$	$\begin{bmatrix} .197-j.321 \\ .712-j.013 \\ .076+j.588 \end{bmatrix}$	$\begin{bmatrix} .2-j.325 \\ .718+j.021 \\ .056+j.579 \end{bmatrix}$

Table 6.18. Root locus results for example.

Note the formulation is natural in the component connection model context; it is numerically implementable through a continuations approach as above or through an iterative calling of a *QR-algorithm*; it identifies components giving rise to particular composite system eigenvalues by tracing the eigenvalue locus of $F(r)$ (potentially useful in state model order reduction); and finally by replacing L_{11} by $(L_{11} + rP)$ where P is of low rank, the technique offers a means for investigating *coupling gains* on composite system eigenvalue locations as follows.

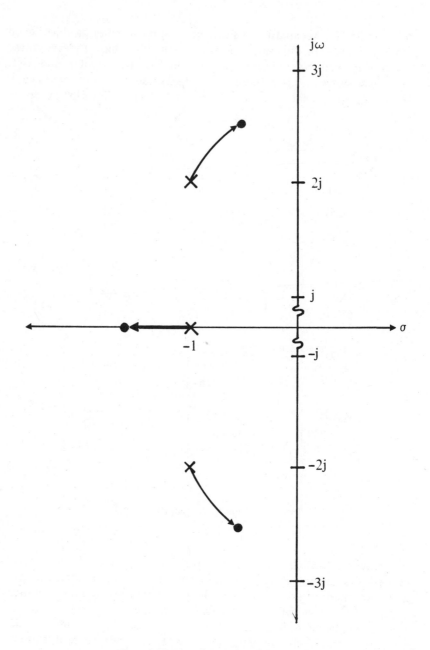

Figure 6.19. Plot of root locus for example.

Assume the composite system eigenvalues/eigenvectors are known and distinct. To investigate the effects of gain changes, replace L_{11} in 6.4 by $(L_{11} + rP)$ where P is a low rank matrix characterizing the gain variation and r varies over some relevant finite interval containing zero. This produces

$$F(r) = A + B(I - L_{11}D - rPD)^{-1}(L_{11} + rP)C \quad (6.20)$$

Again computing the root locus via the continuations approach requires $F'(r)$ which can be derived as

$$F'(r) = B(I - L_{11}D - rPD)^{-1}P[I^- D(1 - L_{11}D - rPD)^{-1}$$

$$(L_{11} + rPD)]C \quad (6.21)$$

Of course, if $D = 0$, $F(r) = BPC$ which is constant, sparse, and of low rank. Although 6.20 seems formidable, view rPD as a low rank perturbation on $(I - L_{11}D)$ and use Householder's formula (III.6.15) to compute the necessary inverses during each step of the integration.

Of course, stability or instability is known simply by noting the position of the terminal points of the root locus.

The technique is also useful in accounting for effects of stray capacitances/inductances. Consider that the A matrix in 6.1 is block diagonal (predominately diagonal for circuits) describing *only* component information. Suppose a typical entry is $1/C$ characterizing a capacitance. Clearly dA/dC is zero except for the particular C-dependent diagonal entry. If C_0 is the nominal C, using $dA/dC|C_0$ in 5.10 gives the approximation $d\lambda_i/dC|C_0$. Considering $\Delta C \; d\lambda_i/dC|C_0$ for each i gives a good first order approximation to the direction and magnitude of eigenvalue movement relative to small perturbations in C.

7. INTERNAL AND EXTERNAL STABILITY— FREQUENCY DOMAIN

External stability means input-output stability; i.e., BIBO system stability. Analogous to the notions of section 3, the frequency domain characterization of external system stability depends on the composite system transfer function $S(s)$ where

$$S(s) = L_{22} + L_{21}(I - Z(s)L_{11})^{-1}Z(s)L_{12} \quad (7.1)$$

Recapping the derivation and discussion of section III.2, consider the block diagram rendition of the component connection model given in figure 7.2.

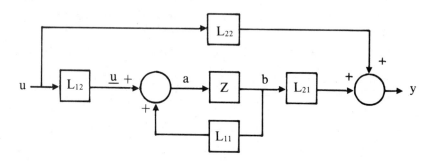

Figure 7.2. Block diagram of component connection model.

Let Z be $Z(s)$, the composite component transfer function matrix discussed in detail in section 3 of this chapter. To distinguish between internal and external stability, we first identify the internal "feedback" structure of the system. All dynamical action occurs between $\underline{u}$ and b. In view of this, define $Q(s)$ to be the *internal composite transfer function matrix* from $\underline{u}$ to b. To derive Q (s), observe

$$b = Z(s)a = Z(s)(\underline{u} + L_{11}b) \tag{7.3}$$

which simplifies to

$$b = (I - Z(s)L_{11})^{-1}Z(s)\underline{u} \tag{7.4}$$

By inspection

$$Q(s) = (I - Z(s)L_{11})^{-1}Z(s) \tag{7.5}$$

Expressing the composite system transfer function in terms of $Q(s)$ yields

$$S(s) = L_{22} + L_{21}Q(s)L_{12} \tag{7.6}$$

Note that theorem III.2.32 guarantees the existence of $(I - Z(s)L_{11})^{-1}$ for almost all connections.

DEFINITION 7.7: A system is *internally stable* if and only if $Q(s)$ has all poles in the open left half complex plane and no poles at s $= \infty$. This notion is akin to uniform boundedness of all state trajectories.

DEFINITION 7.8: A system is *externally stable* if and only if S(s) has all poles in the open left half plane with no poles at s = ∞.
Clearly external stability coincides with classical BIBO stability. The following proposition relates external and internal stability.

PROPOSITION 7.9: Internal stability implies external stability.
Since the L_{ij} matrices are real, they cannot create poles. Thus, equation 7.6 makes proposition 7.9 obvious. Note that by introducing certain structural zeros in L_{21} or L_{12}, the system may be internally unstable but externally stable. Such refinements were adequately discussed at the component level in sections 2 and 3. For reasons discussed therein, inclusion here would be superfluous.
The immediate task is to characterize internal stability of an interconnected system. As a working hypothesis, suppose the composite component transfer function matrix is stable; i.e., Z(s) has all poles in the open left half plane. Inspecting 7.5 clearly indicates that Q(s) has all its poles in the open left half plane if and only if $(I - Z(s)L_{11})^{-1}$ does. Note that in general neither Q(s) nor Z(s) are square. The following theorem characterizes internal stability through $\det(I - Z(s)L_{11})$.

THEOREM 7.10: Suppose each component of an interconnected dynamical system is stable in the sense that all poles of Z(s) are in the open left half plane. The system is internally stable if and only if $\det(I - Z(s)L_{11})$ has no finite zeros with non-negative real parts and no zeros at s = ∞.

PROOF: The key to the forward direction of the theorem is the following equality:

$$I + Q(s)L_{11} = (I - Z(s)L_{11})^{-1} \qquad (7.11)$$

To derive 7.11, observe

$$I + Q(s)L_{11} = (I - Z(s)L_{11})^{-1}(I - Z(s)L_{11}) + (I - Z(s)L_{11})^{-1} Z(s)L_{11}$$

$$= (I - Z(s)L_{11})^{-1} \qquad (7.12)$$

From 7.11 the poles of $(I - Z(s)L_{11})^{-1}$ are seen to be a subset of the poles of Q(s). On the other hand, if p is *not* a pole of $(I - Z(s)L_{11})^{-1}$, then every entry of the fixed complex matrix $(I - Z(p)L_{11})^{-1}$ is bounded. Hence

$$| \det[(I - Z(p)L_{11})^{-1}]| = \frac{1}{\det(I - Z(p)L_{11})} < \infty \qquad (7.13)$$

Obviously then p cannot be a zero of $\det[1 - Z(s)L_{11}]$. Thus, if Q(s) has no poles in the closed right half plane ($\mathbf{C}_+$), including $s = \infty$, then $\det(I - Z(s)L_{11})$ has no finite zeros in $\mathbf{C}_+$ and no zeros at $s = \infty$.

To establish the converse, expand Q(s) via *Cramer's rule* to obtain

$$Q(s) = \frac{\text{Adj}(I-Z(s)L_{11})}{\det(I-Z(s)L_{11})} Z(s) \qquad (7.14)$$

Since Z(s) is stable, any poles of Q(s) in $\mathbf{C}_+$ must appear as zeros of $\det(I - Z(s)L_{11})$. In other words, if $\det(I - Z(s)L_{11})$ has no zeros in $\mathbf{C}_+$ and no zeros at $s = \infty$, then the system is internally stable. The proof is complete.

The restriction on the stability of Z(s) is fundamental to the validity of the above theorem. The following example demonstrates why.

EXAMPLE 7.15: Suppose an interconnected system has Z(s) and L_{11} given as

$$Z(s) = \begin{bmatrix} \dfrac{-1}{s-1} & 0 \\ 0 & \dfrac{(s^2+1)}{s(s+1)} \end{bmatrix} \qquad (7.15)$$

and

$$L_{11} = \begin{bmatrix} 1 & 1 \\ 0 & 1 \end{bmatrix} \qquad (7.16)$$

A little arithmetic yields

$$Q(s) = \frac{1}{s(s-1)} \begin{bmatrix} -(s-1) & s^2+1 \\ 0 & s(s^2+1) \end{bmatrix} \qquad (7.17)$$

Clearly, the poles of $Q(s)$ are $s = 0$ and $s = 1$. Now observe that

$$(I - Z(s)L_{11}) = \begin{bmatrix} \dfrac{s}{s-1} & \dfrac{-1}{s-1} \\ 0 & \dfrac{s-1}{s(s+1)} \end{bmatrix} \tag{7.18}$$

and that

$$\det[1 - Z(s)L_{11}] = \frac{1}{s+1} \tag{7.19}$$

The determinant in 7.19 has no finite zeros and thus provides no information on the location of the finite poles of $Q(s)$.

It is a simple matter to show for any $Z(s)$, if p is a pole of $Q(s)$ then p is a zero of $\det(I - Z(s)L_{11})$ or p is a pole of $Z(s)$. Generalizing theorem 7.10 to allow for an unstable $Z(s)$ requires some sophisticated algebraic machinery in which $Z(s)$ is factored as

$$Z(s) = D(s)^{-1}N(s) \tag{7.20}$$

where $D(s)$ and $N(s)$ are *left co-prime polynomial matrices*.[4] One can then show that p is a pole of $Q(s)$ if and only if p is a zero of the polynomial, $\det[D(s) + N(s)L_{11}]$. Further elaboration of this point would be prohibitively long and the reader is directed to the literature.[4,7]

For the special case where $Z(s)$ is stable, theorem 7.10 sets up the groundwork for the *Nyquist theory* of the next section. Since in this case $\det(I - Z(s)L_{11})$ is a rational function, classical Nyquist theory furnishes a convenient means of determining internal stability.

8. NYQUIST THEORY

Graphical techniques such as *Bode plots*, *Nyquist plots*, *Nichols charts*, etc., underlie all classical scalar control systems design. Of particular importance is the Nyquist plot which classically is a graph of the frequency response of a scalar system.[2] The Nyquist plot renders necessary and sufficient conditions for *stability* as well as *gain and phase margin* information within a unity feedback configuration. The actual test is simple. A rigorous justification of the test entails subtle applications of some sophisticated mathematics. Trends in the literature indicate that *homotopy theory*, a branch of *algebraic topology*, is the foundation on which the classical *graphical stability tests* (Nyquist, Circle, Popov) are built. The first meaningful generalizations of the scalar Nyquist test were the *Circle and Popov criterions*, over thirty years later.[4] Generalizations to multi-input, multi-output linear systems appeared in the early seventies. This section delineates a smidgeon of the recent *multivariable Nyquist theory* in the context of the *component connection model*. To simplify the exposition, we require that the *rational matrix* $Z(s)$ be stable; i.e., $Z(s)$ has all poles in the open left half plane. Furthermore, $Z(s)$ must be a proper rational matrix; i.e., all entries of $Z(s)$ are rational and bounded at $s = \infty$, see section I.3. The mathematics required to relax these assumptions more than offsets any gain in generality.

Basic to all *Nyquist theory* are the concepts of a *curve, a closed curve, and the index of a curve* with respect to a point. The definitions to follow are somewhat restrictive but sufficient for the goals of this text.

DEFINITION 8.1: A curve, γ, is a continuous function of *bounded variation* such that

$$\gamma: \quad [0,1] \rightarrow \mathbf{C} \tag{8.2}$$

The *trace*, $[\gamma]$, *of a curve*, γ, is the image of the *closed unit interval*, $[0,1]$ in $\mathbf{C}$. Note that trace refers to a subset of $\mathbf{C}$ whereas the term curve refers to an actual function. The curve, γ, is said to be *closed* if $\gamma(0) = \gamma(1)$. As an example consider that

$$\gamma_1(t) \quad = \quad \exp(j\pi t) \tag{8.3}$$

is a curve which is not closed, whereas

$$\gamma_2(t) \quad = \quad \exp(j2\pi t) \tag{8.4}$$

is a closed curve. (Plot the traces of each curve). Henceforth the word *contour* (e.g., Nyquist contour) will mean (in a generic sense) the trace of some curve. Observe the positive direction of the unit interval induces a direction or orientation along the associated trace of contour in $\mathbf{C}$ as illustrated in figure 8.5.

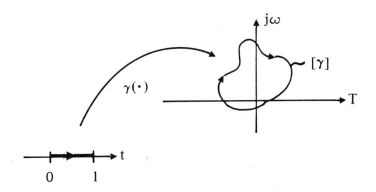

Figure 8.5. Illustration of the induced direction of a curve.

DEFINITION 8.6: If γ is a curve and f is a function defined and continuous on the trace of γ, then the *line integral* of f along γ is

$$\int_{\gamma} f(z)dz \;\triangleq\; \int_{0}^{1} f[\gamma(t)]d\gamma(t) \qquad (8.7)$$

For example, suppose $\gamma(t) = \exp(2\pi t)$ and $f(z) = 1/z$. From 8.7 we have

$$\int_{\gamma} f(z)dz \;=\; \int_{0}^{1} \frac{[2\pi \exp(2\pi t)dt]}{\exp(2\pi t)} \;=\; 2\pi \qquad (8.8)$$

The definition of a line integral permits definition of the notion of degree or index of a closed curve with respect to a point z_0 in $\mathbf{C}$.

DEFINITION 8.9: If γ is a *closed curve* and $z_0 \notin [\gamma]$, then the *index* or *degree* of γ with respect to z_0 is

$$d(\gamma;z_0) \;=\; \frac{1}{2\pi j} \int_{\gamma} (z - z_0)^{-1}dz \qquad (8.10)$$

Observe that under appropriate assumptions

$$\int_{\gamma} (z-z_0)^{-1}dz \;=\; \int_{\gamma} d[\ln(z-z_0)] \;\triangleq\; \int_{\gamma} d[\ln|z-z_0|] + j \int_{\gamma} d[\arg(z-z_0)]$$

$$=\; j \int d[\arg(z-z_0)] \tag{8.11}$$

These definitions interpret graphically as follows. Consider figure 8.12.

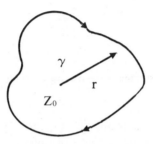

Figure 8.12. Trace of a closed curve, γ.

Equation 8.11 means that the line integral

$$\int_{\gamma} (z-a)^{-1}dz \tag{8.13}$$

measures j times the net increase in angle that the ray, r, accumulates as its tip traverses $[\gamma]$ in the indicated direction. For example, define two curves according to

$$\gamma_1(t) \;=\; \exp(j2\pi t) \tag{8.14}$$

and

$$\gamma_2(t) \;=\; \exp(j4\pi t) \tag{8.15}$$

A straightforward evaluation of 8.10 yields

$$d(\gamma_1;0) \;=\; 1 \tag{8.16}$$

and

$$d(\gamma_2;0) = 2 \qquad (8.17)$$

By plotting the traces, $[\gamma_1]$ and $[\gamma_2]$, it becomes clear that the number of *encirclements* of "0" by the contours, $[\gamma_1]$ and $[\gamma_2]$, is one and two respectively.

The development of Nyquist theory rests on the *analytic* properties of the *Laplace transforms* of the associated time dependent functions.

DEFINITION 8.18: A function, $f:C \to C$, is *analytic* at a point z_0 if f is defined at z_0 and if f is differentiable at every point in *some* open neighborhood of z_0.

A function is *differentiable* at a point $z = z_0$ if the following limit exists:

$$f(z_0) \underset{=}{\triangle} \lim_{\Delta z \to 0} \frac{f(z+\Delta z) - f(z)}{\Delta z} \qquad (8.19)$$

A function is analytic (differentiable) in an open region if it is analytic (differentiable) at every point in the open region. A function is analytic on a closed region, K, if there exists an open region, $G \supset K$, such that the function is analytic on G.

The definition of differentiability is powerful enough to imply that every analytic function is *smooth* and has a power series expansion about each point of its domain. In addition, let γ be a curve and f a function analytic on $[\gamma]$. Then $f \circ \gamma$ is a curve itself.

ARGUMENT PRINCIPLE 8.20: Let $f:C \to C$ be a rational function with poles $\lambda_1, \lambda_2, ..., \lambda_n$ and zeros $\xi_1, \xi_2, ..., \xi_n$. If γ is a closed curve such that $\lambda_i, \xi_k \notin [\gamma]$ for all i and k, then

$$\frac{1}{2\pi j} \int_{\gamma} \frac{f'(z)}{f(z)} dz = \sum_{k=1}^{m} d(\gamma;\xi_k) - \sum_{l=1}^{n} d(\gamma;\lambda_i) \qquad (8.21)$$

Since f is analytic on $[\gamma]$ we may rewrite 8.21 as

$$\frac{1}{2\pi j} \int_{fo\gamma} \frac{dz}{z} = \sum_{k=1}^{m} d(\gamma;\xi_k) - \sum_{i=1}^{n} d(\gamma;\lambda_i) \qquad (8.22)$$

This esoteric looking formula has a simple yet far reaching meaning: with due regard to multiplicity, the number of zeros of f encircled by $[\gamma]$ minus the number of poles of f encircled by $[\gamma]$ equals the number of times the contour $[fo\gamma]$ encircles "0". More importantly, if $f(z)$ has no poles encircled by $[\gamma]$ and if $d(\gamma,\xi_k) = 0$ for $k = 1,..., m$ then the function $1/f(z)$ is analytic in the region encircled by $[\gamma]$.

What is the connection between Nyquist theory and the Argument Principle? Consider figure 8.23 which shows a frequency domain component model in a negative unity feedback loop.

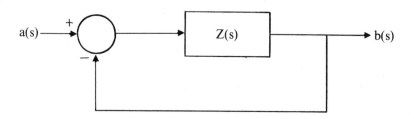

Figure 8.23. Single input, single output component in feedback loop.

Assuming $Z(s)$ is a *proper rational function* in the complex variable, s, the input output relation is given by

$$b(s) = \frac{Z(s)}{1 + Z(s)} a(s) \qquad (8.24)$$

The problem Nyquist solved was to graphically determine stability from the open loop gain and the feedback structure. Consistent with previous discussions in this chapter, assume $Z(s)$ has no poles in the closed right half complex plane, $\mathbf{C}_+$. Clearly then the closed loop system, 8.24, has no poles in $\mathbf{C}_+$ if and only if $(1 + Z(s))$ has no zeros in $\mathbf{C}_+$; i.e., the function $1/(1 + Z(s))$ is analytic in $\mathbf{C}_+$.

Let [Γ] be a semicircular contour enclosing **C**₊ in the clockwise direction as illustrated in figure 8.25. "[Γ]" is called the *Nyquist contour.*

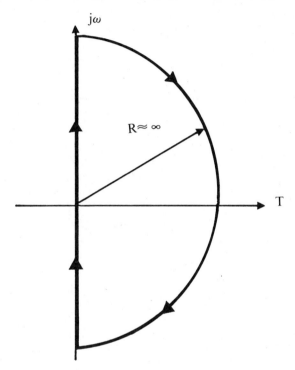

Figure 8.25. Sketch of Nyquist contour.

Since Z is a proper rational function analytic on [Γ], ZoΓ is a valid closed curve. The trace of Z composed with Γ, is called the *Nyquist plot* of the open loop gain, Z.

Under the above assumptions, the argument principle a la 8.22 implies that $(1 + Z(s))$ has no zeros in **C**₊ if and only if the Nyquist plot of Z, [ZoΓ], does not encircle or pass through "–1" where encirclements are taken in the clockwise direction. This is clear since $1 + Z(s) = 0$ is equivalent to $Z(s) = -1$. Thus, one determines stability by inspection of the Nyquist plot. No clockwise encirclements of "–1" implies closed loop stability.

Strictly speaking, Γ is not a curve. However, by adjoining [∞] to **C** to form the *extended complex plane*, this mathematical imprecision disappears. Observe Γ is an intermediary whereas ZoΓ is the important entity and under the above assumptions is always well defined. This brief derivation leads to the following restricted Nyquist theorem.

NYQUIST THEOREM 8.26: Let f:$\mathbf{C} \rightarrow \mathbf{C}$ be analytic on $\mathbf{C}_+$; then f has no zeros on $\mathbf{C}_+$ if and only if the Nyquist plot, [foΓ], does not encircle or pass through zero.

Let us return now to determining *internal stability* as developed in section 7. Recall theorem 7.10 stated that if the composite component transfer function matrix $Z(s)$ has no poles in $\mathbf{C}_+$, then the interconnected system is *internally stable* if and only if det $(I-Z(s)L_{11})$ has no finite zeros with non-negative real parts and no zeros at s= ∞. In view of 7.10, we have the following corollary to the Nyquist theorem.

COROLLARY 8.27: The above characterized interconnected system is internally stable if and only if the Nyquist plot of $\det(I-Z(s)L_{11})$ does not encircle or pass through zero.

Upon observing that $\det(I-Z(s)L_{11})$ is analytic in $\mathbf{C}_+$, whenever $Z(s)$ has no poles in $\mathbf{C}_+$ (as assumed) the proof is immediate. Unfortunately, for large systems $\det(I-Z(s)L_{11})$ becomes a numerically unwieldy object in that numerically computing a symbolic form of $\det(I-Z(s)L_{11})$ is extremely difficult. Consequently, we propose on analogous Nyquist-like test in terms of the *eigenvalue loci* of $(I-Z(j\omega)L_{11})$. Again $Z(s)$ will be analytic in $\mathbf{C}_+$. In particular, for each fixed ω, the nxn complex matrix $Z(j\omega)$ has exactly n eigenvalues, not necessarily distinct. For convenience denote them as $\lambda_i(j\omega)$ for $i = 1,...,$ n. Basically as ω climbs from $_{-\infty}$ to ∞, the location of the eigenvalues envolves continuously in the complex plane. This permits plotting of *eigenvalues loci*.

There are a number of distinctive features associated with the loci. First, the eigenvalue loci, $\lambda_i(j\omega)$, are not uniquely defined since, whenever two particular loci cross, the identity of the correct continued path is lost. Furthermore, each loci, $\lambda_i(j\omega)$, does not in general form a closed contour. Fortunately, it is possible to concatenate the n eigenvalue loci of $(I-Z(j\omega)L_{11})$ to form one or more closed contours.[1] The underlying functions, $\lambda_i(\cdot)$, are piecewise analytic. Let us assume that the concatenation of the loci form m $\leq$ n closed contours in $\mathbf{C}$. Label them $C_1,...,C_m$. In other words, $[\lambda_i o \Gamma]_i = 1,...,m$, are curves in $\mathbf{C}$. When viewed as a group, they form m closed contours, $C_1,...,C_m$.

DEFINITION 8.28: The *spectral Nyquist plot* of $(I-Z(j\omega)L_{11})$ is the set of closed contours $[C_1,...,C_m]$.

The *degree* of the *set* of curves giving rise to this spectral Nyquist plot with respect to a point z_0 is the sum of the clockwise encirclements of the C_i $(i = 1,...,m)$ each with respect to z_0. Informally speaking, the *degree* of the *spectral Nyquist plot* is the sum of the clockwise encirclements of z_0 by the C_i.

THEOREM 8.29: Let each component of an interconnected dynamical system be characterized by a stable rational transfer function matrix. Then the system is internally stable if and only if the spectral Nyquist plot of $(I-Z(j\omega)L_{11})$ does not pass through zero and the degree of this spectral Nyquist plot with respect to "0" is zero.

The proof of this theorem requires further development of complex variable theory so the following is only a cursory sketch. Details appear in reference 1.

The basic goal of a proof to this theorem is to show that the degree of the Nyquist plot of $\det(I-Z(s)L_{11})$ with respect to "0" is equal to the degree of the spectral Nyquist plot with respect to "0". In general, for points other than zero, the degrees of the respective plots are different as the examples at the end of this section demonstrate.

In showing that the degrees of the two plots are equal relative to the point "0", one uses the identity

$$\det(I + Z(j\omega)L_{11}) = \prod_{i=1}^{n} \lambda_i(j\omega) \tag{8.30}$$

where $\lambda_i(j\omega)$ are the aforementioned eigenvalue loci of $(I-Z(j\omega)L_{11})$. Relative to the point zero, one may define appropriate *branch cuts* in $\mathbf{C}$ so that (i) $\log[\lambda_i(j\omega)]$ exists and (ii) the degrees relative to zero of the two plots are equal.

As such, we may define $f(s) = \det(I-Z(s)L_{11})$ and

$$\lambda(s) = \prod_{i=1}^{n} \lambda_i(s) \tag{8.31}$$

so that

$$\int_{f o \Gamma} \frac{ds}{s} = \int_{\lambda o \Gamma} \frac{ds}{s} = \int_{\Gamma} \frac{d\lambda(s)}{\lambda(s)} \tag{8.32}$$

Under the appropriate hypotheses then

$$\int_{\Gamma} \frac{d\lambda(s)}{\lambda(s)} = \int_{\Gamma} d[\log \lambda(s)] \tag{8.33}$$

Using the properties of the log results in

$$\int_{\Gamma} \frac{d\lambda(s)}{\lambda(s)} = \prod_{i=1}^{n} \int_{\Gamma_0\lambda_i} \frac{ds}{s} \tag{8.34}$$

Thus, the degrees of the Nyquist plot and the spectral Nyquist plot with respect to zero coincide. Again, for points other than zero, the degrees are, in general, different. This is due to the impossibility of defining branch cuts so that 8.33 and 8.34 are well defined. This ends the sketch of the proof.

To further simplify theorem 8.29, define the eigenvalue loci of $Z(j\omega)L_{11}$ as $\underline{\lambda}_i(j\omega)$. Then the eigenvalue loci $\lambda_i(j\omega)$ as above and $\underline{\lambda}_i(j\omega)$ satisfy

$$\lambda_i(j\omega) = 1 - \underline{\lambda}_i(j\omega) \tag{8.35}$$

COROLLARY 8.36: Under the hypotheses of theorem 8.29, a system is internally stable if and only if the spectral Nyquist plot of $Z(j\omega)L_{11}$ (i.e., the eigenvalue loci $\underline{\lambda}_i(j\omega)$) does not pass through or encircle the point "1".

To cement these notions in a practical setting, let us illustrate them with two examples.

EXAMPLE 8.36: Consider again the system whose block diagram is given in figure 6.10. Frequency domain descriptions of components I and II are

$$b_1 = \left[\frac{1}{s+1} \right] a_1 \tag{8.37}$$

and

$$\begin{bmatrix} b_2^1 \\ b_1^2 \end{bmatrix} = \frac{1}{s^2+2s+5} \begin{bmatrix} s+1 & -2 \\ 2 & s+1 \end{bmatrix} \begin{bmatrix} a_2^1 \\ a_2^2 \end{bmatrix} \tag{8.38}$$

For convenience, define

$$Z_1 = \left[\frac{1}{s+1} \right] \tag{8.39}$$

and

$$Z_2 = \frac{1}{s^2+2s+5} \begin{bmatrix} s+1 & -2 \\ 2 & s+1 \end{bmatrix} \qquad (8.40)$$

Thus, the composite component transfer function matrix becomes

$$b = Za = \left[\begin{array}{c|c} Z_1 & 0 \\ \hline 0 & Z_2 \end{array}\right] a \qquad (8.41)$$

As in equation 6.16, the connection equations are

$$\begin{bmatrix} a_1 \\ a_2^1 \\ a_2^2 \\ \hline y_1 \\ y_1 \end{bmatrix} = \left[\begin{array}{ccc|cc} 0 & -1 & -1 & 1 & 0 \\ 2 & 0 & 0 & 0 & 0 \\ 0 & 0 & 0 & 0 & 1 \\ \hline 0 & 1 & 0 & 0 & 0 \\ 0 & 0 & 1 & 0 & 0 \end{array}\right] \begin{bmatrix} b_1 \\ b_2^1 \\ b_2^2 \\ u_1 \\ u_2 \end{bmatrix} \qquad (8.42)$$

Computing $Z(s)L_{11}$ leads to

$$Z(s)L_{11} = \begin{bmatrix} 0 & -\dfrac{1}{s+1} & -\dfrac{1}{s+1} \\ \dfrac{2(s+1)}{s_2+2s+5} & 0 & 0 \\ \dfrac{4}{s_2+2s+5} & 0 & 0 \end{bmatrix} \qquad (8.43)$$

Construction of the eigenvalue loci is possible via the *continuations approach* discussed in section 5. However, for the moment observe that the eigenvalue loci of $Z(j\omega)L_{11}$ is given by the solutions of equation 8.44.

$$\underline{\lambda}(j\omega)\left[\underline{\lambda}^2(jw) + \frac{2(j\omega+3)}{(j\omega+1)\,(5-\omega^2+2j\omega)}\right] = 0 \qquad (8.44)$$

Solving this equation for its zeros and plotting the resultant loci produces the following spectral Nyquist plot.

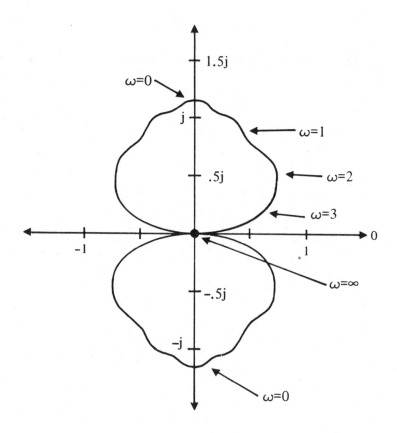

Figure 8.45. Spectral Nyquist plot of $Z(jw)L_{11}$.

Since the resultant contour does not encircle or pass through "1", the system is internally stable. Consequently, it is externally stable. For comparison purposes, figure 8.46 is the Nyquist plot of $\det(I - Z(j\omega)L_{11})$. Observe the radically different shapes of the two plots. Since the degree of the Nyquist plot with respect to zero is zero, the results of the two tests, internal stability, coincide.

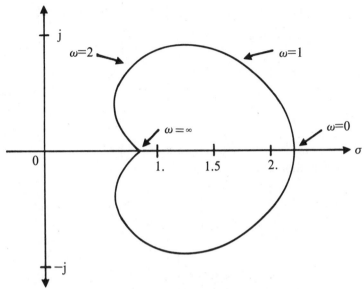

Figure 8.46. Nyquist plot of det $[I-Z(j\omega)L_{11}]$.

EXAMPLE 8.47: As a second example, consider the system block diagram given in figure 8.48.

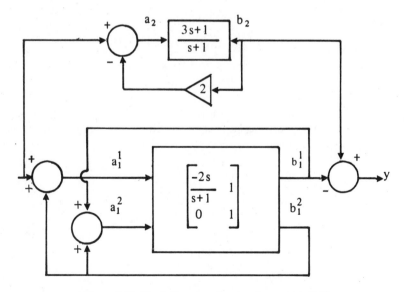

Figure 8.48. Block diagram of system in example 8.47.

The composite component frequency domain model is given by

$$
\begin{bmatrix} b_1^1 \\[2mm] b_1^2 \\[2mm] b_2 \end{bmatrix} = \begin{bmatrix} \dfrac{-2s}{s+1} & 1 & 0 \\[3mm] 0 & 1 & 0 \\[3mm] 0 & 0 & \dfrac{3s+1}{s+1} \end{bmatrix} \begin{bmatrix} a_1^1 \\[2mm] a_1^2 \\[2mm] a_2 \end{bmatrix} \tag{8.49}
$$

The corresponding connection equations are

$$
\begin{bmatrix} a_1^1 \\[1mm] a_1^2 \\[1mm] a_2 \\ \hline y \end{bmatrix} = \left[\begin{array}{ccc|c} 0 & 1 & 0 & 1 \\ 1 & 1 & 0 & 0 \\ 0 & 0 & -2 & 1 \\ \hline -1 & 0 & 1 & 0 \end{array} \right] \begin{bmatrix} b_1^1 \\[1mm] b_1^2 \\[1mm] b_2 \\ \hline u \end{bmatrix} \tag{8.50}
$$

Note that we have embedded the feedback gain around component two into the connection equations. The quantity considered in this example is $(1-Z(s)L_{11})$. A straightforward manipulation yields

$$
(I-Z(s)L_{11}) = \begin{bmatrix} 0 & \dfrac{s-1}{s+1} & 0 \\[3mm] -1 & 0 & 0 \\[3mm] 0 & 0 & \dfrac{7s+3}{s+1} \end{bmatrix} \tag{8.51}
$$

Again, to clearly illustrate the above Nyquist theory, we analytically compute the spectral Nyquist plot of $(I-Z(j\omega)L_{11})$ by setting $s = j\omega$ in the following three equations.

$$
\lambda_1(s) = \sqrt{-\frac{s-1}{s+1}} \tag{8.52}
$$

$$
\lambda_2(s) = -\sqrt{-\frac{s-1}{s+1}} \tag{8.53}
$$

and

$$
\lambda_3(s) = \frac{7s+3}{s+1} \tag{8.54}
$$

Also observe that $\det(I-Z(s)L_{11})$ is given according to

$$\det[I-Z(s)L_{11}] = \frac{7s+3}{s+1} \frac{s-1}{s+1} \tag{8.55}$$

With the aid of equations 8.52 through 8.55, we present the spectral Nyquist plot of $(I-Z(j\omega)L_{11})$ and the Nyquist plot of $\det(I-Z(s)L_{11})$ in figures 8.56 and 8.57 respectively.

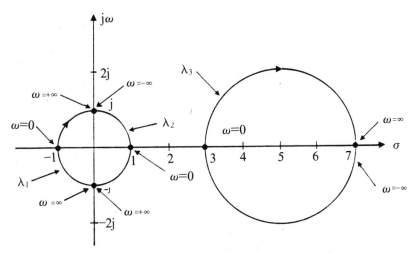

Figure 8.56. Spectral Nyquist plot of $(I-Z(j\omega)L_{11})$.

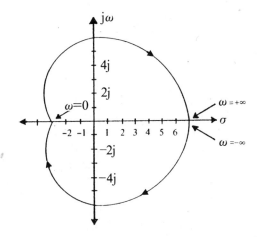

Figure 8.57. Nyquist plot of $\det(I-Z(s)L_{11})$.

Both plots encircle zero indicating an internally unstable system.

To clarify earlier remarks, notice that the degrees of the respective contours relative to points other than zero are, in general, different. For example, in 8.57 the degree of the spectral plot relative to the point "2" is zero, whereas the Nyquist plot encircles "2" once.

Numerical implementation of the spectral Nyquist test is a straightforward application of equations 5.10, 5.11, and 5.12. Recall that these equations govern the eigenvalue/eigenvector movement of a matrix F with respect to a variable parameter, r. Applied to finding the spectral Nyquist plot of, say $Z(j\omega)L_{11}$, define $F = Z(j\omega)L_{11}$ and $r = \omega$. Clearly, $F = Z(j\omega)L_{11}$ where the derivative is with respect to ω. By initializing the equations (5.10, 5.11, and 5.12) at the eigenvalues/eigenvectors of $Z(j\omega)L_{11}$ for some fixed ω ($\omega = 0$ seems to be an optimum choice), we may integrate to obtain the desired plot. In particular, the solution trajectory of 5.10 gives the desired spectral Nyquist plot. Remember that whenever two loci intersect or cross, the equations become singular and one must reinitialize the algorithm.

Finally we wish to point out that the spectral Nyquist plot of $(I-Z(j\omega)L_{11})$ is the true generalization of the scalar Nyquist criterion as opposed to the Nyquist plot of $\det[1-Z(s)L_{11}]$. The scalar Nyquist plot sketches the frequency response or *spectrum* of the return difference operator of the feedback loop. Similarly, the spectral Nyquist plot of $(I-Z(j\omega)L_{11})$ sketches the spectrum of the "return difference operator" of the internal system structure.

9. SENSITIVITY OF STABILITY

A *stability analysis* of a real world system will lack completeness and may instill false intuition unless one accounts for *parameter variations* in the analysis.[10] For example, an *eigenvalue* location of the composite system may be highly sensitive to an amplifier gain. Potential instability could be present even with all eigenvalue locations in the open left half plane. With this in mind, the aim here is to introduce the notion of "*sensitivity of stability*" into the analyses discussed in previous sections. For the *root locus* technique we investigate the eigenvalue motion toward the imaginary axis as a function of a parameter, r. With regard to the *Nyquist analysis*, we set up a measure of the proclivity of the spectral Nyquist plot to cross the point "0" or "1" depending, of course, on the specific test. The Nyquist-theoretic ideas are akin to the classical notion of *gain and phase margin*. Again, the equations of section 5 are useful for numerical implementation.

To develop the notion of sensitivity of stability via the *Nyquist plot*, consider the eigenvalue loci of $(I-Z(j\omega)L_{11})$. Assuming I is nxn, there are n segments to this loci, denoted as $\lambda_1(j\omega,r),...,\lambda_n(j\omega,r)$ where r indicates the possible r-dependence of the loci. The derivative of λ_i with respect to r gives a measure of the motion of the eigenvalue loci with respect to small r-variations. The goal is to fabricate a normalized measure of the motion towards or away from the origin. Implicitly, we are assuming the system is stable for nominal parameter values.

Regarding both the derivative of $\lambda_i(\omega,r)$ and $\lambda_i(\omega,r)$ as vectors in the complex plane, for each fixed r_0 and ω_0, the component towards the origin of the "r-motion" of the eigenvalue-vector, $\lambda_i(\omega_0,r_0)$, is given by

$$\text{Re} < \frac{d\lambda_i}{dr} ,\lambda^i > \tag{9.1}$$

If r increases and 9.1 is positive, the motion is away from the origin; if 9.1 is negative, the motion is toward the origin. Adding the obvious normalization, the length of λ_1, yields

$$\frac{\text{Re} < \frac{d\lambda_i}{dr} ,\lambda_i >}{|\lambda_i|} \tag{9.2}$$

Of course, all pertinent quantities are evaluated at the specific nominal value of r, say r_0. Since we are looking for the greatest proclivity of motion toward the origin, the following equation provides the desired measure of sensitivity of stability relative to the spectral Nyquist plot of $(I-Z(j\omega))L_{11}$.

$$\mathbf{S_r} = \min_i \inf_\omega \left[\frac{\text{Re} < \frac{d\lambda_i}{dr} , \lambda_i>}{|\lambda_i|} \right] \tag{9.3}$$

To illustrate this definition, again consider the block diagram of figure 6.9. Let the potentially variable parameter be the gain, r, of the amplifier connected to the output of component I. The nominal value of r is, of course, 2 and is denoted by r_0. In this case, the r-dependence of $Z(s)L_{11}$ is given by equation 9.4.

$$Z(s)L_{11} = \begin{bmatrix} 0 & -\frac{1}{s+1} & -\frac{1}{s+1} \\ \frac{r(s+1)}{s^2+2s+5} & 0 & 0 \\ \frac{2r}{s^2+2s+5} & 0 & 0 \end{bmatrix} \tag{9.4}$$

Numerical considerations not withstanding, the *r-dependent eigenvalue loci* is given by the zeros of equation 9.5 below.

$$\underline{\lambda}\,(j\omega,r) \left[\underline{\lambda}^2(j\omega,r) + \frac{r(j\omega + 3)}{(j\omega + 1)\,(5-\omega^2 + 2j\omega)} \right] = 0 \quad (9.5)$$

Since one of the eigenvalue loci-functions is identically equal to zero and not r-dependent, it can be eliminated from further consideration. For convenience, define the function $g(j\omega)$ as given in 9.6.

$$g(j\omega) \;=\; -\,\frac{(j\omega + 3)}{(j\omega + 1)\,(5-\omega^2 + 2j\omega)} \quad (9.6)$$

As such, the two remaining eigenvalue loci are given by equations 9.7 and 9.8.

$$\underline{\lambda}_1(j\omega,r) \;=\; \sqrt{r\ g(j\omega)} \quad (9.7)$$

and

$$\underline{\lambda}_2(j\omega,r) \;=\; -\,\sqrt{r\ g(j\omega)} \;=\; -\,\underline{\lambda}_1(j\omega,r) \quad (9.8)$$

Recall that the eigenvalue loci of $Z(j\omega)L_{11}$ and $(I-Z(j\omega)L_{11})$ can be related by

$$\lambda\,(j\omega,r) \;=\; 1\,-\,\underline{\lambda}\,(j\omega,r) \quad (9.9)$$

From this equation it is easy to see that

$$\frac{d\underline{\lambda}\,(j\omega,r)}{dr} \;=\; -\,\frac{d\lambda\,(j\omega,r)}{dr} \quad (9.10)$$

Therefore, the definition of sensitivity of stability as per equation 9.3 reduces to the following:

$$S_r \;=\; \min_{i}\ \inf_{\omega}\ \left[\frac{Re < -\dfrac{d\underline{\lambda}}{dr}\,,\ (1-\underline{\lambda}) >}{|\,1 - \underline{\lambda}\,|} \right] \quad (9.11)$$

Doing the necessary minimization we arrive at the answer

$$\mathbf{S_r} = -.1251163 \qquad (9.12)$$

where $\omega = 3.1245$ is the minimizing frequency, i.e. the frequency at which the spectral Nyquist plot is "most likely" to cross the origin due to variations in r.

Numerical implementation of the above measure initially requires calculation of dF/dr. In the above example define $F = Z(s)L_{11}$ as in 9.4. Consequently

$$\frac{dF}{dr} = \frac{1}{s^2+2s+5} \begin{bmatrix} 0 & 0 & 0 \\ s+1 & 0 & 0 \\ 2 & 0 & 0 \end{bmatrix} \qquad (9.13)$$

Using this in equation 5.10 permits one to compute the desired derivatives at each fixed ω, as

$$\frac{d\lambda_i}{dr} = \frac{<\dot{F}e_i, \underline{e}_i>}{<e_i, \underline{e}_i>} \qquad (9.14)$$

Of course, it is also necessary to compute the eigenvectors e_i and $\underline{e}_i$ at each fixed ω. The next step is to evaluate 9.2. One embeds this calculation within a standard numerical minimization routine as per section V.6. After finding the "inf" for each i, choose the smallest. The "smallest" is then the solution to 9.3.

Clearly such a numerical procedure is grossly inefficient. Hopefully the future will reveal a means for computing the eigenvalues of a parameterized matrix in a simple efficient way. Intuitively such an algorithm should entail a *continuations method* which utilizes sparse matrix techniques.

Characterization of potential instability is also possible through the *root locus* technique. Suppose the composite system eigenvalues are known. To in some sense complete the stability analysis account must be made for *eigenvalue movement* with regard to some potentially variable, r. The instantaneous eigenvalue motion is given by $d\lambda_i/dr$. Clearly the eigenvalue motion toward the imaginary axis is

$$\text{Re}\left(\frac{d\lambda_i}{dr}\right) \tag{9.15}$$

Assuming r increases, then if 9.15 is positive, the eigenvalue r-motion is towards the imaginary axis; the motion is oppositely directed if 9.15 is negative. To obtain a uniform measure for the system of eigenvalues, one normalizes by the distance from the $j\omega$-axis. Thus, for each eigenvalue we may define the sensitivity of stability with respect to a parameter r according to

$$\mathbf{S}_r^i = \frac{\text{Re}\left(\frac{d\lambda_i}{dr}\right)}{\text{Re}[\lambda_i]} \tag{9.16}$$

Basically, the motion is akin to the *small change sensitivity analysis* of the previous chapter.

$$\frac{d\lambda_i}{dr} = \frac{\langle \frac{dF}{dr} e_i, \underline{e}_i \rangle}{[e_i, \underline{e}_i]} \tag{9.17}$$

As an example, again consider the block diagram given in figure 6.9. The gain of the amplifier connected to the output of component I is the potentially variable parameter having nominal value 2. As such

$$F(r) = \begin{bmatrix} -1 & -1 & -1 \\ r & -1 & -2 \\ 0 & 2 & -1 \end{bmatrix} \tag{9.18}$$

Thus, the derivative of $F(r)$ is

$$\frac{dF(r)}{dr} = \begin{bmatrix} 0 & 0 & 0 \\ 1 & 0 & 0 \\ 0 & 0 & 0 \end{bmatrix} \tag{9.19}$$

The table below enumerates the appropriate unnormalized eigenvectors.

i	e_i	$\underline{e}_i$
1	$\begin{bmatrix} -3.508 \\ 1.0 \\ -3.196 \end{bmatrix}$	$\begin{bmatrix} -3.196 \\ 1.0 \\ -1.91 \end{bmatrix}$
2	$\begin{bmatrix} .509+j.939 \\ 2.0 \\ .196-j1.57 \end{bmatrix}$	$\begin{bmatrix} .196+j1.57 \\ 2.0 \\ .410-j.172 \end{bmatrix}$
3	$e_3 = \underline{e}_2$	$e_3 = \underline{e}_3$

Table 9.20. Unnormalized eigenvectors e_i and $\underline{e}_i$ of F and F* respectively.

Computing the sensitivity of $\lambda_1 = -1.625$ with respect to r at the nominal value of $r = 2$, we have

$$S_r^1 = -.11780 \tag{9.21}$$

This number indicates that the eigenvalue is somewhat insensitive to r-variations and its motion is away from the imaginary axis for increasing r. Computing the sensitivity of $\lambda_2 = -.687+j2.5$ (and by symmetry) λ_3 we have

$$S_r^2 = S_r^3 = .13937 \tag{9.22}$$

Whenever the gain, r, increases, the eigenvalue motion is toward the imaginary axis. Since the sensitivity of stability values are small, movement toward $j\omega$–axis would not be rapid. Knowledge of these numbers then in some sense completes the stability analysis.

10. PROBLEMS

1. A system is said to be *causal* if whenever two input vectors, $a_1(t)$ and $a_2(t)$, coincide for $t \leq T$ then the corresponding outputs coincide for $t \leq T$. Show that the system of equation 2.1 is causal if and only if $Z(t) = 0$ for $t < 0$.

2. Give examples of functions $f(t)$ and $g(t)$ such that $||f(t)||_1 < ||f(t)||_\infty$ and $||g(t)||_\infty < ||g(t)||_1$; i.e., neither the L_1 *norm* nor the *sup norm* dominate each other. Note, for sequences or functions defined on a finite interval domination does occur.

3. Let a system be characterized by the state model

$$\dot{x} = Ax + Ba$$

$$y = Cx$$

Show that it is BIBO stable if it is BIBS stable.

4. Give an example of a system defined by the state equation of problem 3 which is BIBO stable but not BIBS stable.

5. Show that if a complex number, λ, is an *eigenvalue* of matrix A then the $(\lambda I - A)$ is not invertible.

6. Show that the eigenvalues of the matrix A^t are the complex conjugates of the eigenvalues of A.

7. Compute the eigenvalues and *eigenvectors* of the matrix

$$A = \begin{bmatrix} 1 & 2 \\ -2 & 0 \end{bmatrix}$$

and its transpose and show that they satisfy 2.47.

8. Expand the rational function

$$Z(s) = \frac{s^2(s+1)(s-1)}{(s+2)(s^2+6s+9)}$$

as per equation 3.2.

9. Determine whether or not the system described by the transfer function of problem 8 is BIBO stable.

10. Derive equations 3.24 and 3.25.

11. Give a complete proof for the validity of properties i), ii), and iii) of theorem 3.26.

12. For the A matrix of problem 7 compute $(sI - A)^{-1}$ via the recursive formula of theorem 3.36.

13. Let the matrix A have eigenvalue λ with corresponding eigenvector e and let $B = TAT^{-1}$. Then show that B has eigenvalue λ with eigenvector $T^{-1}e$.

14. Show that the diagonal entries of a *triangular matrix* are its eigenvalues.

15. Apply the *LU-factorization* algorithm of theorem 4.11 to compute the eigenvalues of the matrix

$$A = \begin{bmatrix} 5 & 2 \\ -2 & 0 \end{bmatrix}$$

16. Show that a matrix is orthogonal if and only if $Q^t = Q^{-1}$.
17. Apply the *QR algorithm* to compute the eigenvalues of the matrix of problem 15.
18. Find a similarity transformation which transforms the following matrix into *upper Hessenberg form.*

$$A = \begin{bmatrix} 1 & 0 & 0 & 1 & 0 & 0 \\ 1 & 0 & 1 & 0 & 0 & 0 \\ 0 & 0 & 0 & 0 & 0 & 1 \\ 0 & 1 & 0 & 0 & 1 & 0 \\ 0 & 0 & 0 & 0 & 1 & 1 \\ 0 & 1 & 0 & 0 & 0 & 0 \end{bmatrix}$$

19. Let A be a matrix with distinct eigenvalues and let T be the matrix whose columns are made up of the eigenvectors of A. Then show that the eigenvectors of A* are the columns of $[T^{-1}]$. Here, "*" denotes the complex conjugate transpose of a matrix (which is the natural generalization of the transpose operation to complex matrices).
20. Compute the eigenvalues and eigenvectors of the matrix

$$F(r) = \begin{bmatrix} r & 1 \\ 1 & r \end{bmatrix}$$

for $0 \leq r \leq 1$.
21. Reformulate the root *locus technique* of section 6 by letting

$$F(r) = A + rB[1 - L_{11}D]^{-1}L_{11}C$$

and compute F(r). Discuss the advantages and disadvantages of this formulation vis-a-vis the formulation of section 6.

11. REFERENCES

1. Barmen, J. F., and J. Katznelson, "A Generalized Nyquist-type Stability Criterion for Multivariable Systems", *Inter. Jour. on Cont.*, Vol. 20, pp. 593-622, (1974).
2. Chen, C. T., *Introduction to Linear System Theory*, Holt, Rinehart and Winston, New York, 1970.
3. Desoer, C. A., *Notes for a Second Course on Linear Systems*, Van Nostrand, New York, 1970.
4. Desoer, C. A., and M. Vidyasagar, *Feedback Systems: Input-Output Properties*, Academic Press, New York, 1975.
5. Faddeev, D. K., V. N. Faddeeva, *Computational Methods of Linear Algebra*, Freeman, San Francisco, 1963.
6. Michel, A. N., and R. K. Miller, *Qualitative Analysis of Large-Scale Dynamical Systems*, Academic Press, New York, 1977.
7. Rosenbrock, H. H., *State Space and Multivariable Theory*, Nelson, London, 1970.
8. Rudin, W., *Real and Complex Analysis*, McGraw-Hill, New York, 1967.
9. Tewerson, R. P., *Sparse Matrices*, Academic Press, New York, 1972.
10. Van Ness, J. E., Boyle, J. M., and F. Imad, "Sensitivities of Multiloop Control Systems", *IEEE Trans. on Auto. Cont.*, Vol. XX AC-10, pp. 308-314, (1965).

VII. CONTROL

1. INTRODUCTION:

While discussing the importance of control theory, a friend of the authors' aptly remarked, "If it ain't got controls on it, it's no damn good." This facetious comment is no understatement. The material of this chapter provides a bit of an introduction to this fascinating area. Hopefully the interconnection viewpoint will motivate the student to pursue further work.

Schematically we may view the general linear deterministic control problem as in figure 1.1.

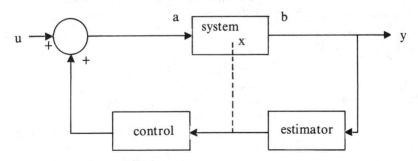

Figure 1.1. State space control configuration.

The problem is to control the system state vector, x, by applying an appropriate input, a. In particular, the *controllability* problem requires the existence of an input vector, a(t), defined over a finite interval [t_0, t_f], driving the initial state of the system, x(t_0), to some desired final state, x(t_f). Practically speaking, the problem is to find the appropriate input which makes the system state behave in a prescribed fashion.

A second problem is *stabilizability*, which is an intriguing controllability problem. Given a component with state model

$$\dot{x} = Ax + Ba$$
$$b = Cx + Da \tag{1.2}$$

Construct a state feedback gain matrix, K, so that the resultant component is stable. That is to say, the dynamics of the controlled component are characterized by the stable differential equation

$$\dot{x} = (A + BK)x \tag{1.3}$$

in which all state trajectories decay exponentially to zero. If the component is *completely controllable*, then the addition of a suitable bias term, say v(t), produces an input which asymptotically drives any initial state to a v-dependent final state. If the component, 1.2, has unstable directed modes, then a stabilizing feedback matrix, as in 1.3, exists if and only if the unstable modes are controllable.

Computation of the controlling input underscores both of the above problems. In turn, this requires knowledge of the state vector. For controllability one needs initial state information whereas for stabilizability continuous measurements of the state are fed back.

Seldom, if ever, does the control engineer have direct access to state information. The output equation

$$b = Cx + Da \tag{1.4}$$

distorts or destroys state information. Of course if C is invertible one may compute x from measurements of the accessible variables, b and a. Generally an *estimation* scheme is necessary. Here one estimates x from the accessible variables, a and b, in conjunction with the knowledge that x must satisfy a differential equation, 1.2.

The estimation problem is the dual of the controllability and stabilizability problems. Using the standard terminology, a component, 1.2, is *observable* if and only if its initial state, $x(t_0)$, can be computed from measurements on $a(t)$ and $b(t)$ over the finite time interval $[t_0, t_f]$; the component is *detectable* if $x(t_0)$ can be asymptotically computed from measurements on $a(t)$ and $b(t)$ for $t_0 \leqslant t < \infty$.

The purpose of the present chapter is to discuss the four problems of system controllability, stabilizability, observability, and detectability in the context of our theory of interconnected dynamical systems. [1,5,8] Sections 2 through 6 constitute a short course on the control of a single system component while sections 7 and 8 deal with the application of these ideas to interconnected dynamical systems. The former sections are devoted to the concepts of controllability, stabilizability, observability, and detectability; respectively. The emphasis in these sections is on the derivation of tests to determine whether or not a system component is controllable, etc., and the formulation of algorithms for constructing appropriate controllers and detectors.

Section 6 centers on the *control of composite interconnected systems* with the respective tests formulated in terms of the connection matrices. Since the "tests" are rank tests on large matrices, section 4 covers some numerical methods for *rank computation*.

The components of a large interconnected system are often geographically dispersed. Thus *global control laws* require an intricate communications net surrounding a central computer. Such a global interplay is bypassed by using a *decentralized control strategy* in which the control input for each component depends only on the state of the component. The advent of the low cost microprocessor is making this strategy most appealing. Section 8 sketches the rudiments of this viewpoint.

2. CONTROLLABILITY

The setting for this topic is the anticipated component state model

$$\dot{x} = Ax + Ba$$
$$b = Cx + Da \tag{2.1}$$

where A is nxn and B is nxm. To unify this exposition with classical ones, and for notational convenience, we define the *state transition matrix* as

$$\Phi(t) = \exp[At] \tag{2.2}$$

An input, $a(t)$, defined over the finite time interval, $[t_0, t_1]$, is said to *transfer* an initial state $x(t_0)$ to $x(t_1)$ if

$$x(t_1) \;=\; \Phi(t_1-t_0)x(t_0) + \int_{t_0}^{t_1} \Phi(t_1-q)Ba(q)dq \qquad (2.3)$$

This follows from theorem I.4.19.

DEFINITION 2.4: Fix $x_0 = x(t_0)$. If for each state $x_1 = x(t_1)$ in $\mathbf{R}^n$, there exists an input which *transfers* x_0 to x_1, then the state x_0 is said to be *controllable*.

DEFINITION 2.5: If every state x_0 in $\mathbf{R}^n$ is controllable on $[t_0,t_1]$, the component is said to be *completely controllable* on $[t_0,t_1]$.

Observe that equation 2.3 is equivalent to

$$0 \;=\; \Phi(t_1-t_0)[x_0-\Phi(t_0-t_1)x_1] + \int_{t_0}^{t_1} \Phi(t_1-q)Ba(q)dq \qquad (2.6)$$

Defining $\underline{x}(t_0) = [x_0-\Phi(t_0-t_1)x_1]$, it becomes clear, after some thought, that complete controllability coincides with the ability to drive every initial state to the zero state.

PROPOSITION 2.7: A component as per 2.1 is completely controllable if and only if every state can be transferred to the zero state.

EXAMPLE 2.8: Suppose the dynamics of a component are given by the scalar differential equation

$$\dot{x} \;=\; \lambda x + \beta a \qquad (2.9)$$

If $\beta = 0$, then by 2.3

$$x(t) \;=\; \exp[\lambda(t-t_0)] \, x(t_0) \qquad (2.10)$$

Thus, $x(t) = 0$ if and only if $x(t_0) = 0$ for any finite time interval $[t_0,t]$. However, if $\beta \neq 0$, the component is completely controllable as follows.   For an arbitrary but specified initial state, x_0, let us exhibit a controlling input. Define the scalar

$$\xi \;=\; \Phi(t_1,t_0)x_0 \;=\; \exp[\lambda(t_1-t_0)]x_0 \qquad (2.11)$$

Define the constant input

$$a(q) \;=\; \frac{\lambda \xi}{\beta[1-\Phi(t_1-t_0)]} \tag{2.12}$$

With these definitions a straightforward calculation verifies that

$$0 \;=\; \Phi(t_1,t_0)x_0 + \int_{t_0}^{t_1} \Phi(t_1-q)\beta a(q)dq \tag{2.13}$$

For each x_0, the input according to 2.12 drives x_0 to zero, so the component is completely controllable.

Upon some close scrutiny of the above discussion, the careful reader will observe that complete controllability is independent of the time interval, whereas the amplitude of the controlling input increases exponentially as (t_1-t_0) goes to zero. This property rests on the time invariant nature of the component and will be properly demonstrated later in the section.

Another important observation is that there are an infinite number of controlling inputs. Indeed, let $a_1(t)$ be any input such that

$$0 \;=\; \int_{t_0}^{t_1} \Phi(t_1-q)\beta a_1(q)dq \tag{2.14}$$

Let $a(q)$ be as in equation 2.12. Then $a_2(q) = a_1(q) + a(q)$ also drives the state x_0 to zero for the component of 2.9. This kind of ambiguity led to the development of *optimal control* where one chooses the "best" input from the set of controlling inputs.

The immediate goal is to fabricate controllability tests. Such tests are bound up with the notions of eigenvalue, rank, and positive definiteness. The road to these controllability tests starts out with the notion of *positive definiteness*. Unless stated otherwise, z and w will be vectors in $\mathbf{C}^n$.

Recall the definition of the *complex inner product* for two n vectors

$$\langle z,w \rangle \;=\; \sum_i z_i \overline{w}_i \tag{2.15}$$

where the overbar indicates complex conjugate. This relationship validates the following

$$\langle Mz,w \rangle \;=\; \langle z,M^*w \rangle \tag{2.16}$$

where M is an arbitrary square complex matrix and the superscript "*" means complex conjugate transpose.

DEFINITION 2.17: A matrix M is *Hermitian* if and only if $M = M^*$.

DEFINITION 2.18: An Hermitian matrix M is *positive semidefinite (positive definite)* if and only if

$$<z,Mz> \geqslant 0 \ (>0) \tag{2.19}$$

for all $z \neq 0$.

Clearly if M is real, then M is Hermitian if and only if $M = M^t$. These definitions bring us to our first set of interesting properties.

LEMMA 2.20: The *eigenvalues* of any Hermitian matrix, M, are real.

PROOF: Let λ and e be an eigenvalue and corresponding *eigenvector* of M respectively. Recall $Me = \lambda e$ and $<e,e> \neq 0$. As such

$$
\begin{aligned}
\lambda <e,e> &= <\lambda e,e> = <Me,e> \\
&= <e,Me> = <e,\lambda e> \tag{2.21} \\
&= \bar{\lambda} <e,e>
\end{aligned}
$$

This forces $\lambda = \bar{\lambda}$ making λ real.

THEOREM 2.22: An Hermitian matrix M is positive semidefinite (positive definite) if and only if all the eigenvalues of M are real and non negative (positive).

PROOF: The preceeding lemma established the realness of the eigenvalues. It remains to show $\lambda \geqslant 0$ for positive semi-definite and $\lambda > 0$ for positive definite. Although unnecessary, we assume the eigenvalues of M are distinct so as to simplify the proof. The structure of the proof is as follows: first factor $M = TDT^{-1}$ where D is a diagonal matrix of the eigenvalues of M and T is the matrix whose columns are the corresponding eigenvectors. Then we show $T^{-1} = T^*$ and use the relationship $M = TDT^*$ in the inner product $<z,Mz>$. This will show an explicit dependence of the inner product on the eigenvalues of M. The proof becomes obvious at that point.

As indicated factor $M = TDT^{-1}$ and show first that $T^{-1} = T^*$. Since $M = M^*$, lemma VI.5.4 guarantees that

$$<e_i,e_j> \;=\; \begin{cases} 0 \; ; \; i \neq j \\ 1 \; ; \; i = j \end{cases} \tag{2.23}$$

provided the e_i's are normalized to be unit vectors
This implies that

$$TT^* \;=\; \begin{bmatrix} <e_1,e_1> & \cdots & <e_1,e_n> \\ & & \\ \cdot & & \cdot \\ \cdot & & \cdot \\ \cdot & & \cdot \\ <e_n,e_1> & \cdots & <e_n,e_n> \end{bmatrix} \;=\; I \tag{2.24}$$

Consequently, $T^{-1} = T^*$ — i.e., T is an *unitary matrix*. As claimed, $M = TDT^*$. Define $w = T^*z$, then

$$<z,Mz> \;=\; <z,TDT^*z> \;=\; <w,Dw>$$
$$=\; \sum_i \lambda_i \mid w_i \mid^2 \tag{2.25}$$

This establishes the crux of the proof by showing the dependence of $<z,Mz>$ on the eigenvalues of M.

If $\lambda_i \geq 0$ for all i, then M is positive definite by 2.25. On the other hand if for some j (say $j = 1$), $\lambda_1 < 0$, set

$$z \;=\; T[1, 0, 0, \cdots, 0]^t \tag{2.26}$$

Since $T^* = T^{-1}$, $w = T^*z$ is

$$w \;=\; [1, 0, 0, \cdots, 0]^t \tag{2.27}$$

By 2.25, then

$$<z,Mz> \;=\; \lambda_1 < 0 \tag{2.28}$$

Therefore, M is not positive semidefinite if any one of its eigenvalues is negative.

For the positive definite case, suppose for each i, $\lambda_i > 0$ and $w = T^*z$. Then $<z,Mz> = 0$ only when $w_i = 0$ for all i. Consequently if $z \neq 0$, $<z,Mz> > 0$.

Conversely, suppose M is positive definite. Choose

$$z \; = \; Tw \; = \; T[0, 0, \cdots, 1, \cdots, 0]^t \qquad (2.29)$$

where the only nonzero entry of w is a "1" in the j-th position. The nonsingularity of T implies $z \neq 0$ whenever $w \neq 0$. Using 2.29

$$0 < <z,Mz> \; = \; <w,Dw> \; = \; \lambda_j \qquad (2.30)$$

verifying that the eigenvalues of a positive definite matrix are positive. The proof is complete.

Note, the above theorem connects the ideas of positive definiteness and invertibility. A positive semidefinite matrix is invertible if and only if it is positive definite.

EXAMPLE 2.31: Consider the real symmetric matrix

$$M \; = \; \begin{bmatrix} 3 & 1 & 2 \\ 1 & 1 & 0 \\ 2 & 0 & 2 \end{bmatrix} \qquad (2.32)$$

Note the inverse of M fails to exist. The eigenvalues of M are the roots of the polynomial

$$\lambda(\lambda^2 - 6\lambda + 6) \qquad (2.33)$$

Specifically, $\lambda = 0$, 1.268, and 4.732. All are non-negative, making M positive semidefinite.

In the sequel we often work with positive semidefinite matrices that have the form $M = N*N$ where N is arbitrary, possibly a rectangular matrix. Assume the dimensions of N are mxn with $n \leq m$. Clearly, M is nxn and symmetric.

PROPOSITION 2.34: $M = N*N$ is Hermitian and positive semidefinite.

PROOF: The Hermitian part is obvious. For positive semidefiniteness let $w = Nz$. Consider,

$$<z,Mz> \; = \; <z,N*Nz> \; = \; <Nz,Nz>$$
$$= \; <w,w> \; = \; \sum_i |w_i|^2 \geq 0 \qquad (2.35)$$

which confirms the semidefiniteness.

It is possible to strengthen the relationship between M and N as follows. The Hermitian matrix M is positive definite if and only if the columns of N (rows of N^t) are *linearly independent* – i.e. have maximal rank. These comments are a special case of the following lemma.

LEMMA 2.36: Suppose the entries of a rectangular matrix $N(t)$ are integrable functions over $[t_0, t_1]$. Then the matrix

$$M = \int_{t_0}^{t_1} N(q)N^*(q)dq \tag{2.37}$$

is positive semidefinite. Moreover, M is positive definite if and only if the columns of $N(t)$ are linearly independent – i.e., $N(t)z = 0$ for all t in $[t_0, t_1]$ if an only if $z = 0$. Note when dealing with a matrix whose entries are functions of time the rows of the matrix can be linearly independent even though the number of rows exceeds the number of columns and conversely. This is in contrast to the case of constant matrices where the maximum possible rank of a matrix equals its smallest dimension.[2]

PROOF: For each q the integrand is positive semidefinite. Thus the *partial sums* used to construct the integral are positive semidefinite since the linearity of the inner product guarantees that the sum of positive semidefinite matrices is positive semidefinite. Accordingly the integral, M, is also positive semidefinite.

To show positive definiteness implies the linear independence of the columns of $N(t)$, assume the opposite. This means that there exists z such that $N(t)z = 0$ over $[t_0, t_1]$. The desired contradiction, that M is not positive definite results from the following string of equalities

$$<z, Mz> = \int_{t_0}^{t_1} <z, N^*(q)N(q)z>dq$$

$$= \int_{t_0}^{t_1} <N(q)z, N(q)z>dq = 0 \tag{2.38}$$

Conversely, let $w(t) = N(t)z$. Assuming M is not positive definite we desire to show that $N(t)$ has linearly dependent columns. Consider that for some $z \neq 0$

$$0 \;=\; \langle z, Mz \rangle \;=\; \int_{t_0}^{t_1} \langle N(q)z, N(q)z \rangle dq$$

$$=\; \sum_i \int_{t_0}^{t_1} |w_i(q)|^2 dq \qquad (2.39)$$

Thus, $w(q) \equiv 0$ over $[t_0, t_1]$. Equivalently $N(t)\, z \equiv 0$ over $[t_0, t_1]$ and the columns of $N(t)$ are linearly dependent.

EXAMPLE 2.40: Let $[t_0, t_1] = [0,1]$ and let

$$N(q) \;=\; \begin{bmatrix} q & 0 & q^2 \\ 0 & q & 0 \end{bmatrix} \qquad (2.41)$$

Computing $N(q)^*N(q)$ yields

$$N(q)^*N(q) \;=\; \begin{bmatrix} q^2 & 0 & q^3 \\ 0 & q^2 & 0 \\ q^3 & 0 & q^4 \end{bmatrix} \qquad (2.42)$$

Consider

$$M \;=\; \int_0^1 N(q)^*N(q)dq \;=\; \begin{bmatrix} .333 & 0 & .25 \\ 0 & .333 & 0 \\ .25 & 0 & .2 \end{bmatrix} \qquad (2.43)$$

It is easily checked that m has positive real eigenvalues and, since it is symmetric, it is positive definite.

The main theorem on controllability requires one further preliminary, the well known *Caley-Hamilton theorem*.

THEOREM 2.44: Suppose $d(s) = \det(sI - A)$ is given by

$$d(s) \;=\; s^n + d_1 s^{n-1} + \cdots + d_n \qquad (2.45)$$

then $d(A) = 0$ (the zero matrix). In particular

$$A^n = - \sum_{i=1}^{n} d_i A^{n-i} \qquad (2.46)$$

PROOF: Theorem VI.3.45 showed that

$$(sI-A)^{-1} = \frac{R_o s^{n-1} + \cdots + R_{n-1}}{d(s)} \qquad (2.47)$$

where

$$
\begin{aligned}
R_o &= I \\
R_k &= R_{k-1} A + d_k I \\
0 &= R_{n-1} A + d_n I
\end{aligned}
\qquad (2.48)
$$

By successively eliminating the R_k's beginning with R_{n-1} the result follows.

EXAMPLE 2.49: Let A be defined as

$$A = \begin{bmatrix} 2 & 1 \\ 1 & 1 \end{bmatrix} \qquad (2.49)$$

It is a simple task to verify

$$A^2 = \begin{bmatrix} 5 & 3 \\ 3 & 2 \end{bmatrix} = 3 \begin{bmatrix} 2 & 1 \\ 1 & 1 \end{bmatrix} - \begin{bmatrix} 1 & 0 \\ 0 & 1 \end{bmatrix} = - d_2 A - d_o I \qquad (2.50)$$

where $A^o = I$ and $\det(sI-A) = s^2 - 3s + 1$.

Finally we come to the main result characterizing controllability of a component whose dynamics satisfy

$$\dot{x} = Ax + Ba \qquad (2.51)$$

THEOREM 2.52: For such a component, the following statements are equivalent:

(i)   The component is controllable.
(ii) $\text{Rank}[(\lambda I - A) \vdots B] = n$ for each eigenvalue λ of A.
(iii) $\text{Rank}[Q] = n$ where Q is the *controllability matrix*:
(iv) $\text{Rank}[\exp[-At]B] = n$ – i.e. there are n linearly independent rows.
(v)   The matrix M is positive definite.

Here,

$$Q = [B \mid AB \mid \cdots \mid A^{n-1}B] \tag{2.53}$$

$$M = \int_{t_0}^{t_1} \Phi(t_1-q)BB^*\Phi^*(t_1-q)dq \tag{2.54}$$

Moreover, the input

$$a(t) = -B^*\Phi^*(t_1-t)M^{-1}[\Phi(t_1-t_0)x_0-x_1] \tag{2.55}$$

drives $x_0 = x(t_0)$ to $x_1 = x(t_1)$.

PROOF: The proof is a sequence of five successive implications.
[(i) $\rightarrow$ (ii)] This verification is equivalent to verifying the contra positive –
i.e. if (ii) fails the component is not controllable. Assuming (ii) is false, there
exists $z \neq 0$ such that

$$z^t[(\lambda I-A) \mid B] = 0 \tag{2.56}$$

Equivalently, $z^t(\lambda I-A) = 0$ and $z^tB = 0$. In particular,

$$z^tA = \lambda z^t \tag{2.57}$$

implying

$$z^t A^2 = \lambda z^tA = \lambda^2 z^t \tag{2.58}$$

Successive multiplications on the right by A yields

$$z^tA^k = \lambda^k z^t \tag{2.59}$$

for all k. A Taylor series expansion for $\exp(At)$ in conjunction with 2.59
implies

$$z^t \exp[A(t_1-q)] = \exp[\lambda(t-q)]z^t \tag{2.60}$$

To show the component is not completely controllable, for an arbitrary
input "a" consider

$$z^t \int_{t_0}^{t_1} \Phi(t_1-q)Ba(q)dq = \int_{t_0}^{t_1} \exp[\lambda(t_1-q)] [z^tB]a(q)dq = 0 \tag{2.61}$$

Therefore the state z is orthogonal to the range of the integral operator. In particular, the system cannot be driven from $x(t_o) = 0$ to $x(t_1) = z$.

[(ii) → (iii)] This phase of the proof also proves the contrapositive. Assume rank$[Q] < n$ where Q is the controllability matrix. Thus there exists $z \neq 0$ such that $z^t Q = 0$. Equivalently, $z^t A^k B = 0$ for $k = 0,1,\ldots,n-1$. By Caley-Hamilton $z^t A^k B = 0$ for all integers $k \geqslant 0$. To establish the contrapositive we show there exists $e \neq 0$ such that $e^t[(\lambda I - A)^! B] = 0$ for some eigenvalue, λ, of A. It turns out that e is an eigenvector of A^t and $\bar{\lambda}$ the corresponding eigenvalue. Let

$$S = \left\{ w \mid w^t A^k B = 0, k \geqslant 0 \right\} \tag{2.62}$$

Clearly, $S \neq \phi$ since z is in S. We claim S is an invariant subspace of A^t - i.e. $A^t S \subset S$. Let w be in S and observe

$$(A^t w)^t A^k B = w^t A^{k+1} B = 0 \tag{2.63}$$

Since S is a non-zero invariant subspace of A^t, it must contain an eigenvector of A^{t}.[2] (In fact, each eigenvector of A^t spans an invariant subspace of A^t and when A has distinct eigenvalues every invariant subspace is spanned by some subset of the eigenvectors).[2] As such there exists a vector e in S such that

$$e^t[(\lambda I - A) \mid B] = 0 \tag{2.64}$$

where $\bar{\lambda}$ is the corresponding eigenvalue.

[(iii) → (iv)] Here is another use of the contrapositive argument: Suppose rank $[\exp[-AT]B] < n$ on $[t_o, t_1]$. There then exits $z \neq 0$ such that

$$z^t \exp[-At]B \equiv 0 \tag{2.65}$$

on $[t_0, t_1]$. However, the analyticity of $\exp[-AT]$ and its inverse guarantees the validity of 2.65 for all real t. Taking the derivative of 2.65 and evaluating at $t = 0$ forces $z^t A^k B = 0$ for all k. In particular

$$z^t Q = z^t[B \mid AB \quad \cdots \quad \mid A^{n-1} B] = 0 \tag{2.66}$$

So the rank $[Q] < n$, verifying the contrapositive.

[(iv) → (v)] If $\exp[-AT]B$ has n linearly independent rows where B is nxm with $m \leqslant n$, the implication follows from lemma 2.36.

[(v) → (i)] Recall M is invertible if and only if it is positive definite. Let $a(t)$ be defined as in 2.55. Initializing the component at $x(t_0)$ and applying $a(t)$ yields

$$\Phi(t_1-t_0)x(t_0)$$

$$- \int_{t_0}^{t_1} \Phi(t_1-q)BB^*\Phi^*(t_1-q)M^{-1}[\Phi(t_1-t_0)x(t_0) - x(t_1)]dq$$

$$= \Phi(t_1-t_0)x(t_0)$$

$$- [\int_{t_0}^{t_1} \Phi(t_1-q)BB^*\Phi^*(t_1-q)dq]M^{-1}[\Phi(t_1-t_0)x(t_0) - x(t_1)]$$

$$= x(t_1) \tag{2.67}$$

Thus given that M is positive definite, for any initial state and any final state, we have exhibited a control, $a(t)$, which will drive the initial state to the desired final state. The component then is completely controllable and the proof is complete.

Conditions (ii) and (iii) both indicate that controllability is independent of the time interval over which the controlling input acts. On the other hand, the control, $a(t)$, explicitly and implicitly depends on the time interval as indicated by the form of 2.55. Note, condition ii actually holds for all complex numbers λ since rank $[\lambda I-A] = n$ if λ is not an eigenvalue.

To justify that controllability is time-independent, suppose a component is controllable on $[t_0,t_1]$, then 2.55 permits construction of an alternate control on the interval $[t'_0,t'_1]$ with $t_0 \leqslant t'_0 \leqslant t'_1 \leqslant t_1$. The time invariant nature of the component description allows the time intervals to be translated arbitrarily. Formally speaking, we have the following corollary.

COROLLARY 2.68: Controllability is independent of a specific time interval for components described by 2.1.

EXAMPLE 2.69: Consider the system

$$\begin{bmatrix} \dot{x}_1 \\ x_2 \end{bmatrix} = \begin{bmatrix} -2 & 1 \\ -3 & 2 \end{bmatrix} \begin{bmatrix} x_1 \\ x_2 \end{bmatrix} + \begin{bmatrix} 1 \\ 1 \end{bmatrix} a \tag{2.70}$$

To use test (ii) of the theorem, observe that the eigenvalues of A are ± 1. The test requires we work with the following two matrices:

$$[(I-A) \mid B] = \begin{bmatrix} 3 & -1 & 1 \\ 3 & -1 & 1 \end{bmatrix} \tag{2.71}$$

and

$$[(-I-A) \mid B] = \begin{bmatrix} 1 & -1 & 1 \\ 3 & -3 & 1 \end{bmatrix} \tag{2.72}$$

The first of these test matrices has rank one so the component is not controllable. The second has rank two. Thus we may conclude that the range of the integral operator of 2.3 is *orthogonal* to the subspace of eigenvectors of A^t satisfying $z^t(I-A) = 0$.

EXAMPLE 2.73: Suppose a component has A and B matrices

$$A = \begin{bmatrix} -3 & 2 & 2 \\ 0 & -1 & 0 \\ -6 & 6 & 4 \end{bmatrix}; \quad B = \begin{bmatrix} 1 & 1 & 2 \\ 0 & 1 & 1 \\ 2 & 0 & 2 \end{bmatrix} \tag{2.73}$$

The controllability matrix Q is

$$\begin{aligned} Q &= [B \mid AB \mid A^2B] \\ &= \begin{bmatrix} 1 & 1 & 2 & 1 & -1 & 0 & 1 & 1 & 2 \\ 0 & 1 & 1 & 0 & -1 & -1 & 0 & 1 & 1 \\ 2 & 0 & 2 & 2 & 0 & 2 & 2 & 0 & 2 \end{bmatrix} \end{aligned} \tag{2.74}$$

The rank of Q is two so the component is not completely controllable. It turns out that the set of controllable states forms a subspace spanned by the columns of Q. In addition any state having a non-zero projection on the null space of Q^t is an uncontrollable state. In particular, the state $x_0 = [2,-2,1]^t$ lies in the null space of Q^t and is uncontrollable. Furthermore any state having a component in the direction of $[2,-2,1]^t$ is uncontrollable. On the other hand, any state which can be expressed as a linear combination of $[1,0,2]^t$ and $[1,1,0]^t$ is controllable. Note that these vectors are the first two columns of the Q matrix. These comments serve as a good lead-in to the following very important theorem.[5,8]

THEOREM 2.75: Suppose a component has controllability matrix Q with rank$(Q) = q < n$. Suppose the component dynamics satisfy $x = Ax + Ba$. Then there exists a nonsingular matrix, T, such that the equivalent component description

$$\dot{z} = [T^{-1}AT]z + [T^{-1}B]a \qquad (2.76)$$

derived from the original component description by letting $x = Tz$, has state matrices of the form

$$T^{-1}AT = \left[\begin{array}{c|c} A_{11} & A_{12} \\ \hline 0 & A_{22} \end{array}\right] \qquad (2.77)$$

and

$$T^{-1}B = \left[\begin{array}{c} B_1 \\ \hline 0 \end{array}\right] \qquad (2.78)$$

where A_{11} is $q{\times}q$, A_{22} is $(n-q){\times}(n-q)$, etc. Moreover, the component described by the q-dimensional state model

$$\dot{z}_1 = A_{11} z_1 + B_1 a \qquad (2.79)$$

is completely controllable, where z_1 represents the first q entries of z.

PROOF: Define the transformation T according to

$$T = [p_1, \cdots, p_q \mid n_1, \cdots, n_{n-q}] \qquad (2.80)$$

where q is the dimension of the range of Q since rank$[Q] = q$ and $(n-q)$ is the dimension of the null space of Q^t. The set $[p_i]$ are a basis for range$[Q]$ chosen as any q columns of Q which are linearly independent. The set $[n_i]$ are a basis for null$[Q]$. Clearly the vectors n_i and p_j are pairwise orthogonal.

The transformation T is nonsingular since the combined set $\{p_i, n_j \mid 1 \leqslant i \leqslant q, \ 1 \leqslant j \leqslant n-q\}$ form a basis for $\mathbf{R}^n$ (the state space). As such, substituting the equality x = Tz into

$$\dot{x} = Ax + Ba \qquad (2.81)$$

yields

$$\dot{z} = [T^{-1}AT]z + [T^{-1}B]a \qquad (2.82)$$

Since the columns of B are vectors in the range of Q, T^{-1} maps these column vectors onto other vectors whose only possible non-zero entries are in the first q places. This verifies 2.78.

To verify 2.77 first observe that the space spanned by p_i, is invariant under A. To see this consider

$$AQ = [AB \vdots A^2B \vdots \cdots \vdots A^nB] \qquad (2.83)$$

and apply Caley-Hamilton to A^nB. Therefore A maps the first q column of T onto vectors which are linear combinations of the elements of the set p_i. Now T^{-1} maps these vectors onto vectors whose only possible non-zero entries are in the first q places. As such $T^{-1}AT$ has the form specified by 2.77.

To complete the proof observe that the controllability matrix (designated Q_T) for the transformed system of 2.76 is

$$Q_T = TQ = \begin{bmatrix} B_1 & \vdots & A_{11} \ B_1 & \vdots & \cdots & \vdots & A_{11}^{n-1} \ B_1 \\ \cdots & & \cdots & & \cdots & & \cdots \\ 0 & \vdots & 0 & \vdots & \cdots & \vdots & 0 \end{bmatrix} \qquad (2.84)$$

Since T is nonsingular, rank Q_T = Q. Clearly then the first q rows of Q_T must have rank q since the remaining rows are zero. Therefore by Caley-Hamilton Rank $[Q_1]$ = 9 where

$$Q_1 = [B_1 \vdots A_{11} \ B_1 \vdots \cdots \vdots A_{11}^{q-1} \ B_1] \qquad (2.85)$$

and the component description of 2.79 is completely controllable. The proof is now complete.

The power of this decomposition theorem is aptly demonstrated by regarding the form of the integral operator

$$\int_{t_o}^{t_f} \exp[T^{-1}AT(t_f-q)]\ T^{-1}Ba(q)dq$$

$$= \int_{t_o}^{t_f} \left[\begin{array}{c} \exp[A_{11}(t_f-q)]B_1 \\ \hline 0 \end{array}\right] a(q)dq \qquad (2.86)$$

Thus the *controllable states* are clearly those in the range of Q, all other states being uncontrollable.

EXAMPLE 2.87: Let us continue example 2.73 in which

$$A = \begin{bmatrix} -3 & 2 & 2 \\ 0 & -1 & 0 \\ -6 & 6 & 4 \end{bmatrix} ; B = \begin{bmatrix} 1 & 1 & 2 \\ 0 & 1 & 1 \\ 2 & 0 & 2 \end{bmatrix} \qquad (2.88)$$

which has controllability matrix given in 2.74. The vector $[2,-2,-1]^t$ spans the null space of Q^t. The transformation matrix is

$$T = \begin{bmatrix} 1 & 1 & 2 \\ 0 & 1 & -2 \\ 2 & 0 & -1 \end{bmatrix} \qquad (2.89)$$

and T^{-1} becomes

$$T^{-1} = \frac{1}{9}\begin{bmatrix} 1 & -1 & 4 \\ 4 & 5 & -2 \\ 2 & -2 & -1 \end{bmatrix} \qquad (2.90)$$

This yields the equivalent system description

$$\left[\begin{array}{c|c} A_{11} & A_{12} \\ \hline 0 & A_{22} \end{array}\right] = \left[\begin{array}{cc|c} 1 & 0 & -14 \\ 0 & -1 & 2 \\ \hline 0 & 0 & 0 \end{array}\right] \qquad (2.91)$$

and

$$\begin{bmatrix} B_1 \\ \hline 0 \end{bmatrix} = \begin{bmatrix} 1 & 0 & 1 \\ 0 & 1 & 1 \\ \hline 0 & 0 & 0 \end{bmatrix} \tag{2.92}$$

Obviously the reduced controllability matrix

$$Q_1 = [B_1 \mid A_{11} B_1]$$

$$= \begin{bmatrix} 1 & 0 & 1 & 1 & 0 & 1 \\ 0 & 1 & 1 & 0 & -1 & -1 \end{bmatrix} \tag{2.93}$$

has rank 2 and the reduced component model is completely controllable.

The component description derived by using the transformation $z = T^{-1}x$ is designated the *controllability canonical form* of the given component model. We may view the form as

$$\begin{bmatrix} \dot{z}_1 \\ \hline \dot{z}_2 \end{bmatrix} = \begin{bmatrix} A_{11} & A_{12} \\ \hline 0 & A_{22} \end{bmatrix} \begin{bmatrix} z_1 \\ \hline z_2 \end{bmatrix} + \begin{bmatrix} B_1 \\ \hline 0 \end{bmatrix} a \tag{2.94}$$

As such, any vector of the form $[z_1,0]^t$ is controllable and any vector of the form $[z_1,z_2]$ for $z_2 \neq 0$ is uncontrollable in that no input will drive the z_2 part to zero. Finally, any vector of the form $[0,z_2]^t$ is *completely uncontrollable*.

3. STABILIZATION

This section tackles the theory of component *stabilization*. Stabilization depends intrinsically on the problem of *spectral assignability* or *pole placement* which in turn is generically equivalent to the problem of *controllability*. A component is *stabilizable* whenever the *directed modes* corresponding to right half plane eigenvalues can be moved to the left half plane through state feedback.

Suppose A has distinct eigenvalues $\lambda_1, \cdots \lambda_n$ and eigenvectors $e_1, \cdots, e_n$, then recall that $\exp(\lambda_i t)e_i$ is a directed mode. Assuming e_i is real, then this mode is controllable if and only if e_i lies in the controllable subspace of the pair (A,B). The controllable subspace of the pair (A,B) is simply the span of the columns of the controllability matrix, $Q = [B \mid AB \mid \cdots \mid A^{n-1}B]$. If e_i and e_j corresponds to a complex conjugate pair of eigenvalues then each mode is controllable if and only if both the real and imaginary parts of e_i lie in the controllable subspace of the pair (A,B).

Using the notation due to Wonham,[8] denote by $<A \mid B>$ the controllable subspace of the pair (A,B) – i.e. $< A \mid B >$ is the span of the columns of the controllability matrix. In the above, $\mathbf{B}$ is the image of the B-matrix. Let $b_1, \cdots, b_m$ be the columns of B, then $A\mathbf{B}$ is the span of the vectors $Ab_1, \cdots, Ab_m$. Similarly we may characterize $A^j\mathbf{B}$ as the span of $A^jb_1, \cdots, A^jb_m$. Using the above notation, $< A \mid \mathbf{B}>$ is precisely defined as

$$< A \mid \mathbf{B} > \ = \ \mathbf{B} + A\mathbf{B} + \cdots + A^{n-1}\mathbf{B} \tag{3.1}$$

where "+" means *subspace addition*. The above description of $A^j\mathbf{B}$ establishes the link between the controllability matrix and $< A \mid \mathbf{B} >$.

Before describing a technique for stabilization, we prove that the controllable subspace $< A \mid \mathbf{B} >$ is invariant under state feedback. In other words, given an appropriately dimensioned state feedback matrix K, then $< A + BK \mid \mathbf{B} > = < A \mid \mathbf{B} >$. In a more meaningful sense, the component model

$$\dot{x} \ = \ Ax + Ba \tag{3.2}$$

is controllable if and only if the component model

$$\dot{x} \ = \ (A + BK)x + Ba \tag{3.3}$$

is controllable.

LEMMA 3.4: For state feedback K

$$< A + BK \mid \mathbf{B} > \ = \ < A \mid \mathbf{B} > \tag{3.4}$$

PROOF: Observe that

$$(A + BK)\mathbf{B} = A\mathbf{B} + BK\mathbf{B} \tag{3.5}$$

Similarly

$$(A + BK)^j\mathbf{B} = \mathbf{B} + A\mathbf{B} + \cdots + A^j\mathbf{B} \tag{3.6}$$

By the Caley-Hamilton theorem

$$(A + BK)^j\mathbf{B} = < A \mid \mathbf{B} > \tag{3.7}$$

for all j. Hence

$$< A + BK\,|\,\mathbf{B} > = < A\,|\,\mathbf{B} > \qquad (3.8)$$

By symmetry the reverse inclusion follows – i.e. define $A_o = A + BK$; the above implies that $< A_o - BK\,|\,\mathbf{B} > = < A_o\,|\,\mathbf{B}>$.

The importance of this theorem is that state feedback neither destroys the property of controllability nor does it change the *original controllable subspace*. Those states controllable without feedback are still controllable with arbitrary state feedback.

Stabilizability requires moving those eigenvalues of A located in $\mathbf{C}_+$ into the left half complex plane. Assuming 3.2 is a completely controllable component model, the goal is to construct a *state* feedback matrix, K, so that the eigenvalues of the model 3.3 lie in the left half plane. This is possible only when the "eigenvectors" (possibly generalized eigenvectors) of the non-negative eigenvalues lie in the controllable subspace, $< A\,|\,\mathbf{B} >$.

Let Λ be a *symmetric set* of n complex numbers where *symmetric* means that non-real numbers appear in complex conjugate pairs. Denote by $\sigma(A)$ the *spectrum* of the matrix A where the terms spectrum, eigenvalue, characteristic value, etc. mean the same thing. Our first task is to solve the *spectral assignability problem* which proceeds in several stages. The first stage is to develop a spectral assignability technique for the special case of a single input component model in which A and B have the *rational canonical form* – i.e., A and B have the form

$$A = \begin{bmatrix} 0 & 1 & 0 & 0 & \cdots & 0 & 0 \\ 0 & 0 & 1 & 0 & \cdots & 0 & 0 \\ \cdot & \cdot & \cdot & \cdot & \cdots & \cdot & \cdot \\ 0 & \cdot & \cdot & \cdot & \cdots & 0 & 1 \\ a_1 & a_2 & a_3 & \cdot & \cdots & \cdot & a_n \end{bmatrix} \qquad B = \begin{bmatrix} 0 \\ \cdot \\ \cdot \\ 0 \\ 1 \end{bmatrix} \qquad (3.9)$$

One tackles the general case by converting the model to the rational canonical form solving the assignability problem and then transforming back.

The *characteristic polynomial*, d (s), of a nxn matrix A is given by $\det(sI-A)$. The polynomial d (s) is *monic* and of n-th degree. The zeros of d (s), $\lambda_1, \ldots, \lambda_n$, are the eigenvalues of A. Two forms of d (s) prove useful:

$$d(s) = (s - \lambda_1)(s - \lambda_2) \cdots (s - \lambda_n) \qquad (3.10)$$

and

$$d(s) = s^n - [a_1 + a_2 s + \cdots + a_n s^{n-1}] \qquad (3.11)$$

LEMMA 3.12: The A-matrix given in 3.9 has the characteristic polynomial given by 3.11.

The proof is straightforward. Simply expand $\det[sI-A]$ along the bottom row.

Now let a component model

$$\dot{x} = Ax + Ba \qquad (3.12)$$

be in the rational canonical form, equation 3.9. It is easily checked that the pair (A,B) is completely controllable. Since the model is single input every state feedback matrix K has the form

$$K = [k_1 \ k_2 \cdots k_n] \qquad (3.13)$$

Thus, A+BK becomes

$$A + BK = \begin{bmatrix} 0 & 1 & 0 & 0 & \cdot & \cdot & 0 & 0 \\ 0 & 0 & 1 & 0 & \cdot & \cdot & 0 & 0 \\ \cdot & & \cdot & \cdot & \cdot & \cdot & \cdot & \cdot \\ 0 & & \cdot & \cdot & \cdot & \cdot & 0 & 1 \\ a_1+k_1 & a_2+k_2 & \cdot & \cdot & \cdot & \cdot & & a_n+k_n \end{bmatrix} \qquad (3.14)$$

This observation is the key to the spectral assignability problem for this special case: find a state feedback matrix K so that the spectrum of A+BK, $\sigma(A+BK)$, is given by a prespecified symmetric set $\Lambda = \{\lambda_1, \ldots, \lambda_n\}$.

The characteristic polynomial of the *desired* A+BK matrix will be

$$(s - \lambda_1)(s - \lambda_2) \cdots (s - \lambda_n)$$

$$= s^n - [\underline{a}_1 + \underline{a}_2 s + \cdots + \underline{a}_n s^n] \qquad (3.15)$$

By defining the feedback matrix K according to

$$K = [\underline{a}_1 - a_1, \ \underline{a}_2 - a_2, \cdots, \ \underline{a}_n - a_n] \qquad (3.16)$$

the resultant A+BK matrix is

$$
A+BK = \begin{bmatrix}
0 & 1 & 0 & 0 & \cdot & \cdot & 0 & 0 \\
0 & 0 & 1 & 0 & \cdot & \cdot & 0 & 0 \\
\cdot & \cdot & \cdot & \cdot & \cdot & \cdot & \cdot & \cdot \\
0 & 0 & \cdot & \cdot & \cdot & \cdot & 0 & 1 \\
\underline{a}_1 & \underline{a}_2 & \cdot & \cdot & \cdot & \cdot & \cdot & \underline{a}_n
\end{bmatrix}
\tag{3.17}
$$

which has the desired spectrum. This completes stage 1.

Stage 2 considers the general single input model. Assume (A,B) is an arbitrary single input pair which is controllable. Via a *non-singular* transformation called a *similarity transformation* we associate a new pair $(\hat{A},\hat{B})$ with (A,B) where $\hat{A}$ and $\hat{B}$ have the form given in 3.9. The spectrum of A is invariant under any similarity transformation. Knowledge of this transformation then permits one to solve the general spectral assignability problem for controllable single input systems.

Since the pair (A,B) is controllable, the controllability matrix $Q = [B \vdots AB \vdots \cdots \vdots A^{n-1}B]$ has rank n implying the existence of Q^{-1}. Define v as the last row of Q^{-1}. Construct the desired transformation matrix V as

$$
V = \begin{bmatrix}
v \\
\hline
vA \\
\hline
\cdot \\
\cdot \\
\cdot \\
\hline
vA^{n-1}
\end{bmatrix}
\tag{3.18}
$$

Now compute V^{-1}. The matrices $\hat{A}$ and $\hat{B}$ are given by

$$
\begin{aligned}
\hat{A} &= V A V^{-1} \\
\hat{B} &= V B
\end{aligned}
\tag{3.19}
$$

The proof of this fact is straightforward and can be found in numerous texts.[1]

To construct a feedback matrix K so that $\sigma(A+BK) = \Lambda$ for a prespecified symmetric set Λ, construct an *intermediate feedback matrix* $\hat{K}$ such that $\sigma(\hat{A}+\hat{B}\hat{K}) = \Lambda$. This may be done by the method outlined in stage one. Applying the inverse similarity transformation which preserves the assigned spectrum of $\hat{A}+\hat{B}\hat{K}$ yields

$$\begin{aligned} A+BK &= V^{-1}[\hat{A}+\hat{B}\hat{K}]V \\ &= A + B\,[\hat{K}V] \end{aligned} \tag{3.20}$$

Hence the required feedback matrix is

$$K = \hat{K}V \tag{3.21}$$

EXAMPLE 3.22: Suppose a component model has the (A,B) pair

$$A = \begin{bmatrix} 1 & 1 \\ 0 & -1 \end{bmatrix}, \quad B = \begin{bmatrix} 0 \\ 1 \end{bmatrix} \tag{3.23}$$

To meet certain design specifications it is required that the component be stable and have eigenvalues $-1 \pm j$.

The controllability matrix for the pair (A,B) is

$$Q = [B \vdots AB] = \begin{bmatrix} 0 & 1 \\ 1 & -1 \end{bmatrix} \tag{3.24}$$

which has rank 2 implying that the pair (A,B) is controllable. Hence the state feedback K exists so that $\sigma(A+BK) = [-1 \pm j]$. To find the transformation V which converts A and B into rational canonical form observe that

$$Q^{-1} = \begin{bmatrix} 1 & 1 \\ 1 & 0 \end{bmatrix} \tag{3.25}$$

Setting v = [1   0] implies

$$V = \begin{bmatrix} v \\ \hline vA \end{bmatrix} = \begin{bmatrix} 1 & 0 \\ \hline 1 & 1 \end{bmatrix} \tag{3.26}$$

Therefore the canonical form $(\hat{A},\hat{B})$ of the original pair (A,B) is

$$\hat{A} = VAV^{-1} = \begin{bmatrix} 0 & 1 \\ 1 & 0 \end{bmatrix}; \ \hat{B} = VB = \begin{bmatrix} 0 \\ 1 \end{bmatrix} \qquad (3.27)$$

From 3.27 the characteristic polynomial of $\hat{A}$ is $s^2 - [1 + 0s]$ whereas the desired characteristic polynomial is

$$(s + 1 + j) \ (s + 1 - j) = s^2 - [-2 - 2s] \qquad (3.28)$$

Consequently the intermediate feedback is

$$\hat{K} = [(-2-1) \quad (-2-0)] = [-3 \quad -2] \qquad (3.29)$$

The required feedback is

$$K = \hat{K} V = [-5 \quad -2] \qquad (3.30)$$

in which case

$$A + BK = \begin{bmatrix} 1 & 1 \\ -5 & -3 \end{bmatrix} \qquad (3.31)$$

has the necessary spectrum.

Stage 3 of our development outlines a pole placement scheme for controllable multi-input components. The technique is to convert the multi-input component to an "equivalent" single input component and then use the above techniques to assign the poles of the "equivalent" model. One applies the inverse conversion to get the desired feedback.

LEMMA 3.32: Let a component have a controllable state model

$$\dot{x} = Ax + Ba \qquad (3.33)$$

Let b be any non-zero column of B; then there exists state feedback K such that the model

$$\dot{x} = (A+BK) x + b a \qquad (3.34)$$

is controllable.

PROOF: Define the family of independent vectors

$$v^{i+1} = Av^i + b \tag{3.35}$$

where $v^1 = b$ and $i = 1, \cdots, p$ and where p is chosen as the largest integer for which the family v^i forms a linearly independent set. Let A be nxn. Either $p = n$ or there is some other column of B, say $\underline{b}$, which is independent of the family $\{v^i \mid i = 1, \cdots, p\}$. Indeed, if no such $\underline{b}$ exists, then every column of the controllability matrix $Q = [B \mid AB \mid \cdots \mid A^{n-1}B]$ is linearly dependent on the family $\{v^i \mid i = 1, \cdots, p\}$.

To see this claim, realize that each column of Q has the form $A^i b_0$ for some column b_0 of B. Since b_0 depends on the v^i

$$b_0 = \sum_{k=1}^{p} \xi_i [A^j v^k] \tag{3.36}$$

for some set of scalars ξ_i. However the vectors $A^j v^k$ depend on the family $\{v^i \mid i=1, \cdots, p\}$ by the assumption of the choice of p. In particular, for each $\{v^k, 1 \leqslant k \leqslant p\}$ there exist scalars ζ_i^k such that

$$A^j b_0 = \sum_{k=1}^{p} \sum_{i=1}^{p} (\xi_k \zeta_i^k v^k) \tag{3.37}$$

Hence if every column of B depends on the v_i, then every column of Q does. However controllability implies $\text{rank}(Q) = n$, so either $p = n$ or there does exist $\underline{b}$ independent of $\{v^i \mid i=1, \cdots, p\}$. If b exists define new vectors independent of all previously constructed v^i's. Again the total number of the v^i's either equals n or from the assumption of controllability another column of B is independent of the new family of v^i's.

By repeating the above process finitely many times one obtains n independent v^i's satisfying

$$v^{i+1} = Av^i + b^i \tag{3.38}$$

where $v^i = b^i$ is that column of B chosen so that v^{i+1} is independent of the previously computed vectors. Note that consecutive b^i's may be identical.

Given the family of vectors $\{v^i \mid i = 1, \cdots n\}$ as computed above and the set $\{b^i \mid i = 1, \cdots, n\}$, define the matrix $K = MT^{-1}$ where M is a matrix whose j–i entry is one if b^i represents the j-th column of B and zero otherwise and

$$T = [v^1 \mid v^2 \mid \cdots \mid v^n] \tag{3.39}$$

This construction yields the equality

$$BKv^i = b^i \tag{3.40}$$

Finally, the above constructions render the pair $(A+BK,b)$ controllable. Specifically, 3.38 and 3.40 imply

$$(A+BK)\, v^i = Av^i + b^i = v^{i+1} \tag{3.41}$$

Iterating 3.42 produces

$$(A+BK)^{i-1}\, b = (A+BK)^{i-1}\, v^1 = v^i \tag{3.42}$$

Therefore the controllability matrix of the pair $(A+BK,b)$ is

$$\begin{aligned} Q &= [b \mid (A+BK)b \mid \cdots \mid (A+BK)^{n-1}b] \\ &= [v^1 \mid v^2 \mid \cdots \mid v^n] \end{aligned} \tag{3.43}$$

which has rank n, completing the proof.

This lemma allows the conversion of a multi-input controllable pair (A,B) to an equivalent single input controllable pair (A+BK, b) for suitable choice of K for each chosen b. Fortunately there is a more numerically amenable way of finding an equivalent single input pair as follows.

Randomly generate the entries of K and take a random linear combination of the columns of B. In particular, define b = Bv for an appropriately dimensioned vector v whose entries are randomly generated. The resultant pair (A+BK,b) is controllable *with probability one* whenever the original pair (A,B) is controllable. Both pairs contain the same information and can be constructed from each other since K and v are known. Note that the phrase "with probability one" is similar to the notion *"for almost all connections"* developed in chapter III. Finally the reader is cautioned against generating random entries of K and v with values between zero and one *only*. Such a limited random number generation will typically result in numerically ill-conditioned matrices and vectors.[8]

Assigning the spectrum of a multi-input controllable component proceeds as follows:

(i) Record dimensions of A and B. Check the controllability of the pair (A,B).

(ii) Randomly generate a feedback matrix, K_o, and define $A_o = A+BK_o$. With probability one, A_o has distinct eigenvalues. Check this. If not, repeat.

(iii) Take a random linear combination of the columns of B according to $b = Bv$. Record v. With probability one the pair (A_o,b) is controllable. Check. If not, generate another b.

REMARK: The pair (A_o,b) represents an "equivalent" single input system.

(iv) Assign the prespecified spectrum, Λ, via the methods already discussed by constructing state feedback k such that the spectrum of $(A_o+bk) = \Lambda$.

(v) The desired feedback matrix for the component is given by $K = K_o + vk$. Check that $(A + BK)$ has the required spectrum.

Finally, note that this algorithm is adaptable to component models not completely controllable. Specifically the controllable subspace of a pair (A,B) admits arbitrary pole assignment up to the dimension of $< A \mid B >$: Given a component model

$$\dot{x} = Ax + Ba \tag{3.44}$$

convert it to *controllability canonical form* as per equation 2.76, i.e., 3.44 is equivalent to

$$\begin{bmatrix} \dot{z}_1 \\ \hline \dot{z}_2 \end{bmatrix} = \begin{bmatrix} A_{11} & A_{12} \\ \hline 0 & A_{22} \end{bmatrix} \begin{bmatrix} z_1 \\ \hline z_2 \end{bmatrix} + \begin{bmatrix} B_1 \\ \hline 0 \end{bmatrix} a \tag{3.45}$$

under the state transformation $x = Tz$ where T is defined by equation 2.80. The reduced system

$$\dot{z}_1 = A_{11} z_1 + B_1 a \tag{3.46}$$

is completely controllable. Thus by using the above algorithm, for any symmetric set Λ of q complex numbers (where A_{11} is qxq) there exists state feedback K_1 such that the eigenvalues of $(A_{11} + B_1K_1)$ are given by Λ.

Clearly the admissible control law for 3.46 takes the form

$$[K_1 \mid K_2] \begin{bmatrix} z_1 \\ -- \\ z_2 \end{bmatrix} + a \tag{3.47}$$

and the "controlled" form of 3.45 is

$$\begin{bmatrix} \dot{z}_1 \\ --- \\ \dot{z}_2 \end{bmatrix} = \begin{bmatrix} A_{11}+B_1K_1 & A_{12}+B_1K_2 \\ ------ & ------ \\ 0 & A_{22} \end{bmatrix} \begin{bmatrix} z_1 \\ --- \\ z_2 \end{bmatrix} + \begin{bmatrix} B_1 \\ --- \\ 0 \end{bmatrix} a \tag{3.48}$$

The eigenvalues of this model consist of the eigenvalues of $(A_{11} + B_1K_1)$ adjoined to those of A_{22}. Now K_2 may be chosen to minimize interaction between the controllable and uncontrollable modes etc. From equations 2.77 and 2.78

$$A + BK = T \begin{bmatrix} A_{11}+B_1K_1 & A_{12}+B_1K_2 \\ ------ & ------ \\ 0 & A_{22} \end{bmatrix} T^{-1} \tag{3.49}$$

$$= A + B [K_1 \mid K_2] T^{-1}$$

Hence the state feedback K which will assign the spectrum Λ to the controllable subspace $< A \mid \mathbf{B} >$ is given by

$$K = [K_1 \mid K_2] T^{-1} \tag{3.50}$$

Clearly then if the eigenvectors of the unstable eigenvalues of A lie in the controllable subspace, they may be reassigned to the left half plane via appropriate state feed back. On the other hand, if the component model is stabilizable there must exist state feedback which assigns right half plane eigenvalues to the left half plane. The following alternate characterization of stabilizability proves useful.

THEOREM 3.51: A component having the usual state model is stabilizable if and only if

$$\text{rank}[\lambda_i I - A \mid B] = n \tag{3.52}$$

for all unstable eigenvalues. Generally speaking, "all unstable eigenvalues" include all those with non-negative real parts and we will take it as such.

PROOF: Let T be the transformation associated with the conversion to controllability canonical form. Since T is nonsingular, for any λ

$$\text{rank}[(\lambda I - A) \mid B] = \text{rank}\left(T^{-1}[(\lambda I - A) \mid B] \begin{bmatrix} T & \vdots & 0 \\ \text{---} & \vdots & \text{---} \\ 0 & \vdots & I \end{bmatrix} \right)$$

$$= \text{rank}\left(\begin{bmatrix} (\lambda I - A_{11}) & \vdots & -A_{12} & \vdots & B_1 \\ \text{-------} & \text{---} & \text{------} & \text{---} & \text{--} \\ 0 & \vdots & \lambda I - A_{22} & \vdots & 0 \end{bmatrix} \right) \qquad (3.53)$$

This rank is less than n if and only if λ is an eigenvalue of A_{22}. Indeed if λ is an eigenvalue of A_{22} the rows of the bottom partition are dependent. Conversely, if λ is not an eigenvalue of A_{22}, the rows of the bottom partition are independent as are those of the top partition since the pair (A_{11}, B_1) is controllable. See theorem 2.52.

For the component to be stabilizable each eigenvalue of A_{22} must be stable. Equivalently the unstable eigenvalues of A are not eigenvalues of A_{22}. Obviously then the component is stabilizable if and only if the rank of the matrix in 3.53 is n. In particular, all unstable eigenvalues are those of A_{11} hence assignable to the left half plane and the eigenvalues of A_{22} are all stable.

EXAMPLE 3.54: Consider the component model

$$\begin{bmatrix} \dot{x}_1 \\ \dot{x}_2 \end{bmatrix} = \begin{bmatrix} 1 & 0 \\ -2 & -1 \end{bmatrix} \begin{bmatrix} x_1 \\ x_2 \end{bmatrix} + \begin{bmatrix} 1 \\ -1 \end{bmatrix} a \qquad (3.55)$$

which has eigenvalues $\lambda_1 = -1$ and $\lambda_2 = 1$. Observe that

$$\text{rank}[\lambda_1 I - A \mid B] = \text{rank}\left(\begin{bmatrix} -2 & 0 & \vdots & 1 \\ 2 & 0 & \vdots & -1 \end{bmatrix} \right) = 1 \qquad (3.56)$$

and

$$\text{rank}[\lambda_2 I - A \mid B] = \text{rank}\left(\begin{bmatrix} 0 & 0 & \vdots & 1 \\ & & \vdots & \\ 2 & 2 & \vdots & -1 \end{bmatrix}\right) = 2 \quad (3.57)$$

This component is not controllable but the unstable mode is controllable, hence the component is stabilizable.

Finally, we close this section with an example illustrating spectral assignment on the controllable subspace.

EXAMPLE 3.58: Let a component have state model

$$\begin{bmatrix} \dot{x}_1 \\ \dot{x}_2 \\ \dot{x}_3 \end{bmatrix} = \begin{bmatrix} -3 & 2 & 2 \\ 0 & -1 & 0 \\ -6 & 6 & 4 \end{bmatrix} \begin{bmatrix} x_1 \\ x_2 \\ x_3 \end{bmatrix} + \begin{bmatrix} 2 \\ 1 \\ 2 \end{bmatrix} a \bigg] \quad (3.59)$$

The controllability matrix is

$$Q = \begin{bmatrix} 2 & \vdots & 0 & \vdots & 2 \\ 1 & \vdots & -1 & \vdots & 1 \\ 2 & \vdots & 2 & \vdots & 2 \end{bmatrix} \quad (3.60)$$

which has rank 2. The state transformation resulting in the controllability canonical form of 3.59 is

$$T = \begin{bmatrix} 2 & 0 & -2 \\ 1 & -1 & 2 \\ 2 & 2 & 1 \end{bmatrix} \quad (3.61)$$

and

$$T^{-1} = \frac{1}{9} \begin{bmatrix} 2.5 & 2 & 1 \\ -1.5 & -3 & 3 \\ -2 & 2 & 1 \end{bmatrix} \quad (3.62)$$

The *controllability canonical form* is

$$T^{-1}AT = \begin{bmatrix} 0 & 1 & \vdots & 6 \\ 1 & 0 & \vdots & 8 \\ \hdashline 0 & 0 & \vdots & 0 \end{bmatrix}; \quad T^{-1}B = \begin{bmatrix} 1 \\ 0 \\ \hline 0 \end{bmatrix} \quad (3.63)$$

For the reduced system we have

$$
A_{11} = \begin{bmatrix} 0 & 1 \\ 1 & 0 \end{bmatrix}; \; B_1 = \begin{bmatrix} 1 \\ 0 \end{bmatrix} \tag{3.64}
$$

Suppose the controllable subspace is to have eigenvalues $[-1,-1]$. The eigenvalues of A_{11} are $\lambda_1 = 1, \lambda_2 = -1$. The matrices of 3.64 can be put into the rational canonical form by the transformations

$$
V = \begin{bmatrix} 0 & 1 \\ 1 & 0 \end{bmatrix}; \; V^{-1} = \begin{bmatrix} 0 & 1 \\ 1 & 0 \end{bmatrix} \tag{3.65}
$$

which results in

$$
\hat{A}_{11} = V \, A_{11} \, V^{-1} = \begin{bmatrix} 0 & 1 \\ 1 & 0 \end{bmatrix}; \; B_1 = \begin{bmatrix} 0 \\ 1 \end{bmatrix}
$$

The characteristic polynomial associated with the spectrum $[-1,-1]$ is $s^2 - [-1-2s]$. Hence we choose $\hat{K}_1 = [-2 \; -2]$. Thus

$$
K_1 = \hat{K}_1 \, V^{-1} = [-2 \; -2] \tag{3.66}
$$

Let us choose $K_2 = -6$. Hence,

$$
\begin{bmatrix} A_{11} & A_{12} \\ 0 & A_{22} \end{bmatrix} + \begin{bmatrix} B_1 \\ 0 \end{bmatrix} [K_1 \,|\, K_2]
$$

$$
= \begin{bmatrix} -2 & -1 & 0 \\ 1 & 0 & 8 \\ 0 & 0 & 0 \end{bmatrix} \tag{3.67}
$$

which has the desired spectrum. The feedback matrix relative to the original (A,B) pair is

$$K = [K_1 \mid K_2]T^{-1} = \frac{1}{9} [10 \ -10 \ -14] \qquad (3.68)$$

In this example, the controlled system has eigenvalues $[-1, -1, 0]$. The directed mode corresponding to the zero eigenvalue is completely uncontrollable. Hence the controlled component is BIBO stable even with the eigenvalue at zero.

4. SINGULAR VALUE DECOMPOSITION

The variety of tests and algorithms presented in the preceding sections build on the algebraic properties of certain matrices, specifically the *controllability matrix* Q. The following two sections mold analogous routines and tests which rely on the corresponding algebraic properties of the *observability matrix* R. A sound accurate means for fabricating these tests and transformations is the *singular value (or Schmidt) decomposition* (SVD) of a rectangular matrix Q. The SVD of Q dispells such nasty problems as *rank computation* and finding bases for the image of Q and the null space of Q.[4]

Recall the *rank* of Q designates the dimension of the controllable subspace of the pair (A,B). From the theory of *positive semidefinite matrices* developed in section 2, the rank of Q is simply the number of non-zero eigenvalues of QQ^t which is symmetric positive semidefinite There are several excellent routines available at most computing centers for finding the eigenvalues, etc., of such matrices. The problem, however, is the matrix QQ^t. Suppose

$$Q = \begin{bmatrix} 1 & \epsilon \\ 1 & -\epsilon \end{bmatrix} \qquad (4.1)$$

making

$$QQ^t = \begin{bmatrix} 1+\epsilon^2 & 1-\epsilon^2 \\ 1-\epsilon^2 & 1+\epsilon^2 \end{bmatrix} \qquad (4.2)$$

If ϵ is sufficiently small relative to the word length of the computer then QQ^t will have rank 1 whereas Q has rank 2. Hence forming the matrix QQ^t as an intermediate step in determining the rank of Q is numerically dangerous. the SVD finds the number of non-zero eigenvalues outside a prespecified neighborhood of zero without ever constructing QQ^t. Moreover the SVD provides a measure of the robustness of the rank of Q, i.e., is the rank of Q fixed under small perturbations? Although the rank of Q in 4.1 is two, an ϵ-perturbation on column two makes the rank one. When determining the dimension of the controllable subspace, the "robust rank of Q" commonly referred to as the *effective rank* is the important dimension since a component model approximates the physical process only in a nominal sense. This begs the question: when are two vectors *"almost" dependent* or when are they *maximally independent*? Again the SVD provides a sound characterization.

THEOREM 4.3: For all nxm matrices Q, there exists orthogonal (unitary) matrices U and V where U is nxn and V is mxm such that the *singular value decomposition* of Q is given as

$$Q = U \Sigma V = [U_1 \mid U_2] \begin{bmatrix} S & 0 \\ \hline 0 & 0 \end{bmatrix} \begin{bmatrix} V_1^t \\ \hline V_2^t \end{bmatrix} \qquad (4.4)$$

where $S = \mathrm{diag}(\sigma_1, \cdots, \sigma_q)$, $\sigma_1 \geqslant \cdots \geqslant \sigma_q$, and σ_i^2 is a positive eigenvalue of the matrix QQ^t.

The numbers σ_i are the non-zero *singular values* of Q which are the positive square roots of the non-zero eigenvalues of QQ^t. Note that the positive eigenvalues of QQ^t and Q^tQ coincide.[4] The columns of U form an orthonormal set of eigenvectors of QQ^t corresponding to appropriate eigenvalues of QQ^t. The columns of V are an orthonormal set of eigenvectors of Q^tQ corresponding to the eigenvalues of Q^tQ. In particular, for $1 \leqslant i \leqslant q$, the i-th column of U is the eigenvector associated with the eigenvalue σ_i^2 and similarly for V. For $i \geqslant q$ the columns of U are an orthonormal basis for the null space of QQ^t and correspondingly for V. Since the matrices U and V are orthogonal (unitary) $U^{-1} = \overline{U}^t$ and $V^{-1} = \overline{V}^t$.

It turns out that the image of Q (the controllable subspace of the pair (A,B)) equals the span of the columns of U_1. Denote the span of the columns of a matrix M by **M**. The null space of Q^t is the orthogonal complement of U_1 in R^n. Null (Q^t) has an orthonormal basis given by the columns of U_2, hence $\mathrm{null}(Q^t) = U_2$.

These properties of the submatrices of U find immediate application when converting a component model to controllability canonical form. Using the above properties in conjunction with equation 2.80 the *controllability canonical form* of a pair (A,B) with controllability matrix Q is

$$
\begin{bmatrix} A_{11} & \vdots & A_{12} \\ \text{---} & \vdots & \text{---} \\ 0 & \vdots & A_{22} \end{bmatrix} = T^{-1}AT = U^tAU \tag{4.5}
$$

and

$$
\begin{bmatrix} B_1 \\ \text{---} \\ 0 \end{bmatrix} = T^{-1}B = U^tB \tag{4.6}
$$

As an example and application of the SVD consider the pair (A,B)

$$
A = \begin{bmatrix} -3 & 2 & 2 \\ 0 & -1 & 0 \\ -6 & 6 & 4 \end{bmatrix}; \quad B = \begin{bmatrix} 0 & 1 \\ -1 & 1 \\ 2 & 0 \end{bmatrix} \tag{4.7}
$$

which produces the controllability matrix

$$
Q = \begin{bmatrix} 0 & 1 & \vdots & 2 & -1 & \vdots & 0 & 1 \\ -1 & 1 & \vdots & 1 & -1 & \vdots & -1 & 1 \\ 2 & 0 & \vdots & 2 & 0 & \vdots & 2 & 0 \end{bmatrix} \tag{4.8}
$$

The singular value decomposition of Q is $Q = U \Sigma V^t$ where accurate to three decimal places;

$$
U = [U_1 \vdots U_2] = \begin{bmatrix} .541 & .512 & \vdots & -.894 \\ .126 & .734 & \vdots & .894 \\ .831 & -.445 & \vdots & .447 \end{bmatrix} \tag{4.9}
$$

$$
\Sigma = \begin{bmatrix} S & \vdots & 0 \\ \text{---} & & \text{---} \\ 0 & \vdots & 0 \end{bmatrix} = \begin{bmatrix} 3.782 & 0 & \vdots & 0 & 0 & 0 & 0 \\ 0 & 3.271 & \vdots & 0 & 0 & 0 & 0 \\ \text{---} & & & \text{---} & & & \\ 0 & 0 & \vdots & 0 & 0 & 0 & 0 \end{bmatrix} \tag{4.10}
$$

and

$$
V^t = \begin{bmatrix} V_1^t \\ \hline V_2^t \end{bmatrix} = \begin{bmatrix} .406 & .176 & .759 & -.176 & .406 & .176 \\ -.497 & .381 & .266 & -.381 & -.497 & .381 \\ \hline .707 & 0 & 0 & 0 & -.707 & 0 \\ 0 & .707 & 0 & 0 & 0 & .707 \\ .267 & .535 & -.535 & 0 & .267 & .535 \\ .130 & -.194 & -.259 & -.907 & .130 & -.194 \end{bmatrix} \quad (4.11)
$$

The actual matrix U Σ V^t does not equal Q but closely approximates Q. This points out potential *round-off error* introduced by the computer's *finite word length*, i.e.,

$$
U \Sigma V^t = \begin{bmatrix} -.002 & .998 & 1.998 & .998 & -.002 & .998 \\ -1.000 & .999 & 1.001 & -.999 & -1.000 & .999 \\ 2.000 & -.002 & 1.998 & .002 & 2.000 & -.002 \end{bmatrix} \quad (4.12)
$$

The controllability canonical form of the pair (A,B) is

$$
U^t A U = \left[\begin{array}{cc|c} .835 & -.983 & 13.19 \\ -.315 & -.83 & -3.78 \\ \hline 0 & 0 & 0 \end{array} \right] \quad (4.13)
$$

The eigenvalues of U^{t}AU are 0, −.999, and 1.0045 which closely approximate the eigenvalues of A which are 0.0, −1.0, and 1.0. Furthermore, we have

$$
U^t B = \left[\begin{array}{cc} 1.536 & .667 \\ -1.624 & -.222 \\ \hline 0 & 0 \end{array} \right] \quad (4.14)
$$

The inverse transformation is equally easy to implement. Observe that there is significant round-off error. For a transformation T which is not an *orthogonal matrix*, the round-off error would be more pronounced. It turns out that orthogonal matrices do not magnify round-off errors making their use highly attractive in numerical applications. This results because multiplication by orthogonal matrices preserves vector length and the angle between vectors.

Unfortunately the SVD algorithm still leaves one with the problem of determining *what constitutes a non-zero singular value*. Suppose the SVD of a specific controllability matrix Q has five potentially non-zero singular values, three of which are, say, greater than .1 and two less than .001. One could legitimately conclude that a *lower bound* on the rank of Q is three. Hence, the *effective rank* may be three whereas the actual rank is five. Physically, this interprets as meaning that the dimension of the controllability subspace is *actually* five but two of the basis vectors for the space will always be *nearly* dependent on the other three. An example of such a set of vectors could be

$$v_1 = \begin{bmatrix} 1 \\ 0 \\ 0 \\ 0 \\ 0 \end{bmatrix} ; v_2 = \begin{bmatrix} 0 \\ 1 \\ 0 \\ 0 \\ 0 \end{bmatrix} ; v_3 = \begin{bmatrix} 0 \\ 0 \\ 1 \\ 0 \\ 0 \end{bmatrix} ; v_4 = \begin{bmatrix} 0 \\ 0 \\ 1 \\ \epsilon \\ 0 \end{bmatrix} ; v_5 = \begin{bmatrix} 0 \\ 1 \\ 1 \\ 0 \\ \epsilon \end{bmatrix} \qquad (4.15)$$

for sufficiently small ϵ. Intuitively, a small perturbation on the component parameters could collapse the dimension of the controllability subspace from five to three. Hence it behooves one to work with the effective rank. These notions motivate the following characterization of effective rank which loosely means strongly independent or nearly orthogonal.

DEFINITION 4.16: The *condition number* of matrix Q having full rank is

$$\text{cond}(Q) = \frac{\sigma_{max}}{\sigma_{min}} \qquad (4.17)$$

where σ_{max} is the largest singular value of Q, σ_{min} is the smallest singular value of Q, and a rectangular matrix of *full rank* is one with either all its columns or all its rows linearly independent.

Since Q has full rank by assumption, all singular values are positive and 4.17 is well defined. If Q does not have full rank, cond (Q) is infinite. Clearly, if cond (Q) is "large" the effective rank of Q is less than the actual rank. If cond(Q) = 1 then the effective rank equals the actual rank and the rows (or columns depending on the dimensions) are maximally independent, i.e., nearly orthogonal. An orthogonal matrix has condition equal to one.

Typically, Q is not full rank. To characterize the *effective* dimension of the controllability subspace, it is necessary to apply the above ideas to the set of singular values of Q which are considered non-zero. Note that these kinds of considerations must influence a practical design since the rank of a matrix is a discontinuous function of its entries: small perturbations cause rank changes.

Unfortunately, at the time of this writing, numerically stable schemes for converting a controllable pair to *rational canonical* form, assigning spectrum, and then converting back, are still under development. Difficulties arise because in many cases the roots of the characteristic polynomial are highly sensitive to small perturbations in the coefficients. The higher the degree of characteristic polynomial, the more critical the sensitivity. Hence the transformations involved must have extended numerical accuracy.

5. OBSERVABILITY

Dual to the notion of *controllability* is *observability*, as we shall see. In addition to this dual kinship, observability plays a fundamental role in *pole placement* (spectral assignability) using *output feedback*.[1,8] Once again the basic component description is

$$
\begin{aligned}
\dot{x} &= Ax + Ba \\
b &= Cx + Da
\end{aligned}
\tag{5.1}
$$

The model structure has the b-vector as the accessible output, not the state vector x. The C-matrix masks the n-dimensional state dynamics. Techniques for unmasking x from b spring naturally, to a large extent, from the definition of observability.

To set the stage, recall from theorem I. 4.19 that the solution to 5.1 is

$$
b(t) = C\Phi(t-t_o)x(t_o) + \int_{t_o}^{t_1} C\Phi(t_1-q)Ba(q)dq + Da(t)
\tag{5.2}
$$

where $\Phi(\cdot)$ is the *state transition matrix* defined in 2.2. Assume the input and output are known functions over the finite interval $[t_o, t_1]$.

DEFINITION 5.3: The state $x_o = x(t_o)$ is observable on the interval $[t_o, t_1]$ if and only if 5.2 is uniquely solvable for x_o.

By hypothesis, $a(t)$ is known and the problem is equivalent to solving

$$b(t) = C \exp[A(t-t_0)]x_0 \tag{5.4}$$

for x_0, $t_0 \leqslant t \leqslant t_1$. The smoothness of both $b(\cdot)$ and the matrix exponential implies

$$\left. \frac{d^i}{dt^i} b(t) \right|_{t_0} = \left. \frac{d^i}{dt^i} C \exp[A(t-t_0)]x_0 \right|_{t_0} \tag{5.5}$$

for all i, which simplifies to

$$b^{(i)}(t_0) \triangleq \frac{d^i b}{dt^i}(t_0) = C A^i x_0 \tag{5.6}$$

These vectors are the respective coefficients corresponding to like powers of $(t-t_0)$ of a Taylor series of both sides of 5.4. Hence, the state x_0 is observable if and only if there exists a *unique solution* to the following algebraic equation.

$$\begin{bmatrix} b^{(0)}(t_0) \\ \hline b^{(1)}(t_0) \\ b^{(2)}(t_0) \\ \hline \cdot \\ \cdot \\ \cdot \end{bmatrix} = \begin{bmatrix} C \\ \hline CA \\ CA^2 \\ \hline \cdot \\ \cdot \\ \cdot \end{bmatrix} x_0] \tag{5.7}$$

By the *Caley-Hamilton theorem*, 2.44, 5.7 is *uniquely solvable* if and only if 5.8 is uniquely solvable.

$$\begin{bmatrix} b^{(0)}(t_0) \\ \hline b^{(1)}(t_0) \\ \hline \cdot \\ \cdot \\ \cdot \\ \hline b^{(n-1)}(t_0) \end{bmatrix} = \begin{bmatrix} C \\ \hline CA \\ \hline \cdot \\ \cdot \\ \cdot \\ \hline CA^{n-1} \end{bmatrix} x_0] \tag{5.8}$$

Without loss of generality we assume the number of component outputs at most equals the number of states. In general 5.8 is uniquely solvable if and only if the *observability matrix,* R, has full rank, where

$$R = \begin{bmatrix} C \\ \hline CA \\ \hline \cdot \\ \cdot \\ \cdot \\ \hline CA^{n-1} \end{bmatrix} \tag{5.9}$$

However, if x_o has *no component in the direction of the null space of R,* then 5.8 can be uniquely solved for x_o. Rather than delineate the mathematics we illustrate the meaning as follows. Suppose a component model is

$$\begin{bmatrix} x_1 \\ x_2 \\ x_2 \end{bmatrix} = \begin{bmatrix} -3 & 2 & 2 \\ 0 & -1 & 0 \\ -6 & 6 & 4 \end{bmatrix} \begin{bmatrix} x_1 \\ x_2 \\ x_2 \end{bmatrix}$$

$$b = \begin{bmatrix} -2 & 1 & 1 \end{bmatrix} \begin{bmatrix} x_1 \\ x_1 \\ x_2 \end{bmatrix} \tag{5.10}$$

The controllability matrix R is

$$R = \begin{bmatrix} -2 & 1 & 1 \\ \hline 0 & 1 & 0 \\ \hline 0 & -1 & 0 \end{bmatrix} \tag{5.11}$$

A basis of the null space of R is $[1,0,2]^t$. The subspace of $\mathbf{R}^3$ *orthogonal to* null (R) is $[-2,1,1]^t$, $[0,1,0]^t$. Every vector x_o in $\mathbf{R}^3$ has the form

$$x_o = \alpha_1 \begin{bmatrix} -2 \\ 1 \\ 1 \end{bmatrix} + \alpha_2 \begin{bmatrix} 0 \\ 1 \\ 0 \end{bmatrix} + \alpha_3 \begin{bmatrix} 1 \\ 0 \\ 2 \end{bmatrix} \tag{5.12}$$

Now every vector x_o for which $\alpha_3 = 0$ is observable in that if x_o has the form of 5.11 and $\alpha_3 = 0$ then 5.8 is uniquely solvable for x_0, on the other hand, if $\alpha_3 \neq 0$ x_0 cannot be recovered from knowledge of a(t) and b(t) over the interval $[t_0, t_1]$. Any state lying in the null space of R is said to be *completely unobservable*. Any state lying in the *orthogonal complement* of null (R) is observable and all other states are "partially" observable.

Obviously if rank (R) = n, i.e., the columns of the observability matrix are linearly independent, 5.8 is uniquely solvable. Specifically, if rank (R) = n, every state x_0 is observable.

DEFINITION 5.13: A component modeled by 5.1 is completely observable if and only if every state x_o is observable.

The above development is independent of the time interval $[t_0, t_1]$. This is not unforseen since once $x(t_o)$ is known

$$x(t) \;=\; \Phi(t{-}t_0)x_0 + \int_{t_0}^{t_1} \Phi(t_1{-}q)Ba(q)dq \tag{5.14}$$

PROPOSITION 5.15: The component of 5.1 is completely observable if and only if rank (R) = n where R is the observability matrix.

This proposition forges the link between observability and *controllability*. The *duality* is explicitly formalized below.

THEOREM 5.16: The component model

$$\begin{aligned} \dot{x} &= Ax \\ b &= Cx \end{aligned} \tag{5.17}$$

is completely observable if and only if the component model

$$\dot{x} \;=\; A^t x + C^t a \tag{5.18}$$

is completely controllable.

PROOF: Since the transpose operation leaves the rank of a matrix unchanged, rank (R) = n if and only if rank $(R^t) = n$ where

$$R^t \;=\; [C^t \mid A^t C^t \mid (A^t)^2 C^t \mid \cdots \mid (A^t)^{n-1} C^t] \tag{5.19}$$

Clearly, R^t is the controllability matrix for the component description of 5.18 verifying the theorem.

Based on this duality, the entire controllability theory sketched in section 2 carries over as

THEOREM 5.20: For the component dynamics of equation 5.1, the following are equivalent conditions:
(i)   The component is completely observable.
(ii) $\text{Rank}[(\lambda I-A)|C] = n$ for each eigenvalue λ of A.
(iii) $\text{Rank}(R) = n$ where R is the observability matrix.
(iv) $\text{Rank}[C \exp[-At]] = n$, i.e., there are n linearly independent columns.
(v)   The matrix N is positive definite. Here

$$N = \int_{t_0}^{t_1} \Phi^*(t_1-q)C^*C\ \Phi\ (t_1-q)dq \qquad (5.21)$$

Moreover, the initial state x_0 may be computed from $b(t)$, $t_0 \leqslant t \leqslant t_1$, as

$$x_0 = N^{-1} \int_{t_0}^{t_f} \Phi^*(q-t_0)C^* b(q)dq \qquad (5.22)$$

EXAMPLE 5.23: Suppose a component has dynamics

$$\begin{bmatrix} x_1 \\ x_2 \end{bmatrix} = \begin{bmatrix} 0 & 0 \\ 1 & 0 \end{bmatrix} \begin{bmatrix} x_1 \\ x_2 \end{bmatrix}$$

$$b \end{bmatrix} = \begin{bmatrix} 1 & 1 \end{bmatrix} \begin{bmatrix} x_1 \\ x_2 \end{bmatrix} \qquad (5.24)$$

The component is completely observable since $\text{rank}(R) = 2$ where

$$R = \begin{bmatrix} 1 & 1 \\ 1 & 0 \end{bmatrix} \qquad (5.25)$$

As an alternate test we compute N and assume $[t_o, t_1] = [0,1]$. Clearly,

$$\exp[At] = \begin{bmatrix} 1 & 0 \\ t & 1 \end{bmatrix} \tag{5.26}$$

which leads to

$$N = \int_0^1 \begin{bmatrix} 1 & q \\ 0 & 1 \end{bmatrix} \begin{bmatrix} 1 \\ 1 \end{bmatrix} [1 \quad 1] \begin{bmatrix} 1 & 0 \\ q & 1 \end{bmatrix} dq$$

$$= \begin{bmatrix} 2.3333 & 1.5 \\ 1.5 & 1 \end{bmatrix} \tag{5.27}$$

The inverse of N is

$$N^{-1} = \begin{bmatrix} 12 & -18 \\ -18 & 28 \end{bmatrix} \tag{5.28}$$

Aided by lemma 2.36, N is positive definite since it is invertible.

Finally suppose $b(t) = 1$, $0 \leq t \leq 1$; assisted by equation 5.22, the initial state is given by

$$x_0 = N^{-1} \int_0^1 \begin{bmatrix} 1 & q \\ 0 & 1 \end{bmatrix} \begin{bmatrix} 1 \\ 1 \end{bmatrix} 1] \, dq = \begin{bmatrix} 0 \\ 1 \end{bmatrix} \tag{5.29}$$

This concludes the example, however, the duality continues.

THEOREM 5.30: Suppose a component has model

$$\dot{x} = Ax$$
$$b = Cx \tag{5.31}$$

and suppose $\text{rank}(R) = r < n$. Then there exists a nonsingular matrix, T, and a corresponding state transformation, $x = Tz$, such that the equivalent component

$$\dot{z} = [T^{-1}AT]z$$
$$b = [CT]z \tag{5.32}$$

has new state matrices

$$T^{-1}AT = \begin{bmatrix} A_{11} & \vdots & 0 \\ --- & + & --- \\ A_{21} & \vdots & A_{22} \end{bmatrix} \tag{5.33}$$

and

$$CT = [C_1 \vdots 0] \tag{5.34}$$

Moreover, the reduced system description

$$\dot{z}_1 = A_{11}z_1$$
$$b = C_1z_1 \tag{5.35}$$

is completely observable.

To derive this new transformation, T, regard the dual component model

$$\dot{x} = A^t x + C^t a \tag{5.36}$$

The controllability matrix for the dual system is

$$R^t = [C^t \vdots A^t C^t \vdots \cdots \vdots (A^t)^{n-1} C^t] \tag{5.37}$$

Let $\hat{T}$ be the transformation resulting in a *controllability canonical form* for the *dual system*. Recall, the first r columns of $\hat{T}$ are the r linearly independent columns of R^t where $r = \text{rank}(R^t)$. The remaining columns span the null space of R. This produces canonical matrices of the form

$$\hat{T}^{-1}A^t\hat{T} = \begin{bmatrix} \hat{A}_{11} & \vdots & \hat{A}_{12} \\ --- & + & --- \\ 0 & \vdots & \hat{A}_{22} \end{bmatrix} \tag{5.38}$$

and

$$\hat{T}^{-1}C^t = \begin{bmatrix} \hat{C}_1 \\ -- \\ 0 \end{bmatrix} \qquad (5.39)$$

Clearly, the transpose of 5.38 and 5.39 correspond to the desired form given in equations 5.33 and 5.34. As such,

$$T^{-1}AT = [\hat{T}^{-1}A^t\hat{T}]^t = \begin{bmatrix} A_{11} & | & 0 \\ ---- & | & ---- \\ A_{21} & | & A_{22} \end{bmatrix} \qquad (5.40)$$

and

$$CT = [\hat{T}C^t]^t = [C_1 \mathrel{|} 0] \qquad (5.41)$$

Clearly, then

$$T = [\hat{T}^{-1}]^t = [\hat{T}^t]^{-1} \qquad (5.42)$$

Component descriptions resulting from the above transformation are in *observability canonical* form. Combining the controllability and observability transformations permits a further decomposition, the *Kalman canonical form*. Here the state transforms into four distinct sectors: the first is completely controllable and completely observable, the second group of entries are controllable but not observable, the third group observable but not controllable, and the last entries correspond to states neither controllable nor observable.

THEOREM 5.43: Let a component model

$$\dot{x} = Ax + Ba$$

$$\qquad (5.44)$$

$$b = Cx$$

have controllability matrix Q with rank $(Q) = q < n$; let the observability matrix, R, have rank r strictly less than n. Then there exists a nonsingular matrix, T, and a state transformation $x = Tz$, resulting in an equivalent component description

$$\dot{z} = [T^{-1}AT]\, z + [T^{-1}B]a$$

$$b = [CT]z$$

(5.45)

where

$$T^{-1}AT = \begin{bmatrix} A_{11} & 0 & A_{13} & 0 \\ A_{21} & A_{22} & A_{23} & A_{24} \\ 0 & 0 & A_{33} & 0 \\ 0 & 0 & A_{43} & A_{44} \end{bmatrix}$$

(5.46)

$$CT = [C_1 \mid 0 \mid C_3 \mid 0]$$

(5.47)

and

$$T^{-1}B = \begin{bmatrix} B_1 \\ B_2 \\ 0 \\ 0 \end{bmatrix}$$

(5.48)

Moreover, the reduced system having state description

$$\dot{z}_1 = A_{11}\, z_1 + B_1\, a$$

$$b = C_1\, z_1$$

(5.49)

is both controllable and observable; the reduced system description

$$\begin{bmatrix} \dot{z}_1 \\ \dot{z}_2 \end{bmatrix} = \begin{bmatrix} A_{11} & 0 \\ A_{22} & A_{22} \end{bmatrix} \begin{bmatrix} z_1 \\ z_2 \end{bmatrix} + \begin{bmatrix} B_1 \\ B_2 \end{bmatrix} a]$$

(5.50)

is controllable; finally

$$
\begin{bmatrix} z_1 \\ -- \\ z_3 \end{bmatrix} = \begin{bmatrix} A_{11} & \vline & A_{13} \\ ----+---- \\ 0 & \vline & A_{33} \end{bmatrix} \begin{bmatrix} z_1 \\ -- \\ z_3 \end{bmatrix}
$$

$$
b = [C_1 \; \vline \; C_3] \begin{bmatrix} z_1 \\ -- \\ z_3 \end{bmatrix} \tag{5.51}
$$

is observable.

The best explanation is an illustration.

EXAMPLE 5.52: Suppose a component has state matrices

$$
A = \begin{bmatrix} -3 & 2 & 2 \\ 0 & -1 & 0 \\ -6 & 6 & 4 \end{bmatrix}; B = \begin{bmatrix} 0 & 2 \\ -1 & 1 \\ 2 & 2 \end{bmatrix}; C = [-2 \; 1 \; 1] \tag{5.53}
$$

The goal is to put these matrices in their respective Kalman canonical form. Step 1 is to find the controllability canonical form. Observe that the controllability matrix, Q, has rank 2 where

$$
Q = \begin{bmatrix} 0 & 2 & \vline & 2 & 0 & \vline & 0 & 2 \\ -1 & 1 & \vline & 1 & -1 & \vline & -1 & 1 \\ 2 & 2 & \vline & 2 & 2 & \vline & 2 & 2 \end{bmatrix} \tag{5.54}
$$

The required transformation T becomes

$$
T = \begin{bmatrix} 0 & 2 & -2 \\ -1 & 1 & 2 \\ 2 & 2 & 1 \end{bmatrix} \tag{5.55}
$$

and

$$
T^{-1} = \frac{1}{18} \begin{bmatrix} -3 & -6 & 6 \\ 5 & 4 & 2 \\ -4 & 4 & 2 \end{bmatrix} \tag{5.56}
$$

As such,

$$\hat{A} = T^{-1}AT = \left[\begin{array}{c|c} A_{11} & A_{12} \\ \hline 0 & A_{22} \end{array}\right] = \left[\begin{array}{cc|c} 0 & 1 & 8 \\ 1 & 0 & 6 \\ \hline 0 & 0 & 0 \end{array}\right] \tag{5.57}$$

and

$$\hat{B} = T^{-1}B = \left[\begin{array}{c} B_1 \\ \hline 0 \end{array}\right] = \left[\begin{array}{cc} 1 & 0 \\ 0 & 1 \\ \hline 0 & 0 \end{array}\right] \tag{5.58}$$

Furthermore, since $x = Tz$ a new C-matrix arises as

$$\hat{C} = CT = [1 \ -1 \ 7] \tag{5.59}$$

the component model (A, B, C) is in controllability canonical form.

Let the new state vector be

$$z = \left[\begin{array}{c} z_1 \\ \hline z_2 \end{array}\right] \tag{5.60}$$

where z_1 is 2x1 and z_2 is 1x1. Moreover let $C = [C_1 \,|\, C_2] = [1 \ -1 \,|\, 7]$. To manipulate these matrices into the Kalman canonical form we consider the controllable and uncontrollable reduced systems separately:

$$\dot{z}_1 = A_{11} z_1 + B_1 a$$
$$\tag{5.61}$$
$$b_1 = C_1 z_1$$

and

$$\dot{z}_2 = A_{22} z_2$$
$$\tag{5.62}$$
$$b_2 = C_2 z_2$$

Obviously the reduced system of 5.62 is completely observable but completely uncontrollable. To check observability of 5.61 we construct the transformation which places it into observability canonical form, in which case

$$T = \begin{bmatrix} .5 & .5 \\ -.5 & .5 \end{bmatrix}; \quad T^{-1} = \begin{bmatrix} 1 & -1 \\ 1 & 1 \end{bmatrix} \qquad (5.63)$$

This leads to another state transformation $\underline{z} = Tz$ which converts 5.57, 5.58, and 5.59 into their Kalman canonical form. In this process

$$T = \frac{1}{2} \begin{bmatrix} 1 & 1 & 0 \\ -1 & 1 & 0 \\ 0 & 0 & 2 \end{bmatrix} \qquad (5.64)$$

and

$$T^{-1} = \begin{bmatrix} 1 & -1 & 0 \\ 1 & 1 & 0 \\ 0 & 0 & 1 \end{bmatrix} \qquad (5.65)$$

The resulting canonical matrices are

$$\begin{bmatrix} A_{11} & 0 & A_{13} \\ A_{21} & A_{22} & A_{23} \\ 0 & 0 & A_{33} \end{bmatrix} = T^{-1}AT = \begin{bmatrix} -1 & 0 & 2 \\ 0 & 1 & 14 \\ 0 & 0 & 0 \end{bmatrix} \qquad (5.66)$$

$$\begin{bmatrix} B_1 \\ B_2 \\ 0 \end{bmatrix} = T^{-1}B = \begin{bmatrix} 1 & -1 \\ 1 & 1 \\ 0 & 0 \end{bmatrix} \qquad (5.67)$$

and finally

$$[C_1 \; | \; 0 \; | \; C_3] = CT = [1 \; | \; 0 \; | \; 14] \qquad (5.68)$$

In practice the design engineer of necessity extracts the controllable and observable subsystem. This justifies the above decomposition. Indeed the controllable and observable subsystem has the same *transfer function matrix* as the original component. To verify this equivalence it suffices to show the equality of the respective *Markov parameters* since as per equation I.4.55 the Markov parameters completely determine the transfer function matrix.

The Markov parameters of the reduced controllable and observable model, (equation 5.49) are

$$M_i = C_1[A_{11}]^{i-1} B_1 \tag{5.69}$$

whereas those of the original component description are

$$M_i = CA^{i-1}B$$

$$= [C_1 \vdots 0 \vdots C_3 \vdots 0]\ T^{-1}T \begin{bmatrix} A_{11} & 0 & A_{13} & 0 \\ A_{21} & A_{22} & A_{23} & A_{24} \\ 0 & 0 & A_{33} & 0 \\ 0 & 0 & A_{23} & A_{44} \end{bmatrix} T^{-1}T \begin{bmatrix} B_1 \\ B_2 \\ 0 \\ 0 \end{bmatrix}$$

$$= C_1[A_{11}]^{i-1} B_1 \tag{5.70}$$

They are, in fact, equal.

This section ends with two theorems characterizing some special properties held by controllable and observable component descriptions.

THEOREM 5.71: A component state model is controllable and observable if and only if the dimension of A is minimal, i.e., the dimension of A equals the *degree* of the state model.

PROOF: If the original component description is either uncontrollable or unobservable, the reduced model, as per 5.49, has dimension (number of rows of A_{11}) strictly lower than n. Both models, however, realize the same transfer function matrix implying that the original component model is not *minimal*.

Conversely, suppose the model of 5.72 is completely controllable-observable, but not minimal.

$$\dot{x} = Ax + Ba$$
$$\tag{5.72}$$
$$b = Cx$$

Let Q and R be the controllability and observability matrices, respectively. Now let the same component have the model of 5.73 assumed controllable, observable, and minimal. Let $\underline{Q}$ and $\underline{R}$ be the controllability and observability matrices.

$$\dot{z} = \underline{A}z + \underline{B}a$$

$$B = \underline{C}z$$

(5.73)

Some arithmetic acrobatics verify that

$$RQ = H_1 = \underline{R}\,\underline{Q}$$

(5.74)

where H_1 is the *Hankel matrix* defined in equation I. 4.49. The equality of 5.74 results since the component description of 5.72 and 5.73 gives rise to the same transfer function matrix and since the Markov parameters depend only on that transfer function matrix. The respective transfer function matrices are equal by hypothesis.

The controllability and observability of 5.72 implies $rank(Q) = rank(R) = n$. Hence, $rank(RQ) = n$. On the other hand, the minimality of 5.73 forces $rank(\underline{R}\,\underline{Q}) = \underline{n} < n$. As such $\underline{R}$ has n columns and $\underline{Q}$ has $\underline{n}$ rows. Since $\underline{n} \neq n$, $\underline{R}\,\underline{Q}$ cannot possibly equal RQ contradicting the hypothesis that 5.72 is observable and controllable but not minimal. Hence, the theorem is true.

The second theorem links different minimal state models having coincident transfer function matrices. By the previous theorem such state models are completely observable and completely controllable.

THEOREM 5.75: Let

$$\dot{x} = Ax + Ba$$

$$b = Cx$$

(5.76)

and

$$\dot{z} = \underline{A}z + \underline{B}a$$

$$b = \underline{C}z$$

(5.77)

be two minimal state models with the same transfer function matrix, then the component model of 5.77 is equivalent to the component model of 5.76 through the *state transformation* $z = Tx$ where

$$T = (\underline{R}^t\underline{R})^{-1}\underline{R}^tR$$

(5.78)

PROOF: Let R, $\underline{R}$, Q, $\underline{Q}$ be the observability/controllability matrices of the indicated models as in the previous theorem $RQ = H_1 = \underline{R}\underline{Q}$. Furthermore,

$$\operatorname{rank}(R) = \operatorname{rank}(Q) = \operatorname{rank}(\underline{R}) = \operatorname{rank}(\underline{Q}) = n \qquad (5.79)$$

By lemma 2.36, R^tR, $\underline{R}^t\underline{R}$, Q^tQ and $\underline{Q}^t\underline{Q}$ are positive definite, hence invertible. Multiplying $\underline{R}\underline{Q} = RQ$ on the left by $(\underline{R}^t\underline{R})^{-1}\underline{R}^t$ yields

$$\underline{Q} = [(\underline{R}^t\underline{R})^{-1}\underline{R}^tR]Q = TQ \qquad (5.80)$$

A similar derivation produces

$$\underline{R} = R[Q \ \underline{Q}^t \ (\underline{Q} \ \underline{Q}^t)^{-1}] \qquad (5.81)$$

Combining 5.80, 5.81, and the equality $RQ = \underline{R}\underline{Q}$ shows that $[\underline{Q}\underline{Q}^t(\underline{Q}\underline{Q}^t)^{-1}] = T^{-1}$, in which case

$$\underline{R} = RT^{-1} \qquad (5.82)$$

Recall that $Q = [B \mathbin{\vert} AB \mathbin{\vert} \cdots \mathbin{\vert} A^{n-1}B]$ and $\underline{Q}$, R, and $\underline{R}$ have corresponding forms. Therefore, 5.80 implies $\underline{B} = TB$ and 5.82 implies $\underline{C} = CT^{-1}$.

To establish the equality $\underline{A} = TAT^{-1}$ observe that $RAQ = H_2 = \underline{R}\underline{A}\underline{Q}$. Aided by 5.80 and 5.82, this yields

$$RAQ = \underline{R}\underline{A}\underline{Q} = RT^{-1}\underline{A}TQ \qquad (5.83)$$

Multiplying on the left by $(\underline{R}^t\underline{R})^{-1}\underline{R}^t$ and on the right by $Q^t(QQ^t)^{-1}$ leaves the desired equivalence proving the theorem.

6. DETECTABILITY

Although equation 5.22 furnishes a definitive formula for computing the initial state of an observable system from inputs and outputs over a finite time interval, hardware implementation of the formula is impractical. The *state estimator* or *dynamic observer* replaces 5.22 with a flexible alternative readily implemented in hardware.[1,8] For the usual component state equation model

$$\dot{x} = Ax + Ba$$
$$b = Cx$$
$$(6.1)$$

we postulate a *dynamic observer* whose state model is

$$\dot{z} = Sz + Ea + Kb$$
$$\hat{x} = z$$
$$(6.2)$$

where in some sense $\hat{x}(t)$ will estimate $x(t)$. Of course justification of 6.2 requires linking S, E, and K to A, B, and C. To construct the link suppose there exists an invertible transformation, T, such that $z = Tx$. Multiplying the dynamical part of 6.1 by T produces

$$\dot{z} = T\dot{x} = TAx + TBa \qquad (6.3)$$

On the other hand, since $b = Cx$ and $z = Tx$, 6.2 simplifies to

$$\dot{z} = (ST + KC)x + Ea \qquad (6.4)$$

A comparison of 6.3 and 6.4 yields

$$S = TAT^{-1} - KCT^{-1}$$
$$E = TB$$
$$(6.5)$$

After some arithmetic manipulations we arrive at

$$\frac{d}{dt}[z - Tx] = S[z - Tx] \qquad (6.6)$$

The solution of this equation is

$$[z(t) - Tx(t)] = \exp[St] [z_0 - Tx_0] \qquad (6.7)$$

If all the eigenvalues of S have negative real parts $z(t)$ converges to $Tx(t)$ as $t \to \infty$. Hence $Tx(t)$ is recoverable from $z(t)$ at least after several time constants. Utilization of the dynamic observer (equation 6.2) requires suitable choice of the spectrum of $S = TAT^{-1} - KCT^{-1}$ which necessitates proper choice of K.

THEOREM 6.8: If $T = I$ and the pair (A,C) is observable, then for any *symmetric set* Λ *of n complex number* there exists K, such that $S = A - KC$ and $\sigma(S) = \Lambda$.

PROOF: If $T = I$, surely $S = A - KC$. If the pair (A,C) is observable, then by the duality theorem (5.16) the pair (A^t, C^t) is controllable. Controllability guarantees the existence of feedback F such that $\sigma(A^t + C^t F) = \Lambda$. By choosing $K = -F^t$ the result follows, i.e.,

$$\sigma(A^t + C^t F) = \sigma(A + F^t C) = \sigma(A - KC) = \Lambda . \qquad (6.9)$$

Note the converse is also true. The converse follows by applying duality to a modified theorem 3.51.

Clearly, observability permits choice of spectrum of S so that the estimated state, $\hat{x}(t) = z(t)$, converges arbitrarily fast to x(t). The *noise bandwidth,* however, limits the speed of convergence in practice.

EXAMPLE 6.10: Build a *dynamic observer* for

$$\begin{bmatrix} \dot{x}_1 \\ x_2 \end{bmatrix} = \begin{bmatrix} -4 & 3 \\ -2 & 1 \end{bmatrix} \begin{bmatrix} x_1 \\ x_2 \end{bmatrix}; \quad b = [1 \ 0] \begin{bmatrix} x_1 \\ x_2 \end{bmatrix} \qquad (6.11)$$

Consider the pair (A^t, C^t) where

$$A^t = \begin{bmatrix} -4 & -2 \\ 3 & 1 \end{bmatrix}; \quad C^t = \begin{bmatrix} 1 \\ 0 \end{bmatrix} \qquad (6.12)$$

This pair is controllable with controllability matrix

$$Q = \begin{bmatrix} 1 & -4 \\ 0 & 3 \end{bmatrix} \qquad (6.13)$$

The *rational canonical form* of the pair (A^t, C^t) is

$$VA^tV^{-1} = \begin{bmatrix} 0 & 1 \\ -2 & -3 \end{bmatrix}; \quad VC^t = \begin{bmatrix} 0 \\ 1 \end{bmatrix} \tag{6.14}$$

where

$$V = \frac{1}{3}\begin{bmatrix} 0 & 1 \\ 3 & 1 \end{bmatrix}; \quad V^{-1} = \begin{bmatrix} -1 & 1 \\ 3 & 0 \end{bmatrix} \tag{6.15}$$

From 6.14, the characteristic polynomial of A^t is $\lambda^2 - [-2-3\lambda]$ resulting in eigenvalues $\lambda_1 = -1$ and $\lambda_2 = -2$. Suppose it is necessary tor the observer eigenvalues to be $-4\pm j$. The characteristic polynomial of S must then be $\lambda^2 - [-17-8\lambda]$. For the canonical system, $-K^t = [-15 \quad -5]$. Thus

$$-K^t = [-15 \quad -5] \frac{1}{3}\begin{bmatrix} 0 & 1 \\ 3 & 1 \end{bmatrix} = [-5 \quad -6.666] \tag{6.16}$$

The resulting S is

$$S = \begin{bmatrix} -9 & 3 \\ -8.666 & 1 \end{bmatrix} \tag{6.17}$$

With S as in 6.17, the model

$$\dot{z} = Sz + Kb$$
$$x = z \tag{6.18}$$

serves as a dynamic observer for the component defined in 6.11.

Since direct access to state variables is seldom possible, the observer provides the state estimate which will drive an appropriate feedback matrix, say $\underline{K}$, which may be used to stabilize the system. As a block diagram, consider the rendition in figure 6.19.

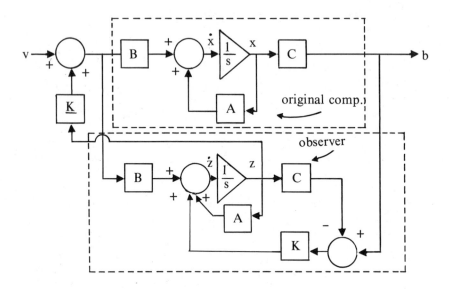

Figure 6.19. Block diagram of, dynamic observer, and state feedback law.

Inspection of this block diagram shows the combined observer and component model to be

$$
\begin{bmatrix} \dot{x} \\ \hline \dot{z} \end{bmatrix} = \begin{bmatrix} A & \vdots & B\underline{K} \\ \hline KC & \vdots & A-KC+B\underline{K} \end{bmatrix} \begin{bmatrix} x \\ \hline z \end{bmatrix} + \begin{bmatrix} B \\ \hline B \end{bmatrix} v \qquad (6.20)
$$

where "v" is now the "new" input variable. Applying the similarity transformation

$$
T = \begin{bmatrix} I & \vdots & 0 \\ \hline I & \vdots & -I \end{bmatrix} \qquad (6.21)
$$

to 6.20 produces

$$
T^{-1} \begin{bmatrix} A & \vdots & BK \\ \hline KC & \vdots & A-KC+BK \end{bmatrix} T = \begin{bmatrix} A+BK & \vdots & -BK \\ \hline 0 & \vdots & A-KC \end{bmatrix} \qquad (6.22)
$$

From 6.22, the eigenvalues of the combined dynamics are simply those of

(A+B$\underline{K}$) and (A–KC). Gratifyingly enough, the eigenvalues of a state feedback control law remain unchanged even in the presence of inaccurate state estimate, $\hat{x}$, introduced by the observer.

As pointed out in the previous section, controllability and observability are dual notions. The observability counterpart to stabilizability is *detectability*. Stabilizability guaranteed the controllability of unstable modes. Via the duality theorem, detectability must guarantee the observability of all unstable modes. As a consequence, we state the dual counterpart of theorem 3.51:

THEOREM 6.23: A component characterized by equation 6.1 is *detectable* if and only if

$$\text{rank} \begin{bmatrix} (\lambda_i I - A) \\ \text{-----} \\ C \end{bmatrix} = n \qquad (6.24)$$

for all unstable eigenvalues, λ_i, of A.

Any eigenvalue λ_j for which the rank in 6.24 is not n corresponds to an *unobservable mode*. States having components in the direction of such an eigenvector (possibly generalized) are unobservable. The *unobservable subspace* of the pair (A,C) consists of all states x_o whose zero input response is identically zero for all time. The unobservable subspace coincides with the null space of the observability matrix

$$R = \begin{bmatrix} C \\ \text{----} \\ CA \\ \text{----} \\ . \\ . \\ . \\ \text{----} \\ CA^{n-1} \end{bmatrix} \qquad (6.25)$$

The observable subspace is the orthogonal complement of null(R) in $\mathbf{R}^n$. Observe that a *singular value decomposition* of R (section 4) provides a convenient algorithm for constructing respective orthonormal bases for these two subspaces. Therefore, the dynamic observer for a pair (A,C) has a blind spot coincident with the null space of R. Theorem 6.23 identifies those modes lying in this unobservable subspace. Hence any state feedback driven by the observer estimate has no effect on null (R). Furthermore, since feedback cannot affect the uncontrollable directed modes of A, feedback driven by an observer influences only the controllable *and* observable subspace of the triple (A,B,C), i.e., the intersection of the controllable and observable subspaces. Again, as per section 4 of this chapter, the SVD algorithm is useful in constructing this intersection.

A component which is both stabilizable and detectable may be stabilized by feedback driven by a dynamic observer. A component which is stabilizable but not detectable cannot be stabilized by feedback driven by a dynamic observer.

To stabilize a detectable and stabilizable component, first design appropriate state feedback which stabilizes the observable subspace. Detectability guarantees that all unstable modes are observable, and stabilizability makes them controllable. Hence, the appropriate state feedback exists. Secondly, build an observer as per theorem 6.8 and use the estimated state to drive the feedback.

At this point the interested reader is encouraged to illustrate the above ideas in the context of the *Kalman canonical form.*

The cascade of a static compensator, K, driven by a dynamic observer (equation 6.11) represents *dynamic compensation.* The dynamic observer has two inputs, "a" and "b". The general *model* of a *dynamic compensator* for this text will be

$$\dot{z} = Sz + Rb$$

$$\tag{6.26}$$

$$a = Qz + Kb + v$$

The dynamic portion of 6.26 is driven only by the component output, b. Technically, a dynamic observer is not a special case of 6.26, although one has the same potential pole assignment capability under the same observability conditions. The *fixed modes*[3] of 6.1 serve to characterize the equivalent *pole placement* capability. One may think of the fixed modes of 6.1 as those *directed modes* which remain invariant under arbitrary output feedback. We index the fixed modes by the associated eigenvalues and view 6.1 as the triple (A,B,C).

DEFINITION 6.27: The *fixed modes* of the triple (A,B,C) are those complex numbers in the set

$$\Phi(A,B,C) = \bigcap_K \sigma(A+BK) \tag{6.28}$$

where $\sigma(A+BKC)$ denotes the set of eigenvalues of $A+BKC$, and the intersection is taken over all conformable feedback matrices K.

Under the static feedback control law

$$a = Kb + v \tag{6.29}$$

equation 6.1 becomes

$$\begin{aligned} \dot{x} &= (A + BKC)x + Bv \\ b &= Cx \end{aligned} \tag{6.30}$$

The fixed modes of the component model are thus those eigenvalues of A which are invariant under the feedback law of 6.29. Alternately the interplay between the uncontrollable and unobservable subspaces isolates the fixed modes. Those eigenvalues for which

$$\text{rank} \begin{bmatrix} \lambda I{-}A \\ \hline C \end{bmatrix} \neq n \tag{6.31}$$

or

$$\text{rank} \, [\lambda I{-}A \mid B] \neq n \tag{6.32}$$

completely determine the set $\Phi(A,B,C)$.

The remainder of this section will show that the fixed modes as determined in 6.28 are invariant under dynamical feedback. On the other hand, we show that the remaining eigenvalues may be arbitrarily reassigned by proper choice of S, R, Q, and K in 6.26.

Combining 6.1 and 6.26 leads to the complete controller-component description

$$\begin{bmatrix} \dot{x} \\ \dot{z} \end{bmatrix} = \begin{bmatrix} A{+}BKC & BQ \\ \hline RC & S \end{bmatrix} \begin{bmatrix} x \\ z \end{bmatrix} + \begin{bmatrix} B \\ 0 \end{bmatrix} v$$

$$b = \begin{bmatrix} C & 0 \end{bmatrix} \begin{bmatrix} x \\ z \end{bmatrix} \tag{6.33}$$

This dynamic compensator design requires proper choice of K, Q, R, and S for placement of the eigenvalues of

$$F = \left[\begin{array}{c|c} A+BKC & BQ \\ \hline RC & S \end{array}\right] \tag{6.34}$$

If for some K, $\sigma(A+BK)$ is stable, then it is permissible to set S = R = 0 in 6.25 which shows that static output compensation is a special form of dynamic compensation. Before proceeding to the main theorem it is worthwhile to interpret the fixed modes of A+BKC via the Kalman canonical form. As per 5.46, 5.47, and 5.48, there exists a nonsingular transformation T such that

$$T^{-1}(A+BKC)T = \left[\begin{array}{c|c|c|c} A_{11}+B_1KC_1 & 0 & A_{13}+B_1KC_3 & 0 \\ \hline A_{21}+B_2KC_1 & A_{22} & A_{23}+B_2KC_3 & A_{24} \\ \hline 0 & 0 & A_{33} & 0 \\ \hline 0 & 0 & A_{43} & A_{44} \end{array}\right] \tag{6.35}$$

The similarity transformation T leaves the eigenvalues of A+BKC unaffected. The block triangularity of $T^{-1}(A+BKC)T$ and its submatrices indicates that the eigenvalues of A+BKC are those of $A_{11}+B_1KC_1$, A_{22}, A_{33}, A_{44}. Hence the fixed modes of the triple (A,B,C) are the fixed modes of the triple (A_{11}, B_1, C_1) and the eigenvalues of the three matrices A_{22}, A_{33}, and A_{44}. Specifically

$$\Phi(A,B,C) = \Phi(A_{11},B_1,C_1) \cup \sigma(A_{22}) \cup \sigma(A_{33}) \cup \sigma(A_{44}) \tag{6.36}$$

THEOREM 6.37: For the component of 6.1, the following are equivalent:
(i)   The spectrum can be arbitrarily assigned by dynamic compensation.
(ii) $\phi(A,B,C) = \phi$, i.e., there are no fixed modes.
(iii) The triple (A,B,C) is controllable and observable.

PROOF: [(i) → (ii)] Let λ be an arbitrary complex number and choose S, Q, R, and K so that λ is not an eigenvalue of F. The hypothesis (i) guarantees the existence of appropriate S, Q, R, and K. The formula for the determinant of a 2x2 partitioned matrix[2] and the fact that λ is not an eigenvalue of F implies

$$0 \neq \det[\lambda I - F] = \det[\lambda I - S] \det[\lambda I - (A+BKC+BQ(\lambda I - S)^{-1}RC)] \tag{6.38}$$

By letting $\underline{K} = K+Q(\lambda I-S)^{-1}R$, 6.38 implies

$$det[\lambda I-(A+B\underline{K}C)] \neq 0 \qquad (6.39)$$

Hence, λ is not an eigenvalue of A+BKC. Since the fixed modes are common to all matrices, A+BKC, λ is not not contained in $\Phi(A,B,C)$. The arbitrariness of λ shows that $\Phi(A,B,C) \neq \phi$.

[(ii) $\rightarrow$ (iii)] From the discussion on the fixed modes of 6.35, $\Phi(A,B,C)$ is null only when each of its constituent parts is null. In particular $\sigma(A_{22}) = \sigma(A_{33}) = \sigma(A_{44}) = \phi$ if and only if each of the matrices is zero dimensional. In consequence the only nontrivial entry in the Kalman canonical form of the original component model is the controllable and observable block. Furthermore, this implies $\Phi(A_{11},B_1,C_1) = \phi$. Hence the original component model is controllable and observable.

[(iii) $\rightarrow$ (i)] Finally, if the component model is controllable and observable one can build a dynamic compensator along lines similar to the dynamic observer and static feedback. For further detail see Wonham.[8] Thus, arbitrary symmetric sets of eigenvalues can be assigned.

This theorem and the Kalman canonical form combine to produce the following two corollaries.

COROLLARY 6.40: In terms of the Kalman canonical form of the triple (A,B,C),

$$\Phi(A,B,C) = \sigma(A_{22}) \cup \sigma(A_{33}) \cup \sigma(A_{44}) \qquad (6.41)$$

COROLLARY 6.42: For the component model 6.1(i.e., the triple (A,B,C)) the following are equivalent:
(i) The triple (A,B,C) is stabilizable by dynamic compensation.
(ii) Re $(\lambda) < 0$ for each λ in $\Phi(A,B,C)$
(iii) The triple (A,B, C) is stabilizable and detectable.

Note that in the above corollaries and theorem, we require the states to be *asymptotically stable*.

7. CONTROL OF COMPOSITE SYSTEMS

The preceding sections present a clear-cut theory for the control of a single component model. In practice the system to be controlled is an interconnection of numerous individual components. This section briefly touches on the control of an interconnected dynamical system characterized by the component connection model

$$\dot{x} = Ax + Ba$$
$$b = Cx + Da \tag{7.1}$$

and

$$a = L_{11}b + L_{12}u$$
$$y = L_{21}b + L_{22}u \tag{7.2}$$

The control properties of the interconnected system are derived by applying the theorems of the preceding sections to the composite state model

$$\dot{x} = Fx + Gu$$
$$y = Hx + Ju \tag{7.3}$$

where

$$F = A + B(I-L_{11}D)^{-1}L_{11}C \tag{7.4}$$

$$G = B(I-L_{11}D)^{-1}L_{12} \tag{7.5}$$

$$H = L_{21}C + L_{21}D(I-L_{11}D)^{-1}L_{11}C \tag{7.6}$$

and

$$J = L_{21}D(I-L_{11}D)^{-1}L_{12} + L_{22} \tag{7.7}$$

In practice, the matrices of the composite system state model are large, non-sparse, and not readily amenable to the various transformations underlying the design of appropriate feedback controllers. One exception to this phenomena is the case of *series parallel connections*. Here, after an appropriate reordering of the components, L_{11} will have *block strictly lower triangular form*. Since A, B, C, and D are *block diagonal* this implies that F takes the form $F = A + E$ where $E = B(I-L_{11}D)^{-1}L_{11}C$ is block strictly lower triangular. As such, 7.3 may be written in the partitioned form

$$
\begin{bmatrix} x_1 \\ x_2 \\ x_3 \\ \vdots \\ x_n \end{bmatrix}
=
\begin{bmatrix}
A_1 & 0 & 0 & \cdots & 0 & 0 \\
E_{21} & A_2 & 0 & & & \\
E_{31} & E_{32} & A_3 & \ddots & & \\
\vdots & & & \cdots & & \\
E_{n1} & E_{n2} & E_{n3} & \cdots & E_{n,n-1} & A_n
\end{bmatrix}
\begin{bmatrix} x_1 \\ x_2 \\ x_3 \\ \vdots \\ x_n \end{bmatrix}
+
\begin{bmatrix} G_1 \\ G_2 \\ G_3 \\ \vdots \\ G_n \end{bmatrix} u
\tag{7.8}
$$

$$
y] = [H_1 \mid H_2 \mid H_3 \mid \cdots \mid H_{n-1} \mid H_n]
\begin{bmatrix} x_1 \\ x_2 \\ x_3 \\ \vdots \\ x_n \end{bmatrix}
+ [D] \ u]
$$

Since F is triangular with the component A_i matrices on the diagonal, it has the same eigenvalues as the individual components and is well suited to the tests for *controllability*, *stabilizability*, *observability* and *detectability*.

THEOREM 7.9: Let a series-parallel system be characterized by equation 7.8. Then it is controllable (stabilizable) if and only if each of the matrices

$$
\begin{bmatrix}
\lambda I - A_1 & 0 & 0 & \cdots & 0 & 0 & G_1 \\
-E_{21} & \lambda I - A_2 & 0 & \cdots & 0 & 0 & G_2 \\
-E_{31} & -E_{32} & \lambda I - A_3 & \ddots & 0 & 0 & G_3 \\
\vdots & & & & \vdots & \vdots & \vdots \\
-E_{k1} & -E_{k2} & -E_{k3} & \cdots & -E_{k,k-1} & \lambda I - A_k & G_k
\end{bmatrix}
\tag{7.10}
$$

has full row rank for each eigenvalue (unstable eigenvalue), λ, of A_k which is not also an eigenvalue of A_i, $i > k$.

PROOF: The proof follows immediately by employing the F and G matrices of equation 7.8 in the eigenvalue test of theorem 2.52 and noting that the eigenvalues of F equal those of A. Moreover, if λ is an eigenvalue of A_k and not one of A_i, $i > k$, the submatrix

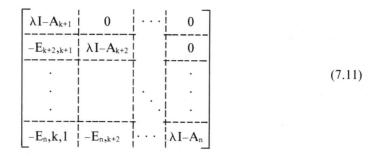

$$\begin{bmatrix} \lambda I - A_{k+1} & 0 & \cdots & 0 \\ -E_{k+2,k+1} & \lambda I - A_{k+2} & & 0 \\ \cdot & & & \cdot \\ \cdot & & \ddots & \cdot \\ \cdot & & & \cdot \\ -E_n,k,1 & -E_{n,k+2} & \cdots & \lambda I - A_n \end{bmatrix} \qquad (7.11)$$

is assured to be non-singular, permitting the corresponding rows and columns of the test matrix to be dropped from the rank test.

In dealing with a *hierarchical system* rather than a series parallel system a similar theorem holds if one replaces the individual components by the *strongly connected sub-systems*.

EXAMPLE 7.12: Consider the cascade system shown in figure 7.13. The component connection model is given by 7.14 and 7.15.

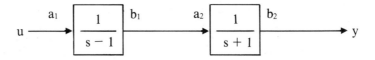

Figure 7.13. Cascade system.

$$\begin{bmatrix} x_1 \\ x_2 \end{bmatrix} = \begin{bmatrix} 1 & 0 \\ 0 & -1 \end{bmatrix} \begin{bmatrix} x_1 \\ x_2 \end{bmatrix} + \begin{bmatrix} 1 & 0 \\ 0 & 1 \end{bmatrix} \begin{bmatrix} a_1 \\ a_2 \end{bmatrix}$$

$$\begin{bmatrix} b_1 \\ b_2 \end{bmatrix} = \begin{bmatrix} 1 & 0 \\ 0 & 1 \end{bmatrix} \begin{bmatrix} x_1 \\ x_2 \end{bmatrix} \tag{7.14}$$

and

$$\begin{bmatrix} a_1 \\ a_2 \\ \hline y \end{bmatrix} = \begin{bmatrix} 0 & 0 & | & 1 \\ 1 & 0 & | & 0 \\ \hline 0 & 1 & | & 0 \end{bmatrix} \begin{bmatrix} b_1 \\ b_2 \\ \hline u \end{bmatrix} \tag{7.15}$$

The composite system has state dynamics

$$\begin{bmatrix} x_1 \\ x_2 \end{bmatrix} = \begin{bmatrix} 1 & 0 \\ 1 & -1 \end{bmatrix} \begin{bmatrix} x_1 \\ x_2 \end{bmatrix} + \begin{bmatrix} 1 \\ 0 \end{bmatrix} u] \tag{7.16}$$

To check controllability, we could use theorem 2.52 which requires computing the rank of two 2x3 matrices, one for $\lambda = 1$ and one for $\lambda = -1$. Using theorem 7.9 it suffices to test the rank of one 2x3 matrix for $\lambda = -1$ and a 1x2 matrix for $\lambda = 1$. For $\lambda = -1$

$$\begin{bmatrix} \lambda - A_1 & | & 0 & | & G_1 \\ \hline -E_{21} & | & \lambda - A_2 & | & G_2 \end{bmatrix} = \begin{bmatrix} -2 & | & 0 & | & 1 \\ \hline -1 & | & 0 & | & 0 \end{bmatrix} \tag{7.17}$$

has rank 2 and for $\lambda = 1$

$$[\lambda - A_1 \ | \ G_1] = [0 \ | \ 1] \tag{7.18}$$

has rank 1. By theorem 7.9 the cascaded system is controllable. Of course, since controllability implies stabilizability the system of figure 7.13 is also stabilizable.

Invoking an argument dual to that used in the proof of theorem 7.9 produces the following theorem.

THEOREM 7.19: Let a series-parallel system be characterized by equation 7.8; then it is observable (detectable) if and only if each of the matrices

$$
\begin{bmatrix}
\lambda I - A_1 & 0 & 0 & \cdots & 0 & 0 \\
-E_{21} & \lambda I - A_2 & 0 & \cdots & 0 & 0 \\
-E_{31} & -E_{32} & \lambda I - A_3 & & 0 & 0 \\
\cdot & & \cdot & & & \cdot \\
\cdot & & & \cdot & & \cdot \\
\cdot & & & & \cdot & \cdot \\
-E_{k1} & -E_{k2} & -k_3 & \cdots & E_{k,k-1} & \lambda I - A_k \\
H_1 & H_2 & H_3 & \cdots & H_{k-1} & H_k
\end{bmatrix}
$$

has full row rank for each eigenvalue (unstable eigenvalue), λ, of A_k which is not also an eigenvalue of A_1, $i > k$.

8. DECENTRALIZED CONTROL

Physical distances as well as other practical snags restrain the engineer in designing global controllers. Electric power system generators have locations distributed over vast geographical areas. Hundreds of miles of high voltage transmission lines connect the various generating stations. Such physical separation makes it impractical and/or uneconomical either to globally monitor or to globally control the interconnected system. This is a prime motivation for the development of *decentralized controllers*.[6,7] A decentralized controller has a control law which feeds the component state or the component measurable output back only to the input of that particular component.

The internal system structure is the setting of our discussion of decentralized control. The *internal system structure* has the representation

$$
\begin{aligned}
\dot{x} &= Ax + Ba \\
b &= Cx
\end{aligned}
\tag{8.1}
$$

and

$$
a = L_{11}b + v \tag{8.2}
$$

where the above symbols take on their usual meaning, and v symbolizes the controlling input. Setting D=0 in 8.1 eliminates direct feedthrough connections which eases the theoretical development.

For v to be a decentralized control law, certain restrictions are imposed on the form of the feedback. Specifically, the output of component, i, may drive through some (possible complex) feedback matrix K_i or through a dynamic compensator only the input of component, i. Suppose v_i is the control input for the i-th component, then the component *output-feedback* control takes the form

$$v_i = K_i C_i x_i \qquad\qquad (8.3)$$

where $C = \text{diag}(C_1, C_2, \cdots, C_r)$, r being the number of components. Define the possible complex feedback matrix, $K_d = \text{diag}(K_1, \cdots, K_r)$, and set, $B = \text{diag}(B_1, \cdots, B_r)$, where the subscript corresponds to a particular component or subsystem. Obviously then the decentralized output feedback control law, v, takes the form

$$v = K_d C \qquad\qquad (8.4)$$

where K_d^C, is the possibly complex feedback to be designed.

The main theorem of this section points to the *existence* of decentralized stabilizing controllers if and only if there exist stabilizing global controllers. The proof uses the above possibly complex output-feedback control law. From an implementation perspective this kind of feedback is impractical. In general decentralized control requires dynamic compensation, i.e., a feedback compensator of the form

$$\begin{aligned} \dot{z} &= Sz + Rb \\ v &= Qz + K_d y \end{aligned} \qquad\qquad (8.5)$$

where S, R, Q, and K_d have the appropriate block diagonal form as determined by the components. In particular $S = \text{diag}(S_1, \cdots, S_r)$, $R = \text{diag}(R_1, \cdots, R_r)$, $Q = \text{diag}(Q_1, \cdots Q_r)$, and K_d is as before.

The need for admitting complex K_d becomes clear after considering the following problem: for the system of figure 8.6 find k_1 and k_2 so that the resultant system has eigenvalues at $-1 \pm j$.

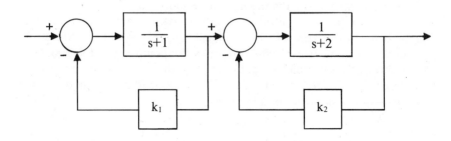

Figure 8.6. Block diagram of cascade system with decentralized output feedback.

For this cascaded system, choosing $k_1 = j$ and $k_2 = -1-j$ will produce an input-output transfer function

$$\frac{y(s)}{u(s)} = \frac{1}{s+1+j} \ \frac{1}{s+1-j} = \frac{1}{s^2+2s+2} \tag{8.7}$$

which has the desired pole locations.

Certainly such a choice is impractical, yet no other choice of real values of k_1 and k_2 would place the system eigenvalues at $-1 \pm j$. On the other hand, choosing the *dynamic compensators*

$$k_1 = \frac{1}{s+1} \quad \text{and} \quad k_2 = -2 \tag{8.8}$$

produces realizable feedback laws which place the eigenvalues in the desired locations. This situation demonstrates the need for dynamic compensation.

The existence of decentralized complex controllers if and only if there exist global controllers depends on the *fixed modes* of the system model. To define the fixed mode concept, note that the internal system structure as defined in 8.1 and 8.2 has the composite model

$$\dot{x} = Fx + Ba$$
$$b = Cx$$
(8.9)

where

$$F = A + BL_{11}C$$
(8.10)

DEFINITION 8.11: For the decentralized control law of 8.4 the *decentralized fixed modes* are given by

$$\theta_d(F,B,C) = \bigcap_{K_d} \sigma(F + BK_dC)$$
(8.12)

where (i) $\sigma(F + BK_dC)$ denotes the spectrum (eigenvalues) of the matrix $F + BK_dC$ and (ii) the intersection is indexed over all appropriately partitioned block diagonal complex matrices K_d resulting in a symmetric spectrum.

Recall that a matrix M has a symmetric spectrum $\sigma(M)$ if and only if all complex eigenvalues occur in complex conjugate pairs. Now for global controllers the associated feedback law is

$$v = KCx$$
(8.13)

hence we define the *fixed modes* of the *globally controlled system* as

$$\theta(F,B,C) = \bigcap_{K} \sigma(F + BKC)$$
(8.14)

where K is an arbitrary complex matrix such that $\sigma(F + BKC)$ is symmetric.

The earlier development of this chapter makes the need for potentially complex K unnecessary. The number of assignable eigenvalues equals the dimension of the intersection of the controllable and observable subspaces, say this dimension is m_0 (see 6.36). In turn this depends on the matrices F, B, and C. In addition any symmetric set of m_0 complex numbers can be assigned *exactly* by some real output feedback matrix K provided the sum of the ranks of B and C equals $m_0 + 1$.

At any rate, since the intersection in 8.12 is indexed over a smaller set of matrices than the intersection of 8.14 we have

$$\theta(F,B,C) \subset \theta_d(F,B,C)$$
(8.15)

Using this notation the existence of possible complex decentralized controllers is guaranteed by

THEOREM 8.16: For the system of 8.1 and 8.2

$$\theta_d(F,B,C) = \theta(F,B,C) = \theta(A,B,C) = \bigcup_i \theta(A_i,B_i,C_i) \quad (8.17)$$

PROOF: Step 1: Show that $\theta(F,B,C) = \theta(A,B,C)$. Consider that

$$\sigma[F + BKC] = \sigma[A+B(K+L_{11})C] = \sigma[A+B\underline{K}C] \quad (8.18)$$

Hence

$$\bigcap_K \sigma[F,B,C] = \bigcap_{\underline{K}} \sigma[A,B,C] \quad (8.19)$$

where K and $\underline{K} = L_{11} + K$ span the same set of matrices. Therefore $\sigma(F,B,C) = \sigma(A,B,C)$.

STEP 2: Now show

$$\theta(A,B,C) = \bigcup_{i=1}^r \theta(A_i,B_i,C_i) \quad (8.20)$$

Observe that A,B, and C are conformable block diagonal matrices. Hence the eigenvalues of $A + BK_dC$ are the union of the sets of eigenvalues of the individual blocks. In particular then 8.20 is valid.

STEP 3: Show that $\theta_d(F,B,C) = \theta(F,B,C)$. Given 8.15, it suffices to show that

$$\theta_d(F,B,C) \subset \theta(F,B,C) \quad (8.21)$$

The approach to follow is inductive. Initially suppose A,B, and C are 2x2 partitioned matrices, i.e., $A = \mathrm{diag}\,(A_1,A_2)$, $B = \mathrm{diag}\,(B_1,B_2)$, and $C = \mathrm{diag}\,(C_1,C_2)$. Let us partition L_{11} accordingly and drop the subscripts as

$$L_{11} = L = \begin{bmatrix} L^{11} & L^{12} \\ L^{21} & L^{22} \end{bmatrix} \quad (8.22)$$

Similarly, let

$$K = \begin{bmatrix} K^{11} & \vdots & K^{12} \\ ---- & + & ---- \\ K^{21} & \vdots & K^{22} \end{bmatrix} \tag{8.23}$$

Recall that the formula for the determinant of a 2x2 partitioned matrix is:

$$\det \begin{bmatrix} X & \vdots & Y \\ --- & + & --- \\ Z & \vdots & W \end{bmatrix} = \det[X]\,\det[W - ZX^{-1}W] \tag{8.24}$$

Now if λ is not a member of $\theta(F,B,C)$, then there exists K dependent on λ such that $\det[\lambda I - (F + BKC)] \neq 0$. The objective is to construct K_d, also λ–dependent, such that $\det[\lambda I - (F + BK_dC)] \neq 0$. For such a λ

$$0 \neq \det[\lambda I - (F + BKC)] = \det[\lambda I - (A+BLC+BKC)]$$

$$= \det \begin{bmatrix} \lambda I - A_1 - B_1 L^{11} C_1 - B_1 K^{11} C_1 & \vdots & -B_1 L^{12} C_2 - B_1 K^{12} C_2 \\ ----------------- & + & ----------- \\ -B_2 L^{21} C_1 - B_2 K^{21} C_1 & \vdots & \lambda I - A_2 - B_2 L^{22} C_2 - B_2 K^{22} C_2 \end{bmatrix}$$

$$\tag{8.25}$$

Define $X = \lambda I - A_1 - B_1 L^{11} C_1 - B_1 K^{11} C_1$ and assume

$$\lambda \in \sigma(A_1 + B_1 L^{11} C_1 + B_1 K^{11} C_1) \tag{8.26}$$

which is always possible by suitable choice of K above. Then

$$0 \neq \det[X]\,\det[\lambda I - A_2 - B_2 L^{22} C_2 - B_2 K^{22} C_2$$

$$- (B_2 L^{21} C_1 + B_2 K^{21} C_1) X^{-1} (B_1 L^{12} C_2 + B_1 K^{12} C_2)]$$

$$= \det[X]\,\det[\lambda I - A_2 - B_2 L^{22} C_2 - B_2 \underline{K}^{22} C_2$$

$$- B_2 L^{21} C_1 X^{-1} B_1 L^{12} C_2] \tag{8.27}$$

where

$$\underline{K}^{22} = K^{22} + K^{21}C_1X + {}^1B_1L^{12} + L^{21}C_1X^{-1}B_1K^{12}$$

$$+ K^{21}C_1X^{-1}B_1K^{12} \tag{8.28}$$

Now, to construct the corresponding decentralized feedback, define

$$K_d = \begin{bmatrix} K^{11} & 0 \\ 0 & \underline{K}^{22} \end{bmatrix} \tag{8.29}$$

Computing $\det[\lambda I - (F + BK_dC)]$ yields

$$\det[\lambda I - (F + BK_dC)] = \det[\lambda I - (A + BLC + BK_dC]$$

$$= \det\begin{bmatrix} \lambda I - A_1 - B_1L^{11}C_1 - B_1K^{11}C_1 & -B_1L^{12}C_2 \\ \hline -B_2L^{21}C_1 & \lambda I - A_2 - B_2L^{22}C_2 - B_2K^{22}C_2 \end{bmatrix}$$

$$= \det[X] \det[\lambda I - A_2 - B_2L^{22}C_2 - B_2\underline{K}^{22}C_2$$

$$- B_2L^{21}C_1X^{-1}B_1L^{12}C_2]$$

$$= \det[\lambda I - (F + BKC)]$$

$$\neq 0 \tag{8.30}$$

Consequently, if λ is not in $\theta(F,B,C)$ then there exists at least one block diagonal feedback matrix, K_d, such that λ is not an eigenvalue of $F + BK_dC$. In particular, λ is not a member of $\theta_d(F,B,C)$.

To extend the above argument from 2x2 partitioned matrices to nxn partitioned matrices one repeats the above construction n–1 times as follows: Let the nxn partitioned matrix below be artificially partitioned into a 2x2 partition as indicated by the double-dashed line. Assume that

$\det[I-(F+BKC)] \neq 0.$

$$K = \begin{bmatrix} K^{11} & K^{12} & K^{13} & \cdots & K^{1n} \\ K^{21} & K^{22} & K^{23} & & K^{2n} \\ K^{31} & K^{32} & K^{33} & & K^{3n} \\ \cdot & & & & \cdot \\ \cdot & & & & \cdot \\ K^{n1} & K^{n2} & K^{n3} & \cdots & K^{nn} \end{bmatrix} \qquad (8.31)$$

Employing the above argument permits formulation of the matrix

$$K_1 = \begin{bmatrix} K^{11} & 0 & 0 & \cdots & 0 \\ 0 & \underline{K}^{22} & \underline{K}^{23} & & \underline{K}^{2n} \\ 0 & \underline{K}^{32} & \underline{K}^{33} & & \underline{K}^{3n} \\ \cdot & & & & \cdot \\ \cdot & & & & \cdot \\ 0 & \underline{K}^{n2} & \underline{K}^{n3} & & \underline{K}^{nn} \end{bmatrix} \qquad (8.32)$$

such that $\det[\lambda I-(F+BK_1C)] \neq 0$. Repartitioning as indicated by the double-dashed line allows one to apply the argument above again. Since the 1-1 entry remains unaffected by the process, one obtains a new matrix of the form

$$K_2 = \begin{bmatrix} K^{11} & 0 & 0 & \cdots & 0 \\ 0 & K^{22} & 0 & & 0 \\ 0 & 0 & K^{33} & & K^{3n} \\ \cdot & & & & \cdot \\ \cdot & & & & \cdot \\ 0 & 0 & K^{n3} & \cdots & K^{nn} \end{bmatrix} \qquad (8.33)$$

such that $\det[\lambda I-(F+BK_2C] \neq 0$. Repeating the process n$-$1 times produces a block diagonal matrix, K_d, such that $\det[\lambda-(F+BK_dC)] \neq 0$. This shows that if λ is not in $\phi(F,B,C)$ then it is also not in $\theta_d(F,B,C)$, thereby verifying that $\theta_d(F,B,C)$ is a subset of $\theta(F,B,C)$. The proof is now complete.

The concept of fixed modes so deftly utilized in the above theorem in some sense generalizes the classical notion of controllability. And although the above theorem says that the fixed modes of the globally controlled system and the locally controlled system coincide, it does not imply that local output feedback, complex or real, will assign an arbitrary prespecified set of eigenvalues. However, if the ranks of C_i and B_i are sufficiently large, then complex local output feedback will suffice to assign an arbitrary prespecified symmetric set of eigenvalues of appropriate number. Still, the utility of real output feedback remains unclear. In the following we will describe an iterative algorithm which will compute real local feedback gains to assign a prespecified set of distinct eigenvalues. The assumptions which underlie the scheme are:

(i) each component is completely controllable,

(ii) all component states are available for feedback,

(iii) the characteristic polynomial of each component is of even order.

The above assumptions provide sufficient conditions for the applicability of the iterative scheme to follow:

For simplicity assume each component is single input. As such, one applies the algorithm on a component by component basis. At each iteration the algorithm will produce a least squares fit to a prespecified set of eigenvalue locations.

STEP 1: Appropriately order the desired set of eigenvalues by associating pairs or groups of desired eigenvalues with particular components. For example if component one has two pairs of complex conjugate eigenvalues moderately close to two pairs of desired complex conjugate eigenvalues then associate these desired eigenvalues with component one. Number these eigenvalues $\lambda_1, \lambda_2, \lambda_3, \lambda_4$. This should be done with each component. This association is important because the decentralized feedbacks readjust the component eigenvalue locations so that when connection information is introduced, the new component eigenvalues migrate to the desired locations. The *root locus* technique of VI.6 is particularly useful in making the association.

STEP 2: Assuming all desired eigenvalues are distinct, compute the matrix

$$E = \begin{bmatrix} \lambda_1^{N-1} & \lambda_1^{N-2} & \cdots & 1 \\ \cdot & \cdot & & \cdot \\ \cdot & \cdot & & \cdot \\ \cdot & \cdot & & \cdot \\ \cdot & \cdot & & \cdot \\ \lambda_N^{N-1} & \lambda_N^{N-2} & \cdots & 1 \end{bmatrix} \qquad (8.34)$$

where N is the number of eigenvalues to be assigned and the set $\{\lambda_i | i=1,N\}$

is the desired set of eigenvalues. Note that complex conjugate pairs must appear in adjacent rows.

STEP 3: Compute the block diagonal transformation T. The blocks of T are 1x1 and have a single entry of one if the associated row of E corresponds to a real eigenvalue. The remaining blocks are 2 x 2 and correspond to conjugate pairs of eigenvalues and have the form

$$\frac{1}{\sqrt{2}} \begin{bmatrix} 1 & 1 \\ -j & j \end{bmatrix} \tag{8.35}$$

Step 4: compute

$$(\lambda I - F)^{-1} = \frac{I\lambda^{N-1} + R_1\lambda^{N-2} + \cdots + R_{N-1}}{d_F(s)} \tag{8.36}$$

by *Leverrier's algorithm* which we describe at the end of the algorithm. Note $d_F(s)$ is the characteristic polynomial of F. Let

$$W = [b, R_1 b, \cdots, R_N b] \tag{8.37}$$

and

$$b = \begin{bmatrix} d_F(\lambda_1) \\ d_F(\lambda_2) \\ \cdot \\ \cdot \\ \cdot \\ d_F(\lambda_N) \end{bmatrix} \tag{8.38}$$

where b is the input vector for the particular iteration.

STEP 6: Compute the matrices TEW^t and TE as

$$Z = TEW^t \tag{8.39}$$

and

$$z = TE \tag{8.40}$$

STEP 7: Compute the inverse of Z.

STEP 8: Compute the pertinent feedback gains as

$$Z^{-1}z = [k_1, \cdots, k_p] \tag{8.41}$$

One repeats the application of the algorithm on a component by component basis. After a suitable number of iterations, the algorithm should converge. Convergence can be determined either by measuring the closeness to the desired eigenvalues or by measuring the closeness to the coefficients of the desired characteristic polynomial or both. Note that F must be updated after each iteration.

In Step 4 it was necessary to use Leverrier's algorithm to compute $(\lambda I - F)^{-1}$. Denote the characteristic polynomial of F by

$$d_F(\lambda) = \lambda^N + a_1\lambda^{N-1} + a_2\lambda^{N-2} + \cdots + a_N \qquad (8.42)$$

Leverrier's algorithm proceeds as follows:

Define $F_1 = F$ and

$$a_1 = -\text{trace}(F_1) \qquad (8.43)$$

and

$$R_1 = F_1 + a_1 I \qquad (8.44)$$

In addition, define

$$F_k = FR_{k-1} \qquad (8.45)$$

$$a_k = -\frac{1}{k}\,\text{trace}(F_k) \qquad (8.46)$$

and

$$R_k = F_k + a_k I \qquad (8.47)$$

These equations set forth Leverrier's algorithm for calculating $(\lambda I - F)^{-1}$.

The preceding eigenvalue assignment scheme has several generalizations. It is possible to relax the restriction of distinct eigenvalues by modifying the construction of W. The scheme can also be adapted to output feedback by modifying step 5 and step 7. In step 5 we have

$$W = C[b, R_1, b, \cdots, R_N b] \qquad (8.48)$$

and in step 7 we compute the *Moore-Penrose pseudo inverse* of Z. Finally the algorithm is adaptable to multi-input components as follows: The scheme is divided into two phases: cycle one, which consists of a single iteration for each component, and the remaining cycles. The division is due to certain numerical considerations as well as generic controllability concepts. In cycle one, for each iteration one sets b (in step 5) equal to a random linear combination of the columns of B which pertain to the appropriate component. With probability one the controllability of the component will be maintained. Hence one can proceed to apply the algorithm. The first cycle consists of doing this procedure for each component once. After completing cycle one the algorithm proceeds as before wherein b is sequentially made equal to successive columns of B.

As an example of the method consider the block diagram of figure 8.49.

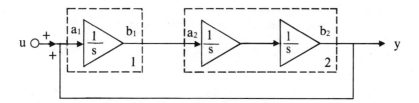

Figure 8.49. System block diagram.

In this system the composite component state model is:

$$
\begin{bmatrix} \dot{x_1} \\ x_2 \\ x_3 \end{bmatrix} = \begin{bmatrix} 0 & 0 & 0 \\ 0 & 0 & 0 \\ 0 & 1 & 0 \end{bmatrix} \begin{bmatrix} x_1 \\ x_2 \\ x_3 \end{bmatrix} + \begin{bmatrix} 1 & 0 \\ 0 & 1 \\ 0 & 0 \end{bmatrix} \begin{bmatrix} a_1 \\ a_2 \end{bmatrix}
$$

(8.50)

The connection equations are

$$
\begin{bmatrix} a_1 \\ a_2 \\ y \end{bmatrix} = \begin{bmatrix} 0 & 0 & 1 \\ 1 & 0 & 0 \\ 0 & 1 & 0 \end{bmatrix} \begin{bmatrix} b_1 \\ b_2 \\ u \end{bmatrix}
$$

(8.51)

Let us find decentralized state feedback to assign the eigenvalues $\lambda_1 = -2$, $\lambda_2 = -1 -j$, $\lambda_3 = \bar{\lambda}_2$. This ordering of the eigenvalues associates λ_1 with component one and λ_2 and λ_3 with component two. Note also that component one does not have an even ordered characteristic polynomial and the eigenvalues will still be assignable by decentralized state feedback. The assignability property is always possible for arbitrary interconnections provided, at most, one component has an odd-ordered characteristic polynomial.

Given the above system model, the composite system F matrix is

$$F = \begin{bmatrix} 0 & 0 & 1 \\ 1 & 0 & 0 \\ 0 & 1 & 0 \end{bmatrix} \tag{8.52}$$

which has eigenvalues at 1, $-.5 \pm j.866$. Application of the root locus technique of VI.6 indicates that component one gives rise to the real eigenvalues and component two to the complex pair.

The structure of the decentralized state feedback matrix is

$$K_d = \left[\begin{array}{c|cc} k_1 & 0 & 0 \\ \hline 0 & k_2 & k_3 \end{array}\right] \tag{8.53}$$

After four cycles of the algorithm (equivalently eight iterations) the following decentralized feedback matrix results.

$$k_d = \left[\begin{array}{c|cc} -2.351 & 0 & 0 \\ \hline 0 & -1.646 & -2.124 \end{array}\right] \tag{8.54}$$

The resulting eigenvalues of the system are $\lambda_1 = -1.997, \lambda_2 = -.9999 - j.998$, and $\lambda_3 = \bar{\lambda}_2$. These are good approximations to the desired ones.

9. PROBLEMS

1. Derive Equation 2.13.
2. Determine which of the following matrices are *postive definite* or *positive semidefinite*.

$$M_1 = \begin{bmatrix} 1 & 1 \\ 1 & 1 \end{bmatrix} \quad M_2 = \begin{bmatrix} 1 & 1 \\ 1 & -1 \end{bmatrix} \quad M_3 = \begin{bmatrix} 1 & 1 \\ 1 & 0 \end{bmatrix}$$

3. Let M be a complex valued matrix which is equal to its own *hermitian conjugate* (i.e., complex conjugate transpose). Show that M has real eigenvalues.

4. Derive equation 2.23 from lemma VI.5.4.

5. Show that a matrix of the form $M = N^tN$ is square and symmetric.

6. Show that a matrix of the form $M = N^tN$ is positive definite if and only if the columns of N are linearly independent.

7. Show that the sum of positive semidefinite matrices is positive semidefinite. Show that the sum of positive semidefinite matrices, at least one of which is positive definite, is positive definite.

8. For the matrix

$$A = \begin{bmatrix} 2 & 1 & 1 \\ 0 & 4 & -1 \\ 0 & 1 & 0 \end{bmatrix}$$

write A^3 as a linear combination of A^2, A, and $A^0 = 1$.

9. Let S be a non-zero subspace of $\mathbf{R}^n$ and M be an n by n matrix which maps vectors in S into vectors in S. Then show that M has at least one eigenvector in S.

10. Determine which of the following pairs of A and B matrices represent *controllable systems*.

$$A = \begin{bmatrix} 1 & 1 \\ 1 & 1 \end{bmatrix} \quad B = \begin{bmatrix} 1 \\ 1 \end{bmatrix}$$

$$A = \begin{bmatrix} 1 & 1 \\ 1 & 1 \end{bmatrix} \quad B = \begin{bmatrix} 1 \\ -1 \end{bmatrix}$$

$$A = \begin{bmatrix} 1 & 1 \\ 1 & 1 \end{bmatrix} \quad B = \begin{bmatrix} 1 \\ 0 \end{bmatrix}$$

$$A = \begin{bmatrix} 0 & 1 & 0 & 0 & 0 & 0 & 0 \\ 0 & 0 & 1 & 0 & 0 & 0 & 0 \\ 0 & 0 & 0 & 1 & 0 & 0 & 0 \\ 0 & 0 & 0 & 0 & 1 & 0 & 0 \\ 0 & 0 & 0 & 0 & 0 & 1 & 0 \\ 0 & 0 & 0 & 0 & 0 & 0 & 1 \\ 2 & 0 & 0 & 1 & 0 & 0 & -1 \end{bmatrix} \quad B = \begin{bmatrix} 0 \\ 0 \\ 0 \\ 0 \\ 0 \\ 0 \\ 1 \end{bmatrix}$$

11. For the system

$$
\begin{bmatrix} x_1 \\ x_2 \end{bmatrix} = \begin{bmatrix} 0 & 2 \\ 1 & 4 \end{bmatrix} \begin{bmatrix} x_1 \\ x_2 \end{bmatrix} + \begin{bmatrix} 1 \\ 1 \end{bmatrix} a
$$

determine a control that will drive the initial state col(1,2) to the zero state.

12. Verify equation 2.86.

13. Transform each of the four systems described in problem 10 into *controllability canonical form*.

14. Let a controllable system be described by the first order state model

$$
\dot{x} = hx + ka
$$

and show that there always exists a feedback control law, a = fx + v, which makes the resultant system stable.

15. Transform the systems characterized by the following A and B matrices to *rational canonical form*.

$$
A = \begin{bmatrix} 1 & 1 \\ 1 & 1 \end{bmatrix} \quad B = \begin{bmatrix} 1 \\ 0 \end{bmatrix}
$$

$$
A = \begin{bmatrix} 1 & 1 \\ 1 & 1 \end{bmatrix} \quad B = \begin{bmatrix} 0 \\ 1 \end{bmatrix}
$$

$$
A = \begin{bmatrix} 1 & 0 & 0 & 0 \\ 2 & 1 & 0 & 0 \\ 3 & 0 & 1 & 0 \\ 4 & 0 & 0 & 1 \end{bmatrix} \quad B = \begin{bmatrix} 1 \\ 0 \\ 0 \\ 0 \end{bmatrix}
$$

16. Find a feedback control law which will stabilize the first system of problem 10.

17. Find a feedback control law for the last system of problem 10 which will place it eigenvalues at $\lambda_1 = -1, \lambda_2 = -2, \lambda_3 = -3, \lambda_4 = -4, \lambda_5 = -5, \lambda_6 = -6$, and $\lambda_7 = -7$.

18. Find feedback control laws which will stabilize the systems of problem 15.

19. Which of the systems of problem 10 are stabilizable?

20. For the systems characterized by the following A and C matrices which are *observable*?

$$A = \begin{bmatrix} 0 & 1 & 1 \\ 1 & -1 & 1 \\ 2 & 0 & 1 \end{bmatrix} \quad C = \begin{bmatrix} 1 & 1 & 0 \end{bmatrix}$$

$$A = \begin{bmatrix} 1 & 0 \\ 2 & 1 \end{bmatrix} \quad C = \begin{bmatrix} 0 & 1 \end{bmatrix}$$

$$A = \begin{bmatrix} 1 & 0 \\ 2 & 1 \end{bmatrix} \quad C = \begin{bmatrix} 0 & 1 \end{bmatrix}$$

21. Let the system

$$\begin{bmatrix} x_1 \\ x_2 \end{bmatrix} = \begin{bmatrix} 0 & 1 \\ 0 & 0 \end{bmatrix} \begin{bmatrix} x_1 \\ x_2 \end{bmatrix}$$

$$\begin{bmatrix} b \end{bmatrix} = \begin{bmatrix} 1 & 1 \end{bmatrix} \begin{bmatrix} x_1 \\ x_2 \end{bmatrix}$$

have output $b(t) = 2 + t$ in the interval $0 \leqslant t \leqslant 1$. Then what is its initial state?

22. Transform the system with the following A, B, and C matrices to *Kalman canonical form*.

$$A = \begin{bmatrix} 1 & 0 & 0 & 1 \\ 0 & 1 & 1 & 1 \\ 1 & 0 & 0 & 0 \\ 0 & 1 & 1 & 0 \end{bmatrix} \quad B = \begin{bmatrix} 1 \\ 0 \\ 0 \\ 0 \end{bmatrix} \quad C = \begin{bmatrix} 1 & 1 & 1 & 1 \end{bmatrix}$$

23. For the two systems described by the matrices

$$A = \begin{bmatrix} -1 & 1 \\ 0 & -1 \end{bmatrix} \quad B = \begin{bmatrix} 0 \\ 2 \end{bmatrix} \quad C = \begin{bmatrix} 1 & 2 \end{bmatrix}$$

$$\underline{A} = \begin{bmatrix} 0 & 1 \\ -1 & -2 \end{bmatrix} \quad \underline{B} = \begin{bmatrix} 0 \\ 2 \end{bmatrix} \quad \underline{C} = \begin{bmatrix} 3 & 2 \end{bmatrix}$$

find a matrix, T, such that $\underline{A} = T^{-1}AT$, $\underline{B} = T^{-1}B$, and $\underline{C} = CT$.

24. For the three systems of problems 22 and 23, which are *detectable*?

25. Compute the *fixed modes* for the systems of problems 22 and 23.

26. For a 2 by 2 partitioned matrix

$$M = \begin{bmatrix} M_{11} & M_{12} \\ M_{21} & M_{22} \end{bmatrix}$$

show that

$$\det(M) = \det(M_{11})\det(M_{22}+M_{21}(M_{11})^{-1}M_{12})$$

$$= \det(M_{22})\det(m_{11}+M_{12}(M_{22})^{-1}M_{21})$$

whenever the appropriate inverse matrices exist.

27. Derive a formula for the inverse of the 2 by 2 partitioned matrix M of problem 26 in terms of the inverses of its blocks.

28. Use the test of theorems 7.9 and 7.19 to determine if the system of figure 8.6 is controllable, observable, stabilizable, or detectable.

29. Which of the following matrices have full column rank?

$$M_1 = \begin{bmatrix} 1 & 1 & 0 \\ 0 & 1 & 0 \\ 1 & 0 & 1 \\ 0 & -1 & 1 \\ 1 & -1 & 2 \end{bmatrix} \quad M_2 = \begin{bmatrix} 0 & 0 & 1 & 2 & 0 \\ 2 & 0 & 1 & 2 & 0 \\ 2 & 0 & 1 & 2 & 0 \\ 1 & 0 & 1 & 2 & 1 \\ 0 & 1 & 1 & 2 & 1 \end{bmatrix}$$

$$M_3 = \begin{bmatrix} 1 & 1 & 1 & 1 & 1 & 1 & 1 & 1 & 1 \\ 2 & 3 & 4 & 5 & 7 & 8 & 9 & 1 & 0 \\ -1 & -1 & -1 & -1 & -1 & -1 & -1 & -1 & -1 \\ 0 & 0 & 1 & 0 & 0 & 0 & 0 & 0 & 0 \\ 2 & 3 & 4 & 0 & 0 & 0 & 0 & 0 & 0 \\ 2 & 3 & 4 & 5 & 0 & 0 & 0 & 0 & 0 \\ 0 & 1 & 3 & 4 & 5 & 7 & 8 & 9 & 1 \\ 1 & 3 & 5 & 7 & 9 & 7 & 5 & 3 & 1 \\ 0 & 0 & 1 & 0 & 0 & 0 & 0 & 1 & 1 \\ 2 & 0 & 2 & 2 & 2 & 2 & 2 & 2 & 2 \end{bmatrix}$$

30. Compute the rank for each of the matrices of the previous problem.
31. Compute the singular values for the matrix

$$
M = \begin{bmatrix}
1 & 1 & 0 & 0 & 0 & 0 & 0 & 0 \\
1 & 2 & 1 & 0 & 0 & 0 & 0 & 0 \\
0 & 1 & -1 & 3 & 0 & 0 & 0 & 0 \\
0 & 0 & 3 & 1 & 2 & 0 & 0 & 0 \\
0 & 0 & 0 & 2 & -2 & 0 & 0 & 0 \\
0 & 0 & 0 & -2 & 1 & 1 & 0 & 0 \\
0 & 0 & 0 & 0 & 1 & 1 & 1 & 0 \\
0 & 0 & 0 & 0 & 0 & 1 & 2 & 2
\end{bmatrix}
$$

32. How many positive eigenvalues does the matrix of problem 31 have?
33. Derive the composite system state equations for a *decentralized system* with component connection model

$$\dot{x} = Ax + Ba$$
$$b = Cx$$

$$a = L_{11}b + L_{12}u$$
$$y = L_{21}b$$

where L_{12} and L_{21} are block diagonal matrices whose partition is conformable with that of A and C.
34. Give an example of a system not having block diagonal B and C matrices for which $\phi_d(F,B,C) \neq \phi(F,B,C)$.
35. Show that theorem 8.16 is valid for the class of decentralized systems formulated in problem 33 where $\underline{B} = BL_{12}$ and $\underline{C} = L_{21}C$ replace B and C in the theorem.
36. Show that a system defined by equation 8.9 and 8.10 (without block diagonal B and C matrices) is stabilizable by a decentralized dynamic controller if and only if $\phi_d(F,B,C)$ contains only stable eigenvalues.
37. Design a decentralized compensator for the system of figure 8.6 which is capable of placing its eigenvalues at $\lambda = -1$ and $\lambda = -5$.
38. Design a dynamic compensator for the system of figure 8.6 which is capable of placing its eigenvalues at $\lambda = -2 \pm j$.

39. Show that if the system characterized by the state model

$$\dot{x} = Ax + Ba$$

is controllable then the *Riccati equation*

$$A^t P + PA - PBB^t P = -1$$

has a unique positive definite symetric solution.

10. REFERENCES

1. Anderson, B.D.O., and J. B. Moore, *Linear Optimal Control*, Prentice-Hall, Englewood Cliffs, 1971.
2. Bellman, R., *Introduction to Matrix Analysis*, McGraw-Hill, New York, 1970.
3. Davidson, E. J., and S. H. Wang, "New Results on the Controllability and Observability of General Systems", IEEE Trans. on Auto. Cont., Vol. AC-20, pp. 123-128, (1975).
4. Forsythe, G. E., Malcolm, M. A., and C. B. Moler, *Computer Methods for Mathematical Computations*, Prentice-Hall, Englewood Cliff, 1977.
5. Kalman, R. E., Ho, Y. C., and K. S. Nerendra, "Controllability of Linear Dynamical Systems" in *Contrib. to Diff. Equations* (ed. J. Hale), Academic Press, New York, pp. 189-213, 1962.
6. Saeks, R., "Decentralized Control of Interconnected Dynamical Systems", IEEE Trans. on Auto. Cont., Vol. AC-24, pp. 269-271, (1979).
7. Wang, S. H., and E. J. Davidson, "On the Stabilization of Decentralized Control Systems" IEEE Trans. on Auto. Cont. Vol. AC-18, pp. 473-478, (1973).
8. Wonham, W. M., *Linear Multivariable Control: A Geometric Approach*, Springer-Verlag, Heidelberg, 1974.

APPENDIX A. PALAIS' THEOREM

The purpose of this appendix is to supply the proofs for lemma III.7.7 and theorem III.7.8 on the existence of a global inverse function. The main theorem is a variation of a result originally due to Palais[2], although Palais' original and highly complex proof, is scrapped in light of a long but elementary argument due to Wu and Desoer.[3]

DEFINITION A.1: A function $f(\cdot):\mathbf{R}^n \to \mathbf{R}^n$ is a *local C^k-diffeomorphism* k $\geqslant 0$, at a point x if there exists an open neighborhood U of x and an open neighborhood V of f(x) such that (1) the restriction of $f(\cdot)$ to U is a one-to-one map of U onto V, and (2) $f(\cdot)$ restricted to U and $f^{-1}(\cdot)$ restricted to V have k continuous derivatives.

Of course, if f is of class C^k ($k \geqslant 1$), the classical *local inverse function theorem*[1] assures that $f(\cdot)$ is a local C^k-diffeomorphism at x if and only if $\det(J_f(x)) \neq 0$ where $J_f(x)$ is the *Jacobian matrix* of partial derivatives for $f(\cdot)$ evaluated at x.

Clearly for k = 0, no such test exists. By default a local C^0-diffeomorphism is a local *homeomorphism*. One final notion necessary for the discussion to follow is the notion of *proper function* by which we mean that for any *compact set* K (i.e., closed and bounded set in $\mathbf{R}^n$) $f^{-1}(K)$ is also compact.

LEMMA A.2: If $f(\cdot)$ is a continuous function mapping $\mathbf{R}^n$ into itself then the following statements are equivalent:
   (i) $f(\cdot)$ is proper,
   (ii) If B is a bounded set then $f^{-1}(B)$ is a bounded set
   (iii) $|f(x)| \to \infty$ as $|x| \to \infty$.

PROOF: First recall two facts of elementary *topology*. One is that a function is *continuous* if and only if the inverse image of *closed sets* are closed. The second is that a set in $\mathbf{R}^n$ is compact if and only if the set is *closed and bounded*.[1]

STEP 1: Show (i) implies (ii). Assume $f(\cdot)$ is proper and that B is a bounded set in $\mathbf{R}^n$. Let $\overline{B}$ be the *closure* of B. The set $\overline{B}$ must be compact since it is closed and bounded. Hence the definition of a proper function guarantees that $f^{-1}(\overline{B})$ is compact. In other words $f^{-1}(\overline{B})$ is closed and bounded. Since B is contained in $\overline{B}$, $f^{-1}(B)$ is contained in $f^{-1}(\overline{B})$. This assures that the inverse of the bounded set B is bounded.

STEP 2: Show (ii) implies (i). Here we must verify that the inverse image of compact sets are compact. Let K be a compact set. As such K is closed and bounded. By (ii), $f^{-1}(K)$ is bounded. Since $f(\cdot)$ is continuous, $f^{-1}(K)$ is also

467

closed. Therefore, $f^{-1}(K)$ is closed and bounded implying compactness. We conclude that $f(\cdot)$ is proper.

STEP 3: This step verifies the equivalence of (iii). First we demonstrate (ii) implies (iii). Assume (ii) is true, but that there exists a sequence of points x_i such that as $i \to \infty$, $x_i \to \infty$, but that $|f(x_i)| < 1M < \infty$. The points $f(x_i)$ then form a bounded set. But by (ii), the inverse image of this set must be bounded, contradicting the hypothesis that the sequence is unbounded. Thus, if (iii) fails then (ii) fails. Therefore, (ii) implies (iii).

Now we show (iii) implies (i). Assume $f(\cdot)$ is *not* proper, i.e., there exists a compact set K such that $f^{-1}(K)$ is not compact. Since K is compact it is closed and bounded. Since $f(\cdot)$ is continuous, $f^{-1}(K)$ is closed. Therefore $f^{-1}(K)$ is *not* bounded, but is closed. Consequently it is possible to pick a sequence $(x_i) \to \infty$ such that (x_i) and the accumulation point are in $f^{-1}(K)$. But this implies that as $|x_i| \to \infty$, then $|f(x_i)| < M < \infty$ which forces condition (iii) to fail. Therefore, (iii) implies (i). The proof is complete.

With the equivalences of this lemma we are more or less equipped to journey through the proof of the C^0-version of *Palais' Theorem.* the proof is long and burdensome but it offers exposure and understanding to many of the topological concepts frequenting system theory.

THEOREM A.3: Suppose $f(\cdot)$ is a continuous function such that $f(\cdot): \mathbf{R}^n \to \mathbf{R}^n$. Then $f(\cdot)$ is a *global homeomorphism* if and only if $f(\cdot)$ is *proper* and a *local homeomorphism* at each point.

PROOF:
STEP 1: The forward direction. Clearly if $f(\cdot)$ is a global homeomorphism it is a local homeomorphism. Moreover since $f(\cdot)$ is a homeomorphism, $f^{-1}(\cdot)$ is continuous. The continuous image of compact sets is compact. Therefore $f^{-1}(\cdot)$ maps compact sets to compact sets implying $f(\cdot)$ is proper.

STEP 2: The reverse direction. To show that $f(\cdot)$ is a global homeomorphism we must show (a) $f(\cdot)$ is onto, (b) $f(\cdot)$ is one-to-one, and (c) $f^{-1}(\cdot)$ is continuous. To show this we use the fact that the only *connected* open and closed sets in $\mathbf{R}^n$ are $\mathbf{R}^n$ itself and the null set. [3] Define $Y = f(\mathbf{R}^n)$. Clearly since $f(\cdot)$ is a local homeomorphism Y cannot be empty. Also Y must be connected since $f(\cdot)$ is continuous. We proceed to show Y is both open and closed implying that $Y = \mathbf{R}^n$.

Let $y = f(x)$ for some x. Since $f(\cdot)$ is locally homeomorphic there exists open neighborhoods, U and V, of x and y respectively such that $f(\cdot)$ maps U homeomorphically onto V. Clearly then V is an open neighborhood of y in the subspace Y under the relative topology induced by $\mathbf{R}^n$. Thus every point y in Y has an open neighborhood about it contained in Y. This says that Y is open.

To verify Y is closed, let y_1 be a sequence of points converging to y such that $y_i \in Y$. We must verify that y, the limit of the sequence, is in Y. This will mean that Y is closed. As such define

$$K = \{y, y_i \mid i = 1,2,..\} \tag{A.4}$$

The set K consists of the sequence y_i and the accumulation point of the sequence y_i. The set K is therefore compact since it consists of a countable sequence unioned with the accumulation point of the sequence.[3] Since $f(\cdot)$ is proper $f^{-1}(K)$ is also compact. Let x_i be a sequence of points in $f^{-1}(K)$. Since $f^{-1}(K)$ is compact x_i has a convergent subsequence, $x_{i_k} \to x$. Again compactness forces x to be in $f^{-1}(K)$.

Continuity of $f(\cdot)$ implies that $f(x_{i_k})$ converges to $f(x)$. By definition $f(x)$ is in Y, since Y is the image of $\mathbf{R}^n$ under $f(\cdot)$. Moreover

$$y = \lim_{k \to \infty} y_{i_k} = \lim_{k \to \infty} f(x_{i_k}) = f(x) \tag{A.5}$$

Thus y is in Y. This says that every convergent sequence whose elements are in Y converges to a point in Y. Therefore Y is closed.

In consequence the connected set Y is both open and closed forcing the equivalence

$$Y = \mathbf{R}^n \tag{A.6}$$

Part (b) will demonstrate that $f(\cdot)$ is one to one. This is the most grueling and, as happenstance, the longest part of the proof. The method of proof is by contradiction. We assume $f(\cdot)$ is not one to one. In other words, there are at least two distinct points, x_1 and x_2, in $\mathbf{R}^n$ such that $f(x_1)=f(x_2)=0$, where no generality is compromised by setting $f(x_1)=f(x_2)=0$ since $\mathbf{R}^n$ is a *topological vector space*. The proof proceeds by showing that the straight line segment connecting x_1 and x_2 must also map to the zero vector under $f(\cdot)$. This will contradict the fact that $f(\cdot)$ is a local homeomorphism. Thus, $f(\cdot)$ must be one to one. The reader is advised to keep the picture of A.7 in mind while reading through the proof.

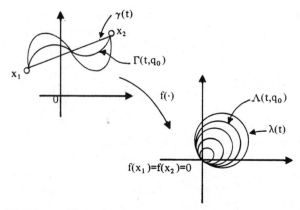

Figure A.7. Diagram illustrating the mappings for the one to one part of proof.

In this vein let $\gamma:[0,1] \to \mathbf{R}^n$ be a continuous map, defined as a curve whose image is the straight line segment between x_1 and x_2. The image of γ, denoted $[\gamma]$, is called the *trace* of γ. In addition, define $I = [0,1]$. Thus for t in I we have

$$\gamma(t) \;=\; (1-t)x_1 + t\, x_2 \tag{A.8}$$

Let us further define another curve, λ, as the composition of $f(\cdot)$ with γ, $\lambda = f \circ \gamma$. In particular, $\lambda:[0,1] \to \mathbf{R}^n$ as

$$\lambda(t) \;=\; f[(1-t)x_1 + t\, x_2] \tag{A.9}$$

Since $f(x_1) = f(x_2) = 0$ we necessarily have that $\lambda(0) = \lambda(1) = 0$. Now let us define another map, commonly called a *homotopy* in the literature, $\Lambda:I \times I \to \mathbf{R}^n$ as

$$\Lambda(t,q) \;=\; (1-q)\lambda(t) \tag{A.10}$$

Intuitively for each q, $\Lambda(\cdot,q)$ is a scaled shrinking of the curve $\lambda(\cdot)$. As q approaches one, (Λ) shrinks to zero.

Now we wish to fabricate a map $\Gamma:I \times I \to \mathbf{R}^n$ such that $\Lambda = f \circ \Gamma$. Obviously, define $\Gamma(t,0) = \gamma(t)$ for all t in I. Since $f(\cdot)$ is a local homeomorphism, there exists for each t_0 in I, open neighborhoods, U of $\Gamma(t_0,0)$ and V of $\Lambda(t_0,0)$, such that $f(\cdot)$ maps U homeomorphically onto V.

Thus it is possible and legitimate to define $\Gamma(t_0,q) = f^{-1}(\Lambda(t_0,q))$ for each q in the interval $[0,q_1]$ where $q_1 > 0$ and q_1 is chosen so that $\Lambda(t,q) \epsilon V$ for all q in the interval $[0,q_1]$. The same arguments permit us to define $\Gamma(t_0,q) = f^{-1}(\Lambda(t_0,q))$ for all q in a new interval $[q_1,q_2]$ since $f(\cdot)$ is also a local homeomorphism at q_1. Clearly then a sequential repetition of the argument admits the above definition of $\Gamma(t_0,q)$ for all q in $[0,1]$.

To substantiate this statement, suppose we could not define $\Gamma(t_0,q)$ over $[0,1]$. Let us reason to a contradiction. Define an increasing sequence of q_k's, $q_k < 1$ with $\Gamma(t_0,q) = f^{-1}(\Lambda(t_0,q))$ for each interval $[q_{k-1},q_k]$. Let

$$T = \lim_{k \to \infty} q_k = \sup_k q_k \leq 1 \qquad (A.11)$$

Thus, T becomes the endpoint of the largest interval over which $\Gamma(t_0,\cdot)$ is defined. As such $\Gamma(t_0,q)$ is well defined over $[0,T]$. The continuity of $\Lambda(\cdot,\cdot)$ implies that

$$\Lambda(t_0,T) = \lim_{k \to \infty} \Lambda(t,q_k) \qquad (A.12)$$

In addition since $f(\cdot)$ is proper the set $K = \{f^{-1}(\Lambda(t_0,q)) | q \text{ in } I\}$ is a compact set. Thus the sequence $\Gamma(t_0,q_k)$ is a sequence contained in the compact set K. Thus there exists a convergent subsequence $\Gamma(t_0,q_{k_j})$. In particular

$$\Gamma(t_0,T) = \lim_{j \to \infty} \Gamma(t_0,q_{k_j}) \qquad (A.13)$$

The continuity of $f(\cdot)$ then implies that

$$\begin{aligned}
\Lambda(t_0,T) &= \lim_{k \to \infty} \Lambda(t_0,q_k) \qquad (A.14) \\
&= \lim_{j \to \infty} \Lambda(t_0,q_{k_j}) \\
&= \lim_{j \to \infty} f(\Gamma(t_0,q_{k_j})) \\
&= f \lim_{j \to \infty} \Gamma(t_0,q_{k_j}) \\
&= f(\Gamma(t_0,T))
\end{aligned}$$

This string of equalities implies that $\Gamma(t_0,\cdot)$ is well defined over the closed interbal $[0,T]$. However by invoking the fact that $f(\cdot)$ is a local homeomorphism at T, there exists an $\epsilon > 0$ such that the domain of $\Gamma(t_0, \cdot)$ may be extended to the open interval $[0, T+\epsilon)$. This contradicts the assumption that T was the endpoint of the largest interval over which $\Gamma(t_0, \cdot)$ was defined. The conclusion is that $\Gamma(t_0, \cdot)$ may be legitimately defined over any finite interval, in particular over $I = [0,1]$.

Moreover, the above argument works for each t_0 in I. This implies that the function $\Gamma(\cdot, \cdot): I \times I \rightarrow \mathbf{R}^n$ is well defined and satisfies

$$\Lambda(t,q) \;=\; f(\Gamma(t,q)) \tag{A.15}$$

In addition, since $\Lambda(0,q) = \Lambda(1,q) = 0 = f(x_2)$, we conclude that $\Gamma(0,q) = x_1$ and $\Gamma(1,q) x_2$ for all q in I. By construction, $\Gamma(t,q)$ is continuous in q for each fixed t. It remains to prove that $\Gamma(t,q)$ is continuous for each fixed q in I.

Let G be an open set in $\mathbf{R}^n$. Let U_t and V_t be open neighborhoods of $\Gamma(t,q_0)$ and $\Lambda(t,q_0)$ respectively such that $f(\cdot)$ maps U_t homeomorphically onto V_t. Now define

$$W = \left[\bigcup_t U_t\right] \cap G = \bigcup_t \left[U_t \cap G\right] = \bigcup_t G_t \tag{A.16}$$

where G_t is open. Clearly the range of $\Gamma(\cdot, q_0)$ is contained in the union of the U_t. This implies that $\Gamma^{-1}(W, q_0) = \Gamma^{-1}(G, q_0) = B$. Hence it suffices to verify that B is open. Observing that $G_t = U_t \cap G$, and that $f(\cdot)$ is a homeomorphism of U_t onto V_t we have

$$f(W) = f\left(\bigcup_t G_t\right) = \bigcup_t f(G_t) \tag{A.17}$$

where $f(G_t)$ is open for each t. Therefore $f(W)$ is open. Since, *locally*, $\Gamma^{-1} = \Lambda^{-1}$ of, we have that $B = \Gamma^{-1}(W, q_0) = \Lambda^{-1}(f(W), q_0)$. Now Λ being continuous and $f(W)$ being open imply that B is open. Consequently $\Gamma(\cdot, q_0)$ is continuous for all q in I.

The above discussion shows that $\Lambda(t,1) = f(\Gamma(t,1) = 0$, for all t in I. But again $\Gamma(t,1)$ is some curve connecting x_1 to x_2. Therefore an entire line segment, $\Gamma(t,1)$, maps to zero. This contradicts the fact that $f(\cdot)$ is a local homeomorphism, i.e., $f^{-1}(0)$ must be a discrete set of points, not the trace of a continuous curve. So $f(\cdot)$ must be one-to-one. (c) Since $f(\cdot)$ is one-to-one and onto, $f^{-1}(\cdot)$ is well defined and onto. Since $f(\cdot)$ is a local homeomorphism, $f^{-1}(\cdot)$ is also a local homeomorphism. Thus $f^{-1}(\cdot)$ is globally continuous. In consequence $f(\cdot)$ is a global homeomorphism. The proof is complete.

REFERENCES

1. Fleming, W., *Functions of Several Variables,* Reading, Addison-Wesley, 1965.
2. Palais, R. S., "Natural Operations on Differential Forms," Trans. of the A.M.S., Vol. 92, pp. 125-141, (1959).
3. Wu, F.F., and C. A. Desoer, "Global Inverse Function Theorem," IEEE Trans. on Circuit Theory, Vol. CT-19, pp. 199-201, (1972).

APPENDIX B. FARKAS' LEMMA

The purpose of this appendix is to give a proof of the generalized Farkas' lemma V.5.49.

LEMMA B.1: Let v_i; $i=1,2, \cdots , m$; and v^i; $i=1,2, \cdots, r$ be two sets of n-vectors. Then if for any x in $\mathbf{R}^n$ the inequalities

$$x^t v_i \geqslant 0 \ ; \ i=1,2, \ldots ,m \tag{B.2}$$

imply at least one of the inequalities

$$x^t v^i \geqslant 0 \ ; \ i=1,2, \ldots ,r \tag{B.3}$$

then there exist non-negative scalars a_i; $i=1,2, \cdots, m$; and a^i; $i=1,2, \cdots, r$; such that

$$\sum_{i=1}^{m} a_i v_i \ = \ \sum_{i=1}^{r} a^i v^i \tag{B.4}$$

where

$$\sum_{i=1}^{r} a^i \ = \ 1 \tag{B.5}$$

PROOF: We will prove the lemma by induction on both n and r. In particular, if $n = 1$ and r is arbitrary we consider four cases. First, if any $v^i=0$, say v^k, B.3 holds for all x and we must verify B.4 and B.5. These equations are, however, trivially satisfied by $a_i=0$; $i=1,2, \cdots , m$, $a^i = 0$; $i=1,2 \cdots ,k-1, k+1, \cdots r$, and $a^k = 1$. As such, we may assume that the v^k are non-zero. Now, if all v_i are zero B.2 is satisfied for every x, positive or negative (remember for $n = 1$ the v^i, v_i, and x's are all scalars and hence may be characterized by sign), and hence for B. 3 to hold there must be at least one v^i which is positive, say v^k, (for B.3 to hold when x, is positive) and at least one v^i which is negative, say v^p, (for B.3 to hold when x is negative). Equations B.4 and B.5 are therefore satisfied by $a_i=0$; $i=1,2, \cdots ,m$; $a^i=0$; $i=1,2, \cdots ,k-1,k+1, \cdots ,p-1,p+1, \cdots ,r$; and

$$a^k \ = \ \frac{|v^p|}{|v^p|+v^k} \ , \quad a^p \ = \ \frac{v^k}{|v^p|+v^k} \tag{B.6}$$

As such, we may assume that the v^i are all non-zero and at least one v_i is non-zero. In particular, if there exist non-zero v^i and v_i with the same sign, say v^k and v_j equations, B.4 and B.5 are satisfied by $a_i=0$; $i=1,2, \cdots ,j-1,j,$ $\cdots,m$; $a^i=0$; $i=1,2, \cdots ,k-1,k+1, \cdots ,r$; $a^k=1$, and

$$a_j \;\; = \;\; \frac{v^k}{v_j} \tag{B.7}$$

Finally, the only remaining possibility is that all of the non-zero v_i have one sign and all of the (non-zero) v^i have the opposite sign in which case the hypotheses of the lemma will not be satisfied. The lemma therefore holds for $n=1$ and all r which we use to initiate an inductive proof for the remaining values of n and r.

Let us assume that the lemma has been verified for $n=1,2, \cdots,p-1$ and all values of r. Using this fact as an inductive hypothesis we will now verify the lemma for the case $n = p$ and $r = 1$ (the case $n = p$ and $r > 1$ will follow). For the case $r = 1$, the hypothesis of the lemma requires that

$$x^t v^1 \;\; = \;\; z \geqslant 0 \tag{B.8}$$

whenever

$$x^t v_i \geqslant 0 \;;i=1,2, \cdots , n \tag{B.9}$$

where the v_i, v^1, and x are p-vectors and z is a scalar. To be able to invoke our inductive hypothesis we desire to make a change of variable which reduces the problem by one dimension. Since the theorem is trivial if $v^1 = 0$(set $a_i=0,i=1,2, \cdots,m$ and $a^1 = 1$) we may assume that $v^1 \neq 0$ in which case B.8 may be expanded as

$$z \;\; = \;\; \sum_{j=1}^{p} x_j v_j^1 \tag{B.10}$$

in which at least one of the v_j^1 are non-zero say v_p^1. As such, we may solve B.10 for x_p via

$$x_p \;\; = \;\; \frac{1}{v_p^1} \, [z - \sum_{j=1}^{p-1} x_j v_j^1] \tag{B.11}$$

which may be substituted into B.9 to express our given inequalities in terms of the p–1 vector $\underline{x}^t = (x_1,x_2, \cdots ,x_{p-1})$ and the scalar z. Of course, these new expressions are still linear in their new variables and hence take on the form

$$\underline{x}^t \underline{v}_i + c_i z = x^t v_i \tag{B.12}$$

where $\underline{v}_1$ is an appropriate p–1 vector constructed from v_i and v^1. Of course, since B.12 is simply a restatement of B.9 our hypothesis implies that $z \geqslant 0$ if

$$\underline{x}^t \underline{v}_i + c_i z \geqslant 0 \; ; \; i=1,2, \cdots ,m \tag{B.13}$$

Now, the expressions in B. 12 may be classified into three groups according to the sign of c_i. In particular, upon reindexing these expressions we may write

$$\underline{x}^t \overline{v}_i = x^t v_i \; ; \; i=1,2, \cdots ,q \tag{B.14}$$

for those indices where $c_i = 0$.

$$\underline{x}^t \overline{v}_i + z = \frac{1}{|c_i|} \underline{x}^t v_i + z \; ; \; i = q+1,q+2, \cdots ,s \tag{B.15}$$

for those indices where $c_i > 0$ and

$$\underline{x}^t \overline{v}_i - z = \frac{1}{|c_i|} \underline{x}^t v_i - z \; ; \; i = s+1,s+2, \cdots ,m \tag{B.16}$$

for those indices where $c_i < 0$. Here $\overline{v}_i = \underline{v}_i$ for $i=,2, \cdots ,q$ and $\overline{v}_i = \underline{v}_i/|c_i|$ $i=q+1,q+2, \cdots ,m$.

Since equations B.14 through B.16 simply represent an alternative expression (and scaling by a positive factor) of B.9 the hypotheses of our lemma may be stated in terms of these expressions as follows: if

$$\underline{x}^t \overline{v}_i \geqslant 0 \; ; \; i=1,2, \cdots ,q \tag{B.17}$$

$$\underline{x}^t \overline{v}_i + z \geqslant 0 \; ; \; i=q+1,q+2, \cdots ,s \tag{B.18}$$

and

$$\underline{x}^t \overline{v}_i - z \geqslant 0 \; ; \; i=s+1,s+2, \cdots ,m \tag{B.19}$$

then $z \geqslant 0$. Note, these hypotheses imply that the set of equations B.18 are non-vacuous. Indeed, if B.17 and B.19 alone implied that $z \geqslant 0$ they would also imply that $\underline{x}^t \bar{v}_i \geqslant 0$; $i=s+1,s+2, \cdots ,m$ and and hence that B.17 and B.19 are satisfied when z was replaced with some arbitrary negative z. This, however, would lead to a contradiction because the hypothesis of the lemma would then imply that $z \geqslant 0$. Note, since v_p^1 is non-zero $\underline{x}$ and z are independent variables and it is possible to arbitrarily change z while holding $\underline{x}$ constant.

For the moment let us also assume that the set of equations B.19 is also non-vacuous and make an additional change of variable by adding equations B.18 and B.19 for all possible combinations of indices. Since the sum of positive numbers is positive this yields the set of inequalities

$$\underline{x}^t \bar{v}_i + \underline{x}^t \bar{v}_j \geqslant 0 \;;i=q+1,q+2, \cdots ,s \; ; j=s+1,s+2, \cdots ,m \quad (B.20)$$

We would like to use B.20 to replace B.19 in the hypotheses of our lemma and hence we desire to show that the inequalities B.17, B.18 and B.20 imply that $z \geqslant 0$.

First let

$$z_0 \;=\; - \; \min_{i=s+1}^{q} [\underline{x}^t \bar{v}_i] \qquad (B.21)$$

Moreover, since this minimum is achieved we may assume that $z_0 = -\underline{x}^t \bar{v}_k$ for some k between $s+1$ and q. Now, by setting $i=k$ in B.18 we obtain

$$z \geqslant -\underline{x}^t \bar{v}_k \;=\; z_0 \qquad (B.22)$$

while for all i between $s+1$ and q, B.18 implies that

$$\underline{x}^t \bar{v}_i + z_0 \;=\; \underline{x}^t \bar{v}_i - \underline{x}^t \bar{v}_k \geqslant 0 \;\; i=s+1,s+2, \cdots q \qquad (B.23)$$

whereas B.20 implies that

$$\underline{x}^t \bar{v}_i - z_0 \;=\; \underline{x}^t \bar{v}_i + \underline{x}^t \bar{v}_k \geqslant 0 \;\; i=q+1,q+2, \cdots ,m \qquad (B.24)$$

As such, B.17, B.18 and B.20 imply the positivity of B.17, B.18, and B.19 when z is replaced by z_0. The hypothesis of our lemma thus implies that z_0 is non-negative and hence B.22 implies that z is non-negative as was to be shown.

As a final step in preparation for the application of our inductive hypothesis we make the observation that if B.17, B.18, and B.20 imply that $z \geq 0$ then B.17 and B.20 alone imply that $\underline{x}^t \bar{v}_i \leq 0$ for some i between s+1 and q. Indeed, if $\underline{x}^t \bar{v}_i > 0$; i=s+1,s+2, $\cdots$,q the three equations would be satisfied using the negative number, z_o, which contradicts the hypothesis of our lemma. As such, we conclude that when the hypotheses of our lemma are satisfied the inequalities

$$\underline{x}^t \bar{v}_i \geq 0 \quad i=1,2, \cdots ,q \tag{B.25}$$

and

$$\underline{x}^t(\bar{v}_i + \bar{v}_j) \;=\; \underline{x}^t \bar{v}_i + \underline{x}^t \bar{v}_j \geq 0 \quad i=q+1, \cdots ,s; \; j=s+1, \cdots ,m \tag{B.26}$$

for any p-1 vector, x, imply that at least one of the inequalities

$$\underline{x}^t \bar{v}_i \leq 0 \quad i=q+1,q+2, \cdots ,s \tag{B.27}$$

holds. Module a minus sign this is, however, precisely our inductive hypothesis for the case n=p–1 and r = s–q. As such, there exists non-negative constants a_i; i=1,2, $\cdots$,q; a_{ij}; i=q+1, $\cdots$,s and j=s+1, $\cdots$,m; and a^i;i=q+1, $\cdots$,s; such that

$$\sum_{i=1}^{q} a_i \bar{v}_i + \sum_{i=q+1}^{s} \sum_{j=s+1}^{m} a_{ij}(\bar{v}_i + \bar{v}_j) \;=\; \sum_{i=q+1}^{s} -a^i \bar{v}_i \tag{B.28}$$

where

$$\sum_{i=q+1}^{s} a^i \;=\; 1 \tag{B.29}$$

Now, multiplying B.28 on the left by $\underline{x}^t$ and adding and subtracting z, appropriately, yields

$$\sum_{i=1}^{q} a_i [\underline{x}^t \bar{v}_i] + \sum_{i=q+1}^{s} \sum_{j=s+1}^{m} a_{ij}[(\underline{x}^t \bar{v}_i + z) + (\underline{x}^t \bar{v}_j - z)]$$

$$+ \sum_{i=q+1}^{s} a^i [\underline{x}^t \bar{v}_i + z] \;=\; z \tag{B.30}$$

with the aid of B.29 for all $\underline{x}$ and z. Now, substituting B.14 through B.16 into B.30 yields

$$\sum_{i=1}^{q} a_i[x^t v_i] + \sum_{i=q+1}^{s} \left[a^i + \sum_{j=s+1}^{m} a_{ij}]/|c_i| \right][x^t v_i] +$$

$$\sum_{j=s+1}^{m} \left[[\sum_{i=q+1}^{s} a_{ij}]/|c_j| \right] [x^t v_j] = [x^t v^t] \tag{B.31}$$

for all x after an appropriate rearrangement of terms. Finally, upon letting $\underline{a}_i = a_i$; $i=1,2, \cdots ,q$;

$$\underline{a}_i = [a_i + \sum_{j=s+1}^{m} a_{ij}]/|c_i| \; ; \; i=q+1,q+2, \cdots s \tag{B.32}$$

and

$$\underline{a}_i = [\sum_{j=q+1}^{s} a_{ji}]/|c_i| \; ; \; i=s+1,s+2, \cdots ,m \tag{B.33}$$

B.31 reduces to

$$\sum_{i=1}^{m} \underline{a}_i[x^t v_i] = x^t v^t \tag{B.34}$$

for all x or equivalently

$$\sum_{i=1}^{m} a_i v_i = v^t \tag{B.35}$$

as was to be shown.

Returning to the case where the set of equations B.19 is vacuous we note that precisely the same conclusion can be reached as in the above case. Indeed, if $\underline{x}^t \overline{v}_i \geqslant 0$; $i=1,2, \cdots ,q$ and each $\underline{x}^t \overline{v}_i > 0$; $i=q+1,q+2, \cdots ,s$; then B.17 and B.18 (with B.19 vacuous) would be satisfied by the negative number z_o which contradicts the hypothesis of our lemma. As such, in this case we conclude that B.17 implies that the conditions for our inductive hypothesis are satisfied and hence the same argument used in the previous case (with the a_{ij} terms deleted) will yield the desired equality for the case $n=p$ and $r=1$.

Given the validity of our lemma for the case $n=p-1$ and all r we have thus verified the lemma for the case of $n=p$ and $r=1$. Now, given the validity of the lemma in this case, we desire to verify the lemma for the case $n=p$ and all r. Note, even if we are only interested in the case $r=1$, as per the classical form of Farkas' lemma (V.5.46) it is necessary to verify the lemma for all r since the previous inductive argument required the validity of the result for the case $n=p-1$ and $r=s-q$ to verify the lemma for the case $n=p$ and $r=1$.

Since the lemma has been verified for the case $n=p$ and $r=1$ let us assume that the lemma has been verified for the case $n=p$ and $r=s-1$ and attempt to inductively verify the lemma for the case $n=p$ and $r=s$. First consider the case where whenever B.2 holds $x^t v^1 \geqslant 0$. Then applying the lemma in the special case where $r=1$ there exist non-negative coefficients $a_i ; i=1,2,\cdots,m$ such that

$$\sum_{i=1}^{m} a_1 v_i = v^1 = \sum_{i=1}^{s} a^i v^i \tag{B.36}$$

where $a^1 = 1$, and $a^1 = 0$; $i=2,3, \cdots ,s$. Since the theorem is trivial in this case we may without loss of generality assume that there exists an $\bar{x}$ such that

$$\bar{x}^t v_i \geqslant 0 \ i=1,2, \cdots ,m \tag{B.37}$$

and

$$\bar{x}^t v^1 < 0 \tag{B.38}$$

Now, consider the new set of $m+1$ inequalities

$$x^t v_i \geqslant 0 ; i=1,2, \cdots ,m \ \text{ and } \ -x^t v^1 \geqslant 0 \tag{B.39}$$

which we would like to show imply the validity of at least one of the $s-1$ inequalities

$$x^t v^i \geqslant 0 ; i=2,3, \cdots ,s \tag{B.40}$$

when the hypotheses of the lemma (i.e., B.2 implies B.3) are satisfied. If the $x^t v_i \geqslant 0 ; i=1,2, \cdots ,m$ and $-x^t v^1 > 0$ then $x^t v^1 < 0$ hence the hypotheses of the lemma imply that $x^t v^i \geqslant 0$ for at least one i between 2 and s (since we know that the inequality holds for some i and that it cannot hold for $i = 1$). In the remaining case where $x^t v_i \leq 0; i=1,2, \cdots, m$; and $x^t v^1 = 0$ we employ a limiting argument working with $x+c\bar{x}$ and taking a limit as c goes to zero from above. Indeed,

$$(x+c\bar{x})^t v_i \;=\; x^t v_i + c\bar{x}^t v_i \geq 0 \; ; \; i=1,2, \cdots, m \tag{B.41}$$

via B.37 and the given property of x. Similarly the fact that $x^t v^1 = 0$ and $\bar{x}^t v^1 > 0$ implies that

$$-(x+c\bar{x})^t v^1 = -x^t v^1 - c\bar{x}^t v^1 > 0 \tag{B.42}$$

since c is positive. The vector $(x+c\bar{x})$ thus satisfies the conditions of the previously discussed case hence the hypotheses of the lemma imply that at least one of the inequalities

$$(x+c\bar{x})^t v^i \;=\; x^t v^i + c\bar{x}^t v^i \geq 0; \; i=2,3, \cdots, s \tag{B.43}$$

must hold for $c \geq 0$. Finally upon taking the limit as c goes to zero from above B.43 reduces to

$$x^t v^i \geq 0 \; ; i=2,3, \cdots, s \tag{B.44}$$

for at least one index. As such, the validity of our given hypothesis of equations B.2 and B.3 implies the validity of the hypothesis that the set of inequalities B.39 implies at least one of the inequalities of equation B.40. Since our lemma is assummed to hold for n=p and r=s–1 this, however, implies the existence of a set of non-negative coefficients $a_i; i=1,2,\cdots,m$; and $a^i; i=1,2,\cdots,s$; such that

$$\sum_{i=1}^{m} a_i v_i - a^1 v^1 \;=\; \sum_{i=2}^{s} a^i v^i \tag{B.45}$$

and

$$\sum_{i=2}^{m} a^i \;=\; 1 \tag{B.46}$$

Finally, rearranging terms in B.45 and scaling yields

$$\sum_{i=1}^{m} \underline{a}_i v_i = \sum_{i=1}^{s} \underline{a}^i v_i \qquad (B.47)$$

and

$$\sum_{i=1}^{s} \underline{a}^i = 1 \qquad (B.48)$$

where

$$\underline{a}_i = a_i / [\sum_{i=1}^{s} a^i] \; ; i=1,2, \cdots ,m \qquad (B.49)$$

and

$$\underline{a}^i = a^i / [\sum_{i=1}^{s} a^i] \; ; i=1,2, \cdots ,s \qquad (B.50)$$

Thus the validity of the lemma for the case n=p and r=s-1 implies its validity for the case n=p and r=s completeing our inductive proof.

BIBLIOGRAPHY

1. SYSTEM THEORY

Anderson, B. D. O., "A System Theory Criterion for Positive Real Matrices", SIAM Jour. Control, Vol. 5, pp. 171-182, (1967).

Barman, J. F., and J. Katznelson, "A Generalized Nyquist-Type Stability Criterion for Multivariable Systems", Inter. Jour. on Cont., Vol. 20, pp. 593-622, (1974).

Brockett, R. W., "The Status of Stability Theory for Deterministic Systems", IEEE Trans. on Auto. Cont., Vol. AC-11, pp. 596-606, (1966).

Brockett, R. W., *Finite Dimensional Linear Systems*, Wiley, New York, 1970.

Chen, C. T., *Introduction to Linear System Theory*, Holt, Rinehart and Winston, New York, 1970.

Cooper, G. R., and C. D. McGillem, *Probalistic Methods in Signal and System Analysis*, Holt, Rinehart, and Winston, New York, 1971.

DeCarlo, R. A., Ph.D. Thesis, Texas Tech University, Aug. 1976.

DeCarlo, R. A., and R. Saeks, "The Encirclement Condition: An approach Using Algebraic Topology", Inter, Jour. on Cont. Vol. 26, pp. 279-287, (1977).

Desoer, C. A., and M. Vidyasagar, *Feedback Systems: Input Output Properties*, Academic Press, New York: 1975.

Desoer, C. A., "A Generalization of the Popov Criterion", IEEE Trans. on Auto. Cont., Vol. AC-10, pp. 182-184, (1965).

Desoer, C. A., *Notes for a Second Course on Linear Systems*, Van Nostrand, New York, 1970.

Ho, B. L., and R. E. Kalman, "Effective Construction of Linear State-Variable Models from Input Output Functions", Regelungstechnik, Vol. 14, pp. 545-548, (1966).

Kalman, R. E., "Canonical Structure of Linear Dynamical Systems", Proc. of the Nat., Acad. Sci., Vol. 48, pp. 596-600, (1962).

Kalman, R. E., "Mathematical Description of Linear Dynamical Systems", SIAM Jour. on Cont., Ser A, Vol. 1, pp. 152-192, (1963).

Koenig, H. E., Tokad, Y., Kesavan, H. K., and H. G. Hodges, *Analysis of Discrete Physical Systems*, McGraw-Hill, New York, 1967.

Kuh, E. S. and R. A. Rohrer, *Theory of Linear Active Networks*, Holden-Day, San Francisco, 1967.

LaSalle, J. P., and S. Lefschetz, *Stability by Liapunov's Direct Method*, Academic Press, New York, 1964.

MacFarlane, A. G. J., and H. H. Rosenbrock, "New Vector-Space Structure for Dynamical Systems", Electronics Letters, Vol. 6, pp. 162-163, (1970).

McMillian, B., "Introduction to Formal Realizability Theory", Bell System Tech. Jour., Vol. 31, pp. 217-279, 541-600, (1952).

Porter, B., *Synthesis of Dynamical Systems*, Nelson, London, 1969.

Rosenbrock, H. H., "On Linear System Theory", Proc. of the IEE (London), Vol. 114, pp. 1353-1359, (1967).

Rosenbrock, H. H., and C. Storey, *Mathematics of Dynamical Systems*, Nelson, London, 1970.

Rosenbrock, H. H., *State Space and Multivariable Theory*, Nelson, London, 1970.

Saeks, R., *Generalized Networks*, Holt, Rinehart, and Winston, New York, 1972.

Sain, M. K., and J. L. Massey, "Invertibility of Linear Time-Invariant Dynamical Systems", IEEE Trans. on Auto. Cont., Vol. AC-14, pp. 141-149, (1969).

Sandberg, I. W., "Some Stability Results Related to Those of V. M. Popov", Bell Syst. Tech. Jour., Vol. 44, pp. 2133-2148, (1965).

Sandberg, I. W., "Stability of Feedback Systems Containing a Single Time-Varying Nonlinear Element", Bell Syst. Tech. Jour., Vol. 43, pp. 1601-1608, (1964).

Silverman, L. M., "Inversion of Multivariable Linear Systems", IEEE Trans. on Auto. Cont., Vol. AC-14, pp. 270-276, (1969).

Willems, J. C., *The Analysis of Feedback Systems*, MIT Press, Cambridge, Mass., 1971.

Wolovich, W. A., *Linear Multivariable Systems*, Springer Verlag, Heidelberg, 1974.

Zadeh, L., and C. A. Desoer, *Linear System Theory*, McGraw-Hill, New York, 1963.

2. SYSTEM MODELING

Anderson, J. H., "Geometrical Approach to the Reduction of Dynamical Systems", Proc. of the IEE (London), Vol. 114, pp. 1014-1018, (1967).

Chen, C. F., and L. S. Shieh, "A Novel Approach to Linear Model Simplifications", Inter. Jour. on Cont., Vol. 8, pp. 561-570, (1968).

Chen, C. F., "Model Reduction of Multivariable Control Systems by Means of Matrix Continued Fractions", Preprints 5th IFAC symposium, pp. 35.1, 35.8, 1972.

Chen, M. S., and W. E. Dillon, "Power System Modeling", IEEE Proc., Vol. 62, pp. 901-905, (1974).

Chien, M. J., "Piecewise Linear Modeling and Simulations", Proc. of the IEEE Inter. Symp. on Circuits and Systems, pp. 81-84, 1977.

Davison, E. J., "A New Method for Simplifying Large Linear Dynamic System", IEEE Trans. on Auto. Cont., Vol. AC-13, pp. 214-215, (1968).

Davison, E. J., "A Method for Simplifying Linear Dynamic Systems", IEEE Trans. on Auto. Cont., Vol. AC-11, pp. 93-101, (1966).

Debs, A. S., and G. Contaxis, "System Identification Approach to External System Equivalents", Proc. of the IEEE Inter. Symp. on Circuits and Systems, pp. 829-838, 1977.

Gibilaro, L. G., and F. P. Lees, "The Reduction of Complex Transfer Function Models Using the Method of Moments", Chem. Eng. Sci., Vol. 24, pp. 85-93, (1969).

Graupe, D., *Identification of Systems*, Van Nostrand Reinhold, New York, 1972.

Hsiah, T. C., "On the Simplification of Linear Systems", IEEE Trans. on Auto. Cont., Vol. AC-17, pp. 372-374, (1972).

Hutton, M. F., Ph.D. Dissertation, Polytech. Inst. of New York, June 1974.

Lal, M., Singh, H., and R. Parthasarathy, "A Minimal Canonical Realization Algorithm for Impulse Response Matrix using Moments", IEEE Proc., Vol. 63, pp. 538-530, (1975).

Lal, M., and R. Mitra, "A Comparison of Transfer Function Simplification Methods", IEEE Trans. on Auto. Cont., Vol. AC-19, pp. 603-604, (1974).

Lal, M., and R. Mitra, "Simplification of Large System Dynamics using a Moment Evaluation Algorithm", IEEE Trans. on Auto. Cont., Vol. AC-19, pp. 603-603, (1974).

Lal, M., and M. E. Van Valkenburg, "Reduced-Order Modeling of Large-Scale Linear Systems", in *Large-Scale Dynamical Systems* (ed. R. Saeks), Point Lobos Press, No. Hollywood, pp. 127-166, 1976.

Lamba, S. S., and S. V. Rao, "A New Frequency Domain Technique for the Simplification of Linear Dynamic Systems", Inter. Jour. on Cont., Vol. 20, pp. 71-79, (1974).

Levy, E. C., "Complex Curve Fitting", IRE Trans. on Auto. Cont., Vol. AC-4, pp. 37-43, (1959).

Liu, R., Chang, Y. C., Kim, K. S., and L. C. Suen, "An Application of the MDR Method to Modeling of Gasoline Consumption of United States", in *Large-Scale Dynamical Systems* (ed. R. Saeks), Point Lobos Press, No. Hollywood, pp. 113-126, 1976.

Liu, R., and L. C. Suen, "The Dynamic Impact on U. S. Economy: Fiscal Policy vs. Monetary Policy", Proc. of the IEEE Inter. Symp. on Circuits and Systems, pp. 135-138, 1975.

Liu, R., Suen, L. C., and C. H. Chiu, "The MDR Method and Its Applications to Socio-Economic Systems", Proc. of the International Computer Symposium, Taipei, Republic of China, pp. 240-253, 1975.

Liu, R., and L. C. Suen, "Minimal Dimension Realization and Identifiability of Input/Output Sequences", EE Memo., Univ. of Notre Dame, April, 1975

Rosenbrock, H. H., "Efficient Computation of Least Order for a Given Transfer Function", Electronics Letters, Vol. 3, pp. 413-414, (1967).

Rosenbrock, H. H., "Computation of Minimal Representations of a Rational Transfer-Function Matrix", Proc. of the IEE (London), Vol. 155, pp. 325-327, (1968).

Sage, A. P., and J. L. Melsa, *System Identification*, Academic Press, New York, 1971.

Shamash, Y., "Model Reduction Using the Routh Stability Criterion and the Pade Approximation Technique", Inter. Jour. on Cont., Vol. 21, pp. 475-484, (1975).

Shamash, Y., "Stable Reduced-Order Models using Pade-Type Approximations", IEEE Trans. on Auto. Cont., Vol. AC-19, pp. 615-616, (1974).

Shamash, Y., "Linear System Reduction Using Pade Approximation to Allow Retention of Dominant Modes", Inter. Jour. on Cont., Vol. 21, pp. 257-272, (1975).

Shamash, Y., Ph.D. Dissertation, Imperial College of Science and Tech., London, U.K., 1973.

Shieh, L. S., and F. F. Gaudiano, "Some Properties and Applications of Matrix Continued Fraction", Proc. of the Twelfth Allerton Conf. on Circuits and Systems, pp. 957-966, 1974

Shieh, L. S., and R. Goldman, "A Mixed Cauer Form for Linear System Reduction", IEEE Trans. on Sys., Man, and Cyber., Vol. SMC-4, pp. 584-588, (1974).

Sinha, N. K., "A New Method of Reduction of Dynamic Systems", Inter. Jour. on Cont., Vol. 14, pp. 111-118, (1971).

Van Ness, J. E., Zimmer, H., and M. Cultu, "Reduction of Dynamic Models of Power Systems", Proc. of the PICA Conf., pp. 105-112, 1973.

Wu, F. F., and N. Narasmihamurthi, "A New Algorithm for Modal Approach to Reduced-Order Modeling", Proc. of the IEEE Inter. Symp. on Circuits and Systems, pp. 583-585, 1977.

3. SYSTEM STRUCTURE

Bailey, F. N., "Decision Processes in Organization", in *Large-Scale Dynamical Systems* (ed. R. Saeks), Point Lobos Press, No. Hollywood, pp. 81-112, 1976.

Branin, F. H., "The Relation between Kron's Method and the Classical Methods of Network Analysis", WESCON Convention Record, Part 2, pp. 1-29, 1959.

Bunch, J. R., and D. J. Rose, "Partitioning, Tearing, and Modification of Sparse Linear Systems", Cornell Univ. Dept. of Computer Science, Tech. Rpt. TR72-149, Nov., 1972.

Frank, H., (ed.), "Special Issue on Large-Scale Networks", IEEE Trans. on Circuit Theory, Vol. CT-20, (1973).

Frank, H., and S. L. Hakimi, *The Workshop on Large-Scale Networks*, NSF Workshop Report, 1974.

Ghausi, M., (ed.), "Special Issue on Large-Scale Systems", IEEE Trans. on Circuits and Systems, Vol. CT-23, (1976).

Happ, W. W., "Flowgraphs for Closed Systems", IEEE Trans. on Aerospace and Electronic Systems, Vol. AES-2, pp. 252-254, (1966).

Harary, F., "Sparse Matrices and Graph Theory", in *Large Sparse Sets of Linear Equations*, (ed. J. K. Reid), Academic Press, New York, pp. 139-150, 1971.

Harrison, B. K., "Large-Scale Linear Systems", in *System Theory*, McGraw Hill, New York, pp. 279-312, 1969.

Higgins, T. J., "Modern Aspects of Large Scale System Science", Jour. of the Franklin Institute, Vol. 286, pp. 553-660, (1968).

Ho, Y. C., and K. C. Chu, "Information Structure in Dynamic Multi-Person Control Problems", Automatica, Vol. 10, pp. 341-351, (1974).

Kazda, L. F., and L. A. Zadeh, *New Directions in System Science and Engineering*, NSF Workshop Report, 1972.

Kevorkian, A. K., "A Decomposition Algorithm for the Solution of Large Systems of Linear Algberaic Equations", Proc. of the IEEE Inter. Symp. on Circuits and Systems, pp. 116-120, 1975.

Kevorkian, A. K., and J. Snoek, "Decomposition in Large-Scale Systems: Theory and Application of Structural Analysis in Partitioning, Disjointing and Constructing Hierarchical Systems", in *Decomposition of Large Scale Problems*, (ed. D. M. Himmelblau), North Holland/ American Elsevier, Amsterdam, 1973.

Kron, G., *Diakoptics*, McDonald, London, 1963.

Liberty, S. R., and R. Saeks, "The Component Connection Model in System Identification, Analysis and Design", Proc. of the 1st Nuclear EMP Meeting, Kirtland AFB, New Mexico, Part 6, pp. 61-76, 1973.

Liu, R. W., "An Approach to Large-Scale Dynamical Systems", Proc. of the 7th Asilomar Conf. on Circuits, Systems and Computers, pp. 89-92, 1973.

Liu, R. W., Singh, S. P., and F. Lu., "Large-Scale Dynamical Systems", Proc. of the 5th Asilomar Conf. on Circuits and Systems, pp. 89-92, 1971.

Mesarovic, M. D., Macko, D., and Y. Takarhara, *Theory of Hierarchical, Multilevel Systems*, Academic Press, New York, 1970.

Marimont, R. B., "System Connectivity and Matrix Properties", Bull. Math. Biophs., Vol. 31, pp. 255-274, (1969).

Ogbuobiri, E. C., Tinney, W. F., and J. W. Walker, "Sparsity-Directed Decomposition for Gaussian Elimination on Matrices", IEEE Trans. on Power Apparatus and Systems, Vol. PAS, 89, pp. 141-155, (1970).

Ozguner, U., and W. R. Perkins, "Structural Properties of Large-Scale Composite Systems", in *Large-Scale Dynamical Systems* (ed. R. Saeks), Point Lobos Press, No. Hollywood, pp. 5-32, 1976.

Ozguner, U., and W. R. Perkins, "Graph Theory in the Analysis of Large Scale Composite Systems", Proc. of the IEEE Inter. Symp. on Circuits and Systems, pp. 121-123, 1975.

Ozguner, U., and W. R. Perkins, "On the Multi-Level Structure of Large Scale Composite Systems", IEEE Trans. on Circuits and Systems, Vol. CAS-22, pp. 618-621, (1975).

Ransom, M. N., and R. Saeks, "The Connection Function—Theory and Application", Inter. Jour. of Circuit Theory and its Applications, Vol. 3, pp. 5-21, (1975).

Ransom, M. N., Ph.D. Thesis, Univ. of Notre Dame, Notre Dame, Ind., 1973.

Ransom, M. N., and R. Saeks, "A Functional Approach to Large-Scale Dynamical Systems", Proc. of the 10th Allerton Conf. on Circuits and Systems, pp. 48-55, 1972.

Rose, D. J., and J. R. Bunch, "The Role of Partitioning in the Numerical Solution of Sparse Systems", in *Sparse Matrices and their Applications*, (eds, D. J. Rose and R. A. Willoughby), Plenum Press, New York, pp. 177-190, 1972.

Rosenbrock, H. H., and A. C. Pugh, "Contributions to a Hierarchical Theory of Systems", Inter. Jour. on Cont., Vol. 19, pp. 845-867, (1974).

Saeks, R., *Large-Scale Dynamical Systems*, Point Lobos Press, No. Hollywood, California, 1976.

Saeks, R., Wise, G., and K. S. Chao, "Analysis and Design of Interconnected Dynamical Systems", in *Large-Scale Dynamical Systems* (ed. R. Saeks), Point Lobos Press, No. Hollywood, pp. 59-80, 1976.

Saeks, R., and S. R. Liberty, "The Component Connection Model in Circuits and Systems", Proc. of the 1st Eurpoean Conf. on Circuit Theory and Design, pp. 141-146, 1973, (IEE London) Conf. Pub. 116).

Sandberg, I. W., "Linear Multiloop Feedback Systems", Bell Systems Tech. Jour., Vol. 42, pp. 116-126, (1961).

Sangiovanni-Vincentelli, A., Chen, L. K., and L. Chua, "A New Tearing Approach—Node-Tearing Nodal Analysis", Proc. of the IEEE Inter. Symp. on Circuits and Systems, pp. 143-147, 1977.

Siljak, D. D., Weissenberger, S., and S. M. Cuk, "Decomposition-Aggregation Stability Analysis", NASA Contract Report, No. 2196, 1973.

Singh, S. P., Ph.D. Thesis, University of Notre Dame, Notre Dame, Ind., 1972.

Singh, S. P., and R. W. Liu, "Irreducibility of a Class of Large-Scale Dynamical Systems", Proc. of the IEEE Inter. Symp. on Circuits and Systems, pp. 124-126, 1975.

Singh, S. P., and R. W. Liu, "Existence of State Equation Representation of Linear Large-Scale Dynamical Systems", IEEE Trans. on Circuit Theory, Vol. CT-20, pp. 239-246, (1973).

Steward, D. V., "Partitioning and Tearing Systems of Equations", SIAM Jour. on Numer. Anal., Series B., Vol. 2, pp. 345-365, (1965).

Steward, D. V., "Tearing Analysis of the Structure of Disorderly Sparse Matrices", in *Sparse Matrix Proceedings*, (ed. R. A. Willoughby), IBM, Yorktown Heights, Rep. No. RA1 (11707), pp. 65-74, 1969.

Tarjan, R., "Depth-First Search and Linear Graph Algorithms", SIAM Jour. on Comput., Vol. 1, pp. 146-160, (1972).

4. SIMULATION

Baty, J. P., and K. L. Stewart, "Organization of Network Equations using Disection Theory", in *Large Sparse Sets of Linear Equations*, (ed. J. K. Reid), Academic Press, New York, pp. 169-190, 1971.

Berry, R., "An Optimal Ordering on Electronic Circuit Equations for Sparse Matrix Solution", IEEE Trans. on Circuit Theory, Vol. CT-18, pp. 40-50, (1971).

Branin, F. H., "A Unifying Approach to the Classical Methods for Formulating Network Equations", Unpublished Notes, Univ. of Waterloo, 1973.

Branin, F. H., "Computer Methods of Network Analysis", IEEE Proc., Vol. 55, pp. 1787-1801, (1967).

Branin, F. H., Hogsett, C. R., Lund, R. L., and L. E. Kugel, "ECAP II-A New Electronic Circuit Analysis Program", IEEE Jour. on Solid State Circuits, Vol. SC-6, pp. 146-166, (1971).

Bryant, P. R., "The Order of Complexity of Electrical Networks", Proc. of the IEE (London), Vol. 106, pp. 174-188, (1959).

Carre, B. A., "The Partitioning of Network Equations for Block Iterations", Comput. J., Vol. 9, pp. 84-97, (1966).

Chandrashekar, M., and H. K. Kesavan, "Graph Theoretical Models for the Piecewise Analysis of Large-Scale Electrical Networks", Unpublished Notes, Univ. of Waterloo, 1974.

Chua, L. O., and P. M. Lin, Computer-Aided Analysis of Electronic Circuits, Prentice-Hall, Englewood, Cliffs, 1975.

Chua, L. O., and Y. F. Lam, "A Theory of Algebraic n-Ports", IEEE Trans. on Circuit Theory, Vol. CT-20, pp. 370-382, (1973).

Chua, L. O., "Efficient Computer Algorithm for Piecewise Linear Analysis of Resistive Nonlinear Network", IEEE Trans. on Circuit Theory, Vol. CT-18, pp. 73-85, (1971).

Davison, E. J., "An Algorithm for the Computer Simulation of Very Large Dynamic Systems", Automatica, Vol. 9, pp. 665-675, (1973).

Davison, E. J., and W. G. Gross, "An Algorithm for the Simulation of Very Large Dynamic Systems", in Large-Scale Dynamical Systems, (ed. R. Saeks), Point Lobos Press, No. Hollywood, pp. 239-256, 1976.

Dembart, D., and A. M. Erisman, "Hybrid Sparse Matrix Methods", IEEE Trans. on Circuit Theory, Vol. CT-20, pp. 641-649, (1973).

Desoer, C. A., and E. S. Kuh, Basic Circuit Theory, McGraw-Hill, New York, 1969.

Director, S. W., "A New Class of Algorithm for Solving Non-linear Circuit Equations", Proc. of the IEEE Inter. Symp. on Circuits and Systems, pp. 77-80, 1977.

Erisman, A. M., "Sparse Matrix Approach to the Frequency Domain Analysis of Linear Passive Electrical Networks", in Sparse Matrices and Their Application, (eds. D. J. Rose and R. A. Willoughby), Plenum Press, New York, pp. 31-40, 1969.

Erisman, A. M., and G. Spies, "Exploiting Problem Characteristic in Sparse Matrix Approach to Frequency Domain Analysis", IEEE Trans. on Circuit Theory, Vol. CT-19, pp. 260-264, (1972).

Fujisawa, T., and E. S. Kuh, "Some Results on the Existence and Uniqueness of Solutions of Nonlinear Networks", IEEE Trans. on Circuit Theory, Vol. CT-18, pp. 501-506, (1971).

Fujisawa, T., and E. S. Kuh, "Piecewise-Linear Theory of Nonlinear Networks", SIAM Jour. on Appl. Math., Vol. 22, pp. 307-328, (1972).

Fujisawa, T., Kuh, E., and T. Ohtsuki, "A Sparse Matrix Method for Analysis of Piecewise-Linear Resistive Networks", IEEE Trans. on Circuits Theory, Vol. CT-19, pp. 571-584, (1972).

Greenbaum, J. R., "A Library of Circuit Analysis Programs", Circuits and Systems, Vol. 7, pp. 4-10, (1974).

Hachtel, G., Brayton, R. K., and F. Gustavson, "The Sparse Tableau Approach to Network Analysis and Design", IEEE Trans. on Circuit Theory, Vol. CT-18, pp. 101-113, (1971).

Hachtel, G., Gustavson, R., Brayton, R., and T. Grapes, "A Sparse Matrix Approach to Network Analysis", Proc. of the Cornell Conf. Computerized Electron, 1969.

Hajj, I. N., "Updating Methods for LU Factorization", Electronics Letters,Vol. 8, pp. 186-188, (1972).

Katzenelson, J., "An Algorithm for Solving Resistive Networks", Bell, Sys. Tech. Jour., Vol. 44, pp. 1605-1620, (1965).

King, I. P.,"An Automatic Reordering Scheme for Simultaneous Equations Derived from Network Analysis", Inter. Jour. Numer. Methods Eng., Vol. 2, pp. 523-533, (1970).

Knox, B., M.S. Thesis, Texas Tech University, Lubbock, Texas, 1974.

Knox, B., and R. Saeks, "A Componentwise Adaptive Step-size Algorithm for the Simulation of LSDS", Proc. of the 8th Asilomar Conf. on Circuits, Systems, and Computers, pp. 580-583, 1974.

Kou, F. F., and J. F. Kaiser, *Systems Analysis by Digital Computer*, Wiley, New York, 1966.

Kuh, E. S., and I. Hajj, "Nonlinear Circuit Theory—Resistive Networks", IEEE Proc. Vol. 59, pp. 340-355, (1971).

MaCalla, W. J., and D. O. Pederson, "Elements of Computer-Aided Circuit Analysis", IEEE Trans. on Circuit Theory, Vol. CT-18, pp. 14-26, (1971).

McNamme, L. P., and N. Potash, "A Users and Programmers Manual for NSASP", Tech. Report 68-38, UCLA, 1968.

Narin, R. S., and C. Pottle, "Effective Ordering of Sparse Matrices Arising from Nonlinear Networks", IEEE Trans. on Circuit Theory, Vol. CT-18, pp. 139-145, (1971).

Ohtsuki, N., and L. Cheung, "A Matrix Decomposition-Reduction Procedure for Pole-Zero Calculation of Transfer Functions", IEEE Trans. on Circuit Theory, Vol. CT-20, pp. 262-271, (1973).

Pinel, J. F., and M. Blostein, "Computer Techniques for the Frequency Analysis of Linear Electrical Networks", IEEE Proc., Vol. 55, pp. 1810-1819, (1967).

Pottle, C., "Comprehensive Active Network Analysis by Digital Computer-A State-Space Approach", Tech. Rpt. EERL 59, Cornell Univ., Ithaca, New York, 1966.

Prasad, N., "Graph Theoretic and Combinatorial Algorithms in Digital System Simulation", Proc. of the 7th Asilomar Conf. on Circuits, Systems and Computers, pp. 29-32, 1973.

Prasad, N., Smith, J., Reiss, J., and S. Robert, "MARSYAS—A New Language for the Digital Simulation of Systems", Research Rpt., Computer Applications Inc., 1969.

Ransom, M. N., "On the State Equations of RLC Networks", Tech Memo. EE 7214, Univ. of Notre Dame, Notre Dame, Ind., 1972.

Russo, P. M., "On the Time-Domain Analysis of Linear Time-Invariant Networks with Large Time-Constant Spreads by Digital Computer", IEEE Trans. on Circuit Theory, Vol. CT-18, pp. 194-197, (1971).

Schichman, H., "Integration System of a Nonlinear Network Analysis Program", IEEE Trans. on Circuit Theory, Vol. CT-17, pp. 378-386, (1970).

Singh, S. P., "Serpentuator Simulation Using MARSYAS", Proc. of the 7th Asilomar Conf. on Circuits, Systems and Computer, pp. 33-37, 1973.

Trauboth, H., and W. McCallum, MARSYAS Users Manual", Tech. Rpt. Al-34812, Computation Lab., NASA/MSFC, 1973.

Trauboth, H., and S. P. Singh, "MARSYAS I & II", IEEE Circuits and Systems Soc. Newsletter, Vol. 6, pp. 6-11 and pp. 9-12, 1973.

Trauboth, H., and N. Prasad, "MARSYAS—A Software Package for Digital Simulation of Physical Systems", Proc. of the Spg. Joint Computer Conf., pp. 223-235, 1970.

Weeks, W. T., Jiminez, A. J., Mahoney, G. W., Metho, D., Quassemzedeh, H., and T. R. Scott, "Algorithms for ASTP—A Network Analysis Program", IEEE Trans. on Circuit Theory, Vol. CT-20, pp. 623-634, (1973).

Zein, D. A., and C. W. Ho, "Frequency-Domain Analysis of Nonlinear Electronic Circuits in a General Purpose Interactive CAD Program in APL", Proc. of the IEEE Inter. Symp. on Circuits and Systems, pp. 122-125, 1977.

5. SENSITIVITY

Barry, R. F., "Some Analytic Fault Isolation Techniques", in *Lecture Notes of Computer Aided Testing and Fault Identification of State Systems*, The University of Wisconsin, May 1968.

Bastian, J. D., "Fault Isolation of Analog Circuits", Proc. of the 12th Asilomar Conf. on Circuits, Systems, and Computers, Pacific Grove, Ca., pp. 154-164, 1978.

Bedrosian, S. D., and R. S. Berkowitz, "Solution Procedure for Single-Element-Kind Networks", IRE Inter. Conv. Rec., Vol. 10, Part 2, p. 16, 1962.

Bedrosian, S. D., Ph.D. Dissertation, University of Pennsylvania, 1961.

Berkowitz, R. S., and R. Wexelblatt, "Statistical Considerations in Element Value Solutions", IRE Trans. on Military Electronics, Vol. MIL-6, pp. 282-88, (1962).

Berkowitz, R. S., and P. S. Krishnaswamy, "Computer Techniques for Solving Electric Circuits for Fault Isolation", IEEE Trans. on Aerospace Elec. Support, Vol. AS-1, pp. 1090-1099, (1963).

Calahan, D. A., *Computer-Aided Design*, McGraw-Hill, New York, 1968.

Chao, K.-S., and R. Saeks, "Continuation Methods in Circuit Analysis", IEEE Proc., Vol. 65, pp. 1187-1194, (1977).

Chau, T. H., and R. Saeks, "Optimal Circuit Design with a Sensitivity Penalty", Proc. of the 12th Allerton Conf. on Circuits and Systems, pp. 942-947, 1974.

Chau, T. H., M.S. Thesis, Texas Tech University, Lubbock, Texas, 1974.

Chen, H. M. S., and R. Saeks, "A Search Algorithm for the Solution of the Multifrequency Fault Diagnosis Equations", IEEE Trans. on Circuits and Systems, Vol. CAS-26, pp. 589-594, (1979).

Chen, W. K., and F. N. Chan, "On the Unique Solvability of Linear Active Networks", IEEE Trans. on Circuits and Systems, Vol. CAS-21, pp. 26-35, 1974.

Director, S. W., and R. A. Rohrer, "Automated Network Design: The Frequency Domain Case", IEEE Trans. on Circuit Theory, Vol. CT-16, pp. 330-337, (1969).

Director, S. W., and R. A. Rohrer, "The Generalized Adjoint Network and Network Sensitivities", IEEE Trans. on Circuit Theory, Vol. CT-16, pp 318-323, (1969).

Director, S. W., "LU Factorization in Network Sensitivity Calculations", IEEE Trans. on Circuit Theory, Vol. CT-18, pp. 184-185, (1971).

Even, S., and A. Lempel, "On a Problem of Diagnosis", IEEE Trans. on Circuit Theory, Vol. CT-14, pp. 361-4, (1967).

Gadenz, R. N., Rezai-Fakhr, M. G., and G. C. Temes, "A Method for the Computation of Large Tolerence Effects", IEEE Trans. on Circuit Theory, Vol. CT-20, pp. 704-708, (1973).

Garzia, R. F., "Fault Isolation Computer Methods", NASA Contractor Report CR-1758, Computer Sciences Corp., Marshall Space Flight Center, Huntsville, Alabama, July 1970.

Goddard, P. J., Villalaz, P. A., and R. Spence, "Method for the Efficient Computation of the Large-Change Sensitivity of Linear Reciprocal Networks", Electronics Letters, Vol. 7, pp. 112-113, (1971).

Goddard, P. J., Ph.D. Thesis, London University, London, 1971.

Hachtel, G. D., and R. A. Rohrer, "Techniques for the Optimal Design and Synthesis of Switching Circuits", IEEE Proc., Vol. 55, pp. 1864-1877, (1967).

Happ, W. W., and D. E. Moody, "Topological Techniques for Sensitivity Analysis", IEEE Trans. on Aerospace and Navigational Electronics, Vol. ANE-11, pp. 248-254, (1964).

Hayashi, S., Hattori, Y., and T. Saski, "Considerations on Network Element Value Evaluation", Electronics and Communications in Japan, Vol. 50, pp. 118-127, (1967, Transl. 1968).

Hochwald, W., and J. D. Bastian, "Component Failure Simulation Program (CFS) Application Study", Proc. of MIDCON, Session 30, Chicago, 1977.

Leung, K. H., and R. Spence, "Multiparameter Large-Change Sensitivity Analysis and Systematic Exploration", IEEE Trans. on Circuits and Systems, Vol. CAS-22, pp. 796-804, (1975).

Neu, E. C., "A New n-Port Network Theorem", Proc. of the 13th Midwest Symp. on Circuit Theory, Minneapolis, paper iv.- 5, 1970.

Plice, W. A., "Automatic Generation of Fault Isolation Tests for Analog Circuit Boards—A Survey", Proc. of ATEX EAST 78, Boston, 1978.

Ransom, M. N., and R. Saeks, "Fault Isolation Via Term Expansion", Proc. of the 4th Pittsburgh Symposium on Modeling and Simulation, pp. 224-228, 1973.

Ransom, M. N., and R. Saeks, "Fault Isolation with Insufficient Measurements", IEEE Trans. on Circuit Theory, Vol. CT-20, pp. 416-417, (1973).

Ransom, M. N., and R. Saeks, "A Functional Approach to Fault Analysis", in *Rational Fault Analysis*, (ed. R. Saeks, and S. Liberty), Marcel Dekker Inc., New York, 1977.

Saeks, R., Singh, S. P., and R. Liu, "Fault Isolation Via Component Simulation", IEEE Trans. on Circuit Theory, Vol. CT-19, pp. 634-640, (1972).

Saeks, R., and S. R. Liberty, *Rational Fault Analysis*, Marcel Dekker, New York, 1977.

Saeks, R., and K.-S. Chao, "Continuations Approach to Large Change Sensivity Analysis", IEE (London) Jour. on Electronic Circuits and Systems, Vol. 1, pp. 11-16, (1976).

Sen, N., and R. Saeks, "Fault Diagnosis for Linear Systems Via Multifrequency Measurements", IEEE Trans. on Circuits and Systems, Vol. CAS-26, pp. 457-465, (1979)

Seshu, S., and R. Waxman, "Fault Isolation in Conventional Linear Systems-A Feasibility Study", IEEE Trans. on Realiability, Vol. R-15, pp. 11-16, (1966).

Singhal, K., Vlach, J., and P. R. Bryant, "Efficient Computation of Large-Change Multiparameter Sensitivity", Inter. Jour. of Circuit Theory and its Appl., Vol. 1, pp. 237-247, (1973).

Sriyananda, H., Towill, D. R., and J. H. Williams, "Voting Techniques for Fault Diagnosis from Frequency Domain Test Data", IEEE Trans. on Reliability, Vol. R-24, pp. 260-267, (1975).

Van Ness, J. E., Boyle, J. M., and F. Imad, "Sensitivities of Multiloop Control Systems", IEEE Trans. on Auto. Cont., Vol. AC-10, pp. 308-314, (1965).

Villalaz, P. A., Ph.D. Thesis, London University, London, 1972.

Vlach, J., *Computerized Approximation and Synthesis of Linear Networks*, Wiley, New York, 1969,

Wong, E., "Application of Decision Theory to the Testing of Large Systems", IEEE Trans. on Aerospace and Electronic Systems, Vol. AES-7, pp. 378-384, (1971).

Wu, F. F., "Multiple Single-Element Modification of Large-Scale Linear Networks", in *Large-Scale Dynamical Systems* (ed. R. Saeks), Point Lobos Press, No. Hollywood, pp. 259-274, 1976.

6. STABILITY OF INTERCONNECTED SYSTEMS

Araki, M., Ando, K., and B. Kondo, "Stability of Sampled-Data Composite Systems with Many Nonlinearities", IEEE Trans. on Auto. Cont., Vol. AC-16, pp. 22-27, (1971).

Araki, M., "Input-Output Stability of Composite Feedback Systems", Publication No. 75/1, Department of Comp. and Cont., Imperal College of Science and Technology, London, 1975.

Araki, M., and B. Kondo, "Stability and Transient Behavior of Composite Nonlinear Systems", IEEE Trans. on Auto. Cont., Vol. AC-17, pp. 537-541, (1972).

Bailey, F. N., "Applications of Lynapunov's Second Method to Interconnected Systems", SIAM Jour. on Cont., Vol. 3, pp. 443-462, (1965).

Cook, P. A., "On the Stability of Interconnected Systems", Math. Research Publication, No. 31, Loughborough University of Technology, Loughborough, 1973.

de Figueriredo, R. J. P., and C. Ho, "Absolute Stability of a System of Nonlinear Networks Interconnected by Transmission Lines", IEEE Trans. on Circuit Theory, Vol. CT-17, pp. 575-584, (1970).

Grujic, L. T., and D. D. Siljak, "Asympototic Stability and Instability of Large-Scale Systems", IEEE Trans. on Auto. Cont., Vol. AC-18, pp. 636-645, (1973).

Ikeda, M., and S. Kadaina, "Large-Scale Dynamical Systems: State Equations, Lipschitz Conditions and Linearization", IEEE Trans. on Circuit Theory, Vol. CT-20, pp. 193-202, (1973).

Lasley, E. L., and A. N. Michel, "Input-Output Stability of Large-Scale Systems", in *Large-Scale Dynamical Systems* (ed. R. Saeks), Point Lobos Press, No. Hollywood, pp. 195-220, 1976.

Lasley, E. L., and A. N. Michel, "Input-Output Stability of Large-Scale Systems", Proceedings of the Eighth Asilomar Conference on Circuits, Systems, and Computers, pp. 472-482, 1974.

Lasley, E. L., and A. N. Michel, "Input-Output Stability of Interconnected Systems", Proc. of the IEEE Inter. Symp. on Circuits and Systems, pp. 131-134, 1975.

Liu, R. W., and F. Lu, "Passivity, Stability and Large-Scale Dynamical Systems", EE Memo, 7203, Univ. of Notre Dame, 1972.

Lu, F. C., Jenkins, L., and R. Liu, "A Two-Matrix Transformation Method for Stability Problems of Large-Scale Dynamical Systems", in *Large-Scale Dynamical Systems* (ed. R. Saeks), Point Lobos Press, No. Hollywood, pp. 221-238, 1976.

Lu, F., Ph.D. Thesis, National Taiwan Univ., Taipei, 1972.

Michel, A. N., and R. K. Miller, "Qualitative Analysis of Large Scale Dynamical Systems", Academic Press, New York, 1977.

Michel, A. N., and D. W. Porter, "Stability Analysis of Composite Systems", IEEE Trans. on Auto. Cont., Vol. AC-17, pp. 222-226, (1972).

Michel, A. N., "Quantitative Analysis of Simple Interconnected Systems: Stability, Boundedness and Trajectory Behavior", IEEE Trans. on Circuit Theory, Vol. CT-17, pp. 292-301, (1970).

Michel, A. N., "Stability, Transient Behavior and Trajectory Boundedness of Interconnected Systems", Inter. Jour. on Cont., Vol. 11, pp. 703-715, (1970).

Michel, A. N., "Stability Analysis of Stochastic Composite Systems", Inter. Jour. on Cont., Vol. AC-20, pp. 246-250, (1975).

Michel, A. N., "Stability and Trajectory Behavior of Composite Systems", IEEE Trans. on Circuits and Systems, Vol. CAS-22, pp. 305-312, (1975).

Michel, A. N., and W. R. Vitacco, "Qualitative Analysis of Interconnected Dynamical Systems with Algebraic Loops: Well-Posedness and Stability", Proc. of the IEEE Inter. Symp. on Circuits and Systems, pp. 602-605, 1977.

Silijak, D. D., "Stability of Large-Scale Systems", Proc. of the 5th IFAC Cong., Part IV, pp. 1-11, 1972.

Silijak, D. D., "Stability of Large-Scale Systems Under Structural Perturbations", IEEE Trans. on Sys. Man. and Cyber., Vol. SMC-2, pp. 657-663, (1972).

Silijak, D. D., "On Stability of Large-Scale Systems Under Structural Perturbations", IEEE Trans. on Sys. Man. and Cyber., Vol. SMC-3, pp. 415-41, (1973).

Silijak, D. D., "On Large-Scale System Stability", Proc. of the 9th Allerton Conf. on Circuits and Systems, pp. 731-741, 1971.

Thompson, W. E., "Exponential Stability of Interconnected Systems", IEEE Trans. on Auto. Cont., Vol. AC-15, pp. 504-506, (1970).

Weissenberger, S., "Stability Regions of Large-Scale Systems", Automatica, Vol. 9, pp. 653-663, (1973).

Weissenberger, S., "Stability Regions of Large-Scale Systems", Proc. of 10th Annual Allerton Conference on Circuit and System Theory, pp. 29-38, 1972.

7. CONTROL

Anderson, B. D. O., and J. B. Moore, *Linear Optimal Control*, Prentice-Hall, Englewood Cliffs, 1971.

Astrom, K. J., *Introduction to Stochastic Control Theory*, Academic Press, New York, 1970.

Bellman, R., *Introduction to the Mathematical Theory of Control Processes*, Vol. 1, Linear Equations and Quadratic Criteria, Academic Press, New York, 1967.

Bellman, R. E., Glicksberg, I., and O. A. Gross, "Some Aspects of the Mathematical Theory of Control Processes", R.A.N.D. Report R-313, 1958.

Chang, S. S. L., *Synthesis of Optimum Control Systems*, McGraw-Hill, New York, 1961.

Cruz, J. B., *Feedback Systems*, McGraw-Hill, New York, 1972.

Fel'baum, A., *Optimal Control Systems*, Academic Press, New York, 1965.

Gilbert, E. G., "Controllability and Observability in Multivariable Control Systems", SIAM Jour. on Cont., Vol. 1, pp. 128-151, (1963).

Gilbert, E. G., "The Decoupling of Multivariable Systems by State Feedback", SIAM Jour. on Cont., Vol. 7, pp. 50-63, (1969).

Heymann, M., "Pole Assignment in Multi-input Linear Systems", IEEE Trans. Auto. Cont., Vol. AC-13, pp. 748-749, (1968).

Horowitz, I. M., *Synthesis of Feedback Systems*, Academic Press, New York, 1963.

Kalman, R. E., "On the General Theory of Control Systems, in Automatic and Remote Control", Proc. of the 1st IFAC Cong., Butterworth, London, Vol. 1, pp. 481-492, 1961.

Kalman, R. E., "Contributions to the Theory of Optimal Control", Bol. Soc. Mathem. Mexicana, pp. 102-119, (1960).

Kalman, R. E., "When is a Linear Control System Optimal?", Jour. of Basic Engrg., Vol. 86, pp. 1-10, (1964).

Kalman, R. E., Ho, Y. C., and K. S. Narendra, "Controllability of Linear Dynamical Systems", in *Contrib. to Diff. Equations* (ed. J. Hale), Academic Press, New York, pp. 189-213, 1962.

Lee, E. B., and L. Markus, *Foundations of Optimal Control Theory*, Wiley, New York, 1967.

Luenberger, D. G., "Observing the State of a Linear System", IEEE Trans. on Military Electronics, Vol. MIL-8, pp. 74-80, (1964).

Morgan, B. S., "The Synthesis of Linear Multivariable Systems by State Variable Feedback", IEEE Trans. on Auto. Contr., Vol. AC-9, pp. 405-411, (1964).

Morse, A. S., and W. M. Wonham, "Decoupling and Pole Assignment by Dynamic Compensation", SIAM Jour. on Cont., Vol. 8, pp. 317-337, (1970).

Newton, G. C., Gould, L. A., and J. F. Kaiser, *Analytical Design of Linear Feedback Controls*, Wiley, New York, 1957.

Porter, B., and R. Crossley, *Modal Control: Theory and Applications*, Taylor and Francis, Barnes and Noble, 1972.

Smith, O. J. M., *Feedback Control Systems*, McGraw-Hill, New York, 1958.

Wonham, W. M., and A. S. Morse, "Decoupling and Pole Assignment in Linear Multivariable Systems: A Geometric Approach", SIAM Jour. on Cont., Vol. 8, pp. 1-18, (1970).

Wonham, W. M., *Linear Multivariable Control: A Geometric Approach*, Springer-Verlag, Heidelberg, 1974.

Wonham, W. M., Algebraic Methods in Linear Multivariable Control", in *System Structure*, (ed. A. S. Morse), IEEE Control Systems Society, IEEE Catalog No. 71C61-CSS, 1971.

Wonham, W. M., "Dynamic Observers: Geometric Theory", IEEE Trans. on Auto. Cont., Vol. AC-15, pp. 258-259, (1970).

Wonham, W. M., "On Pole Assignment in Multi-input Controllable Linear Systems", IEEE Trans. on Auto. Cont., Vol. AC-12, pp. 660-665, (1967).

8. DECENTRALIZED CONTROL

Aoki, M., "On Decentralized Linear Stochastic Control Problems with Quadratic Cost", IEEE Trans. on Auto. Cont., Vol. AC-18, pp. 243-249, (1973).

Aoki, M., "On the Feedback Stabilizability of Decentralized Dynamic Systems", Automatica, Vol. 8, pp. 163-173, (1972).

Aoki, M., and M. T. Li, "Controllability and Stabilizability of Decentralized Dynamic Systems", Proc. of the JACC, pp. 278-286, 1973.

Bhandarkar, M. V., and M. M. Fahmy, "Controllability of Tandem Connected Systems", IEEE Trans. on Auto. Cont., Vol AC-17, pp. 150-151, (1972)

Chen, C. T., and C. A. Desoer, "Controllability and Observability of Composite Systems", IEEE Trans. on Auto. Cont., pp. 402-409, (1967).

Chen, C. T., and C. A. Desoer, "New Results on the Controllability & Observability of General Composite Systems", IEEE Trans. on Auto. Cont., Vol. AC-20, pp. 123-128, (1975).

Cormat, J. P., and A. S. Morse, "Stabilization with Decentralized Feedback Control", IEEE Trans. on Auto. Cont., Vol. AC-18, pp. 679-682, (1973).

Cormat, J. P., "Decentralized Control of Linear Multivariable Systems", Becton Center Technical Report CT-67, Dept. Engineering and Applied Sci., Yale Univ., 1974.

Davison, E. J., and S. G. Chow, "An Algorithm for the Assignment of Closed-Loop Poles using Output Feedback in Large Linear Multivariable Systems", IEEE Trans. on Auto. Cont., Vol. AC-18, pp. 74-75, (1973).

Davison, E. J., "The Stabilizability of General Composite Systems", in *Large-Scale Dynamical Systems* (ed. R. Saeks), Point Lobos Press, No. Hollywood, pp. 183-194, 1976.

Davison, E. J., "The Decentralized Stabilization and Control of a Class of Unknown Non-Linear Time-Varying Systems", Automica, Vol. 10, pp. 309-316, 1974.

Davison, E. J., and S. H. Wang, "New Results on the Controllability & Observability of General Composite Systems", IEEE Trans. on Auto. Cont., Vol. AC-20, pp. 123-128, (1975).

Grasseilli, D. M., "Controllability and Observability of Series Connections of Systems", Richerche di Automatica, Vol. 3, pp. 44-53, (1972).

Hautus, M. L. J., "Input Regularity of Cascaded Systems", IEEE Trans. on Auto. Cont., Vol. AC-20, pp. 120-123, (1975).

Kokotovic, P. V., Perkins, W. R., Cruz, J. B., and G. D'Ans., "De-Coupling Method for Near-Optimum Design of Large-Scale Linear Systems", IEEE Proc., Vol. 116, pp. 889-892, (1969).

Laub, A., Ph.D. Thesis, University of Minnesota, Minneapolis, 1974.

Levine, W. S., and M. Athans, "On the Optimal Error Regulation of a String of Moving Vehicles", IEEE Trans. on Auto. Cont., Vol. AC-11, pp. 355-361, (1966).

Khalil, H., Kokotovic, P., and J. Medanic, "Control Strategies for Multi-model Representation of Large Scale Systems", Proc. of the IEEE Inter. Symp. on Circuits and Systems, pp. 873-876, 1977.

McFadden, D., "On the Controllability of Decentralized Macroeconomic Systems: The Assignment Problem", in *Mathematical Systems Theory and Economics* (eds. H. W. Kuhn and G. P. Szego), Springer-Verlag, Heidelberg, pp. 221-240, 1967.

Ozguner, U., and W. R. Perkins, "Controllability, Pole Placement and Stabilizability in Large Scale Composite Systems", Proc. of the 12th Allerton Conf. on Circuit and System Theory, pp. 441-449, 1974.

Ozguner, U., "The Analysis and Control of Large Scale Composite Systems", Report R-680, Coordinated Science Lab., University of Illinois, 1975.

Richardson, M. H., Leake, R. J., and R. Saeks, "A Component-Connection Formulation for Large-Scale Dynamical Discrete-Time System Optimization", Proc. of the 3rd Asilomar Conf. on Circuits and Systems, pp. 665-670, 1969.

Richardson, M. H., Ph.D. Thesis, Univ. of Notre Dame, Notre Dame, Ind., 1969.

Silijak, D. D., and M. B. Vukcevic,, "Multilevel Control of Large-Scale Systems", in *Large-Scale Dynamical Systems* (ed. R. Saeks), Point Lobos Press, No. Hollywood, pp. 33-58, 1976.

Silijak, D. D., and M. K. Sundareshan, "On Hierarchic Optimal Control of Large-Scale Systems", Proc. of the 8th Asilomar Conference on Circuits, Systems, and Computers, pp. 495-502, 1974.

Silijak, D. D., and M. B. Vukcevic, "On Hierarchic Stabilization of Large-Scale Linear Systems", Proc. of the 8th Asilomar Conf. on Circuits, Systems, and Computers, pp. 503-507, 1974.

Wang, S. H., and E. J. Davison, "On the Stabilization of Decentralized Control Systems", IEEE Trans. on Auto. Cont., Vol. AC-18, pp. 473-478, (1973).

Wang, S. H., and E. J. Davison, "On the Controllability and Observability of Composite Systems", IEEE Trans. on Auto. Cont., Vol. AC-18, pp. 74-75, (1973).

Weissenberger, S., "On the Insensitivity of Large, Decentrally Optimal Control Systems to Modeling Errors", in *Large-Scale Dynamical Systems* (ed. R. Saeks), Point Lobos Press, No. Hollywood, pp. 167-182, 1976.

Weissenberger, S., "Tolerance of Decentrally Optimal Controllers to Nonlinearity and Coupling", Proc. of the 12th Annual Allerton Conference on Circuits and Systems, pp. 87-95, 1974.

Wolovich, W. A., and H. L. Huang, "Composite-System Controllability & Observability", Automatica, Vol. 10, pp. 209-212, (1974).

9. POWER SYSTEMS

Bauman, R., "Sparseness in Power Systems Equations", in *Large Sparse Sets of Linear Equations* (ed. J. K. Reid), Academic Press, New York, pp. 105-126, 1971.

Bergen, A. R., "Analytical Methods for the Problem of Dynamic Stability", Proc. of the IEEE Inter. Symp. on Circuits and Systems, pp. 864-872, 1977.

Chang, A., "Application of Sparse Matrix Methods in Electric Power System Analysis", in *Sparse Matrix Proceedings* (ed. R. A. Willoughby), IBM, Yorktown Heights, Rep. No. RA 1 (No. 11707), pp. 113-121, 1969.

Churchill, M. E., "A Sparse Matrix Procedure of Power System Analysis Programs", in *Large Sparse Sets of Linear Equations* (ed. J. K. Reid), Academic Press, New York, pp. 127-138, 1971.

Davison, E. J., and N. S. Rau, "The Optimal Output Control of a Power System Consisting of a Number of Interconnected Synchronous Machines", Inter. Jour. on Cont., Vol. 18, pp. 1313-1328, (1973).

El-Abiad, A. H., and K. Nagappan, "Transient Stability Regions of Multi-machine Power Systems", IEEE Trans. on Power App. and Sys., Vol. PAS-85, pp. 169-179, (1966).

El-Abiad, A. H., "Advances in Power System Dynamics", Proc. on Research in Electric Transmission and Distribution-The University's Role (Workshop), State Univ. of New York, Buffalo, pp. 84-99, 1974.

Elgerd, O. I., *Electric Energy System Theory*, McGraw-Hill, 1971.

Galiana, F. D., "Analytic Properties of the Load-Flow Problem", Proc. of the IEEE Inter. Symp. on Circuits and Systems, pp. 802-816, 1977.

Haneda, H., "Covergence Analysis of Load-Flow Solution Techniques Based on Componentwise Inequalities", Proc. of the IEEE Inter. Symp. on Circuits and Systems, pp. 817-820, 1977.

Jenkins, L., Ph.D. Dissertation, University of Notre Dame, Notre Dame, Indiana, 1976.

Luders, G. A., "Transient Stability of Multimachine Power Systems via the Direct Method of Liapunov", IEEE Trans. on Power App. and Sys., Vol. PAS-90, pp. 23-36, (1971).

Saeks, R. and R. Hebert, 'An LSDS Model for Power Systems", Proc. of the 7th Asilomar Conf. on Circuits, Systems and Computers, pp. 43-47, 1973.

Sato, N., and W. Tinney, "Techniques for Exploiting the Sparsity of the Nodal Admittance Matrix", IEEE Trans. on Power App. and Sys., Vol. PAS-82, pp. 944-940, (1963).

Semlyn, A., "Step-by-Step Calculation of Disturbances in Interconnected Systems", Proc. of the Inter. Symp. on Circuit Theory, pp. 368-371, 1973.

Smith, O., Makani, K., and L. Krishan, "Sparse Solutions using Hash Storage", IEEE Trans. on Power App. and Sys., Vol. PAS-91, pp. 1396-1407, (1972).

Stagg, G. W., and A. H. El-Abiad, *Computer Methods in Power System Analysis*, McGraw-Hill, New York, 1968.

Stott, B., and O. Alsac, "Fast Decoupled Load Flow", IEEE Trans. Power App. Sys., Vol. PAS-93, pp. 859-869, (1974).

Takahashi, K., Fagan, J., and M. Chen, "Formation of a Sparse Bus Impedance Matrix and its Application to Short Circuit Study", Proc. of the PICA Conf., pp. 63-69, 1973.

Tinney, W. F., and N. M. Peterson, "Steady-state Security Monitoring", in *Real-Time Control of Electric Power Systems* (ed. E. Handschin), Elsevier, New York, pp. 191-214, 1972.

Tinney, W., "Compensation Methods for Network Solutions by Optimally Ordered Triangular Factorization", IEEE Trans. on Power App. and Sys., Vol. PAS-91, pp. 123-127, (1972).

Tinney, W., and J. Walker, "Direct Solution of Sparse Network Equations by Optimally Order Triangular Factorization", IEEE Proc., Vol. 55, pp. 1801-1809, (1967).

Tinney, W. F., "Comments on Sparsity Techniques for Power System Problems", in *Sparse Matrix Proceedings* (ed. R. A. Willoughby), IBM, Yorktown Heights, Report No. RA1 (11707), 1969.

Tinney, W. F., "REI Network Reduction for Power System Application", Proc. of the IEEE Inter. Symp. on Circuits and Systems, pp. 821-828, 1977.

Van Ness, J. E., "Methods of Reducing the Order of Power System Models in Dynamic Studies", Proc. of the IEEE Inter. Symp. on Circuits and Systems, pp. 858-863, 1977.

Willems, J. L., and J. C. Willems, "The Application of Liapunov's Method to the Computation of Transient Stability Regions of Multimachine Power Systems", IEEE Trans. on Power App. and Sys., Vol. PAS-89, pp. 795-801, (1970).

10. NUMERICAL LINEAR ALGEBRA

Branin, F. H., "Poles and Zeroes, Eigenvalues of Matrices, and Roots of Polynominals by the Method of Signatures", Proc. of the IEEE Inter. Symp. on Circuits and Systems, pp. 89-94, 1977.

Brayton, R. K., Gustavson, F. G., and R. A. Willoughby, "Some Results on Sparse Matrices", IBM Res. Rpt. No. RC 2332, 1969.

Collatz, L., *Functional Analysis and Numerical Mathematics*, New York, Academic Press, 1966.

Crout, P. D., "A Short Method for Evaluating Determinants and Solving Systems of Linear Equations with Real or Complex Coefficients," Trans. of the AIEE, Vol. 60, pp. 1235-1241, (1941).

Curtis, A. R., Powell, M. J. D., and J. K. Reid, "On the Estimation of Sparse Jacobian Matrices", Rep. TP 476, Atomic Energy Res. Establishment, Harwell, 1972.

Curtis, A. R., and J. K. Reid, "On Automatic Scaling of Matrices for Gaussian Elimination", Rep. TP-444, Atomic Energy Res. Establishment, Harwell, 1971.

Curtis, A. R., and J. K. Reid, "The Solution of Large Sparse Systems of Linear Equations", Rep. TP-450, Atomic Energy. Res. Establishment, Harwell, 1971.

Erisman, A. M., and W. F. Tinney, "On Computing Certain Elements of the Inverse of a Sparse Matrix", CACM, Vol. 18, pp. 177-179, (1975).

Faddeev, D. K., and V. N. Faddeva, *Computational Methods of Linear Algebra*, Freeman, San Francisco, 1963.

Forsythe, G. E., "Today's Computational Methods of Linear Algebra", SIAM Rev., Vol. 9, pp. 489-515, (1967).

Forsythe, G. E., and C. B. Moler, "Computer Solution of Linear Algebraic Systems", Prentice-Hall, Englewood, Cliffs, 1967.

Fox, L., *Introduction to Numerical Linear Algebra*, Oxford Univ. Press (Clarendon), London and New York, 1965.

Francis, J. G. F., "The Q-R Transformation I & II", Computer Jour., Vol. 4, pp. 265-271, and pp. 332-345, (1961 and 1962).

Garbow, B. S., et al, *Matrix Eigensystem Routines — EISPACK Guide Extension*, Lect. Notes in Comp. Sci., Vol. 51, Springer-Verlag, New York, 1977.

Gustavson, F. G., Linger, W. M., and R. A. Willoughby, "Symbolic Generation of an Optimal Crout Algorithm for Sparse Systems of Linear Equations", Jour. of the ACM, Vol. 17, pp. 87-109, (1970).

Hachtel, "Vector and Matrix Variability Type in Sparse Matrix Algorithms", in *Sparse Matrices and Their Applications* (Ed. D. J. Rose and R. A. Willoughby), Plenum Press, New York, pp. 53-65, 1972.

Harary, F., "A Graph Theoretic Approach to Matrix Inversion", Numer. Math, Vol. 4, pp. 128-135, (1962).

Harary, F., "Graphs and Matrices", SIAM Rev., Vol. 9, pp. 83-90, (1967).

Harary, F., "A Graph Theoretic Method for Complete Reduction of a Matrix with a View Toward Finding its Eigenvalues", Jour. of Math. Phys., Vol. 38, pp. 104-111, (1959).

Householder, A. S., "A Survey of Some Closed Methods for Inverting Matrices", SIAM Jour. on Appl. Math., Vol. 5, pp. 155-169, (1957).

Householder, A. S., "Unitary Triangularization of a Nonsymmetric Matrix", Jour. of the ACM, Vol. 5, pp. 339-342, (1958).

Hsieh, H. Y., and M. S. Ghausi, "A Probabilistic Approach to Optimal Pivoting and Prediction of Fill-in for Random Sparse Matrices", Tech. Rep. 400-214, Electrical Eng. Dept. New York Univ., Nwe York, 1971.

Hsieh, H. Y., and M. S. Ghausi, "On Sparse Matrices and Optimal Pivoting Algorithms", Tech. Rep. 400-213, Electrical Eng. Dept., New York, Univ., New York, 1971.

Jennings, A., "A Compact Storage Scheme for the Solution of Symmetric Linear Simultaneous Equation", Compt. Jour., Vol. 9, pp. 281-285, (1966).

Jimenez, A. J., "Computer Handling of Sparse Matrices", Rept. No. Tr. 00.1873, IBM Yorktown Heights, 1969.

Lee, H. B., "An Implementation of Gaussian Elimination for Sparse Systems of Linear Equations", in *Sparse Matrix Proceedings* (ed. R. A. Willoughby), IBM, Yorktown Heights, Rep. No. RA 1 (No. 11707), pp. 75.-84, 1969.

Nathan, A., and R. K. Even, "The Inversion of Sparse Matrices by a Strategy Derived from their Graphs", Comput. Jour, Vol. 10, pp. 190-194, (1967).

Reid, J. K., *Large Sparse Sets of Linear Equations*, Academic Press, New York, 1971.

Rose, D. J., and R. A. Willoughby, *Sparse Matrices and Their Applications*, Plenum Press, New York, 1972.

Ruthishauser, H., "Solution of Eigenvalue Problems with LR Transformation", U.S. Bur. of Standards Appl. Math. Ser., Vol. 49, pp. 47-81, (1958).

Smith, B. T., et al., *Matrix Eigensystem Routines* — EISPACK Guide, Second Edition, Lect. Notes in Comp. Sci., Vol. 6, Springer-Verlag, New York, 1976.

Smith, D. M., "Data Logistics for Matrix Inversion", in *Sparse Matrix Proceedings*, (ed. R. A. Willoughby), IBM, Yorktown Heights, Report No. RA1 (11707), pp. 127-137, 1969.

Stewart, G. W., *Introduction to Matrix Computations*, Academic Press, New York, 1973.

Strang, G., *Linear Algebra and Its Applications*, Academic Press, New York, 1976.

Tewarson, R. P., *Sparse Matrices*, Academic Press, New York, 1972.

Tewarson, R. P., "Computations with Sparse Matrices, SIAM Rev., Vol. 12, pp. 527-544, (1970).

Tewarson, R. P., "Sorting and Ordering Sparse Linear Systems", in *Large Sparse Sets of Linear Equations* (ed. J. K. Reid), Academic Press, New York, pp. 151-168, 1971.

Tewarson, R. P., "On the Gaussian Elimination for Inverting Sparse Matrices", Computing (Arch. Elektron, Rechnen), Vol. 9, pp. 1-9, (1972).

Tewarson, R. P., "The Crout Reduction for Sparse Matrices", Comput. Jour., Vol. 12, pp. 158-159, (1969).

Tewarson, R. P., "On the Reduction of a Sparse Matrix to Hessenberg Form", Inter. Jour. on Comput. Math., Vol. 2, pp. 283-295, (1970).

Tinney, W. F., and E. C. Ogubuobiri, "Sparsity Techniques: Theory and Practice", Rpt. of the Bonnville Power Administration, Portland, 1970.

Varga, R. A., *Matrix Iterative Analysis*, Prentice-Hall, Englewood Cliffs, 1962.

Wilkinson, J. H., *Rounding Errors in Algebraic Processes*, Prentice-Hall, Englewood Cliffs, 1963.

Wilkinson, J. H., *The Algebraic Eigenvalue Problem*, Oxford University Press, London, 1965.

Wilkinson, J. H., "The Algebraic Eigenvalue Problem", IBM, Yorktown Heights, Rep. No. RA1 (11707), 1969.

Wilkinson, J. H., and C. Reinsch, *Linear Algebra*, Hiedleberg, Springer-Verlag, 1970.

Willoughby, R. A., "Proceedings of Symposium on Sparse Matrices and Their Applications", Report RA 1, (No. 11707), IBM Watson Research Center, Yorktown Heights, 1969.

Wing, O., and T. Papathomas, "Hessenberg Reduction Methods for Sparse Circuits and Systems", Proc. of the IEEE Inter. Symp. on Circuits and Systems, pp. 85-88, 1977.

11. NUMERICAL ANALYSIS

Bellman, R., and S. Dreyfus, *Applied Dynamic Programming*, Princeton Univ. Press, Princeton, 1962.

Bellman, R., *Dynamic Programming*, Princeton Univ. Press., Princeton, 1957.

Bergland, G. D., "A Guided Tour of the Fast Fourier Transform", IEEE Spectrum, Vol. 6, pp. 41-64, (1969).

Branin, F. H., "Widely Covergent Methods for Finding Multiple Solutions of Simultaneous Nonlinear Equations", IBM Jour. of Res. and Dev., Vol. 6, pp. 504-522, (1972).

Brayden, C. G., "A Class of Methods for Solving Nonlinear Simultaneous Equations", Math. of Comp., Vol. 19, pp. 577-593, (1965).

Chao, K. S., and R. J. P. de Figueriredo, "Optimally Controlled Iterative Schemes for Obtaining the Solution of a Nonlinear Equation", Inter. Jour. on Control, Vol. 18, pp. 377-384, (1973).

Chao, K. S., Liu, D. K., and C. T. Pan, "A Systematic Search Method for Obtaining Multiple Solutions of Simultaneous Nonlinear Equations", Proc. of the IEEE Inter. Symp. on Circuits and Systems, pp. 27-31, 1974.

Chua, L. O., and Y. F. Lam, "Global Homeomorphism of Vector-Valued Functions", Jour. of Math. Anal. and Appl., Vol. 39, pp. 600-624, (1972).

Cooley, J. W., and J. W. Tukey, "An Algorithm for the Machine Computation of Complex Fourier Series", Math. of Comp., Vol. 19, pp. 297-301, (1965).

Cooley, J. W., Leweis, P. A., and P. D. Welch, "The Finite Fourier Transform", IEEE Trans. on Audio and Electroacoustics, Vol. AU-17, pp. 77-85, (1969).

Cowell, W., *Portability of Numerical Software*, Lecture Notes in Computer Science, Vol. 57, Springer-Verlag, New York, 1977.

Davidenko, D. J., "On a New Method of Numerical Solution of Systems of Nonlinear Equations", Doklady Akad. Nauk. SSR., Vol. 88, pp. 601-602, (1963).

Davison, E. J., "A High Order Crank-Nicholson Technique for Solving Differential Equations", Computer Jour., Vol. 10., pp. 195-197, (1967).

Fletcher, R., "Generalized Inverse Methods for Least Squares Solution of Systems of Nonlinear Equations", Computer Jour., Vol. 10, pp. 392-399, (1968).

Fletcher, R., and M. J. D. Powell, "A Rapidly Coverging Decent Method for Minimization", Computer Jour., Vol. 6, p. 163, (1963).

Fletcher, R., and C. M. Reeves, "Function Minimization by Conjugate Gradients", Computer Jour., Vol. 7, pp. 149-154, (1964).

Forsythe, G. E., Malcolm, M.A., and C. B. Moler, *Computer Methods for Mathematical Computations*, Prentice-Hall, Englewood Cliffs, 1977.

Fowler, M. E., and R. M. Warten, "A Numerical Integration Technique for Ordinary Differential Equations with Widely Separated Eignevalues", IBM Jour. of Res. and Dev., Vol. 11, pp. 537-543, (1967).

Gear, C. W., "Simultaneous Numerical Solution of Differential-algebraic Equations", IEEE Trans. on Circuit Theory, Vol. CT-18, pp. 89-95, (1971).

Gear, C. W., "The Automatic Integration of Stiff ODE's", Proc. of the IFIPS Congr. pp. A81-A85, (1968).

Gear, C. W., *Numerical Initial Value Problems in Ordinary Differential Equations* Prentice Hall, Englewood Cliffs, 1971.

Hadley, G. F., *Linear Programming*, Wesley-Addison, Reading, 1962.

Hadley, G. F., *Nonlinear and Dynamic Programming*, Wesley-Addison, Reading, 1964.

Hildebrand, F. B., *Introduction to Numerical Analysis*, McGraw-Hill, New York, 1956.

Holtzman, C. A., and R. Liu, "On the Dynamic Equations of Nonlinear Networks with n-coupled elements", Proc. of the 3rd Allerton Conf. on Circuits and Systems, pp. 536-545, 1965.

Kuhn, H. W., and A. W. Tucker, "Nonlinear Programming", in *Proc. of the 2nd Berkeley Symp. on Math. Stat. and Prob.*, (ed. J. Neyman), Univ. of California Press, Berkeley, pp. 482-492, 1951.

Kunz, K., *Numerical Analysis*, McGraw-Hill, New York, 1957.

Ladson, L., *Optimization Theory for Large Systems*, McMillan, New York, 1970.

Lawson, J. D., "Generalized Runge-Kutta Processes for Stable Systems with Large Lipschitz Constants", SIAM Jour. on Numer. Anal., Vol. 4, pp. 372-380, (1967).

Linger, W., and R. A. Willoughby, "Efficient Numerical Integration of Stiff Systems of Ordinary Differential Equations", SIAM Jour. on Numer. Anal., Vol. 7, pp. 47-66, (1970).

Linger, W., "Stability and Error Bounds for Multistep Solutions of Nonlinear Differential Equations", Proc. of the IEEE Inter. Symp. on Circuits and Systems, pp. 277-280, 1977.

Liu, B., "Effect of Finite Word Length on the Accuracy of Digital Filters", IEEE Trans. on Circuit Theory, Vol. CT-18, pp. 670-677, (1971).

Luenberger, D. G., *Optimization by Vector Space Methods*, Wiley, New York, 1969.

Oppenheim, A. V., and C. J. Weinstein, "Effects of Finite Register Length in Digital Filtering and the Fast Fourier Transform", IEEE Proc., Vol. 60, pp. 957-976, (1972).

Osborne, M. R., "A New Method for the Integration of Stiff Systems of Ordinary Differential Equations", Proc. of the IFIP Congress, pp. 200-204, 1968.

Peled, A., and B. Liu, *Digital Signal Processing*, New York, Wiley, 1976.

Rosenbrock, H. H., "Some General Implicit Processes for the Numerical Solution of Differential Equations", Computer Jour., Vol. 5, pp. 329-330, (1962).

Sandberg, I. W., and H. Schikman, "Numerical Integration on Systems of Stiff Nonlinear Differential Equations", Bell Systems Tech. Jour., Vol. 47, pp. 511-527, (1968).

Schubert, L. K., "Modification of a Quasi-Newton Method for Nonlinear Equations with Sparse Jacobian", Math. of Comput., Vol. 25, pp. 27-30, (1970).

Shampine, L. and M. K. Gordon, *Computer Solution of Ordinary Differential Equations*, Freeman, New York, 1975.

Treanor, C. E., "A Method for the Numerical Integration of the Coupled First Order Differential Equations with Greatly Different Time Constants", Jour. of Math. Comput., Vol. 20, pp. 39-45, (1966).

Tretter, S. A., *Intro. to Discrete-Time Single Processing*, Wiley, New York, 1976.

Varaiya, P. P., *Notes on Optimization*, Van Nostrand Reinhold, New York, 1972.

Welch, P. D., "A Fixed Point Fast Fourier Transform Error Analysis", IEEE Trans. on Audio and Electroacoustics, Vol. AU-17, pp. 151-157, (1969).

Wolfe, P., "The Secant Method for Solving Nonlinear Equations", Comm. of the ACM, Vol. 2, pp. 12-13, (1959).

Zangwill, W. I., *Nonlinear Programming: A Unified Approach*, Prentice-Hall, Englewood Cliffs, 1969.

12. MATHEMATICAL FOUNDATIONS

Apostol, T. M., *Mathematical Analysis*, Addison-Wesley, Reading, 1957.

Bers, L., *Topology*, Courant Inst. of Math. Sci., New York, 1956.

Bellman, R., *Introduction to Matrix Analysis*, McGraw-Hill, New York, 1970.

Canon, M. D., Cullum, C. D., and E. Polak, *Theory of Optimal Control and Mathematical Programming*, McGraw-Hill, New York, 1970.

Fleming, W., *Functions of Several Variables*, Addison-Wesley, Reading, 1965.

Frank, H., and I. T. Frisch, *Communication, Transmission, and Transportation Networks*, Addison-Wesley, Reading, 1970.

Gantmacher, F. F., *The Theory of Matrices*, (2 Vols.), Chelsa, New York, 1959.

Greub, W. H., *Linear Algebra*, Springer-Verlag, New York, 1967.

Hadley, G. F., *Linear Algebra*, Addison-Wesley, Reading, 1961.

Hale, J. K., *Ordinary Differential Equations*, Wiley, New York, 1969.

Jacobson, N., *Lectures in Abstract Algebra, Linear Algebra* (2 Vols.), Van Nostrand, Princeton, 1953.

Kalman, R. E., Falb, P. L., and M. A. Arbib, *Topics in Mathematical System Theory*, McGraw-Hill, New York, 1969.

Lefschetz, S., *Differential Equations—Geometric Theory*, Interscience, New York, 1959.

Loeve, M., *Probability Theory*, Van Norstrand, Princeton, 1954.

Massey, W. S., *Algebraic Topology: An Introduction*, Harcourt, Brace, and World, New York, 1967.

Maclane, S., and G. Birkhoff, *Algebra*, Macmillian, New York, 1967.

Mayeda, W., *Graph Theory*, Wiley, New York, 1971.

Newcomb, R. W., *Linear Multiport Synthesis*, McGraw-Hill, New York, 1966.

Palais, R. S., "Natural Operations on Differential Forms", Trans. of the AMS, Vol. 92, pp. 125-141, (1959).

Pease, M. C., *Methods of Matrix Algebra*, Academic Press, New York, 1965.

Ralston, A., *A First Course in Numerical Analysis*, McGraw-Hill, New York, 1965.

Rao, C. R., and S. K. Mitra, *Generalized Inverse of Matrices and its Applications*, Wiley, New York, 1971.

Rudin, W., *Principles of Mathematical Analysis*, McGraw-Hill, New York, 1964.

Rudin, W., *Real and Complex Analysis*, McGraw-Hill, New York, 1967.

Spivak, M., *Calculus on Manifolds*, Benjamin, New York, 1965.

Titchmarsh, E. C., *The Theory of Functions*, Oxford University Press, London, 1932.

Wall, H. S., *Analytic Theory of Continued Fractions*, Chelsa, New York, 1967.

Van der Waerden, B. L., *Algebra*, Ungar., New York, 1970.

Wu, F. F., and C. A. Desoer, "Global Inverse Function Theorem", IEEE Trans. on Circuit Theory, Vol. CT-19, pp. 199-201, (1972).

INDEX